A Natural History of Oregon's Lake Abert in the Northwest Great Basin Landscape

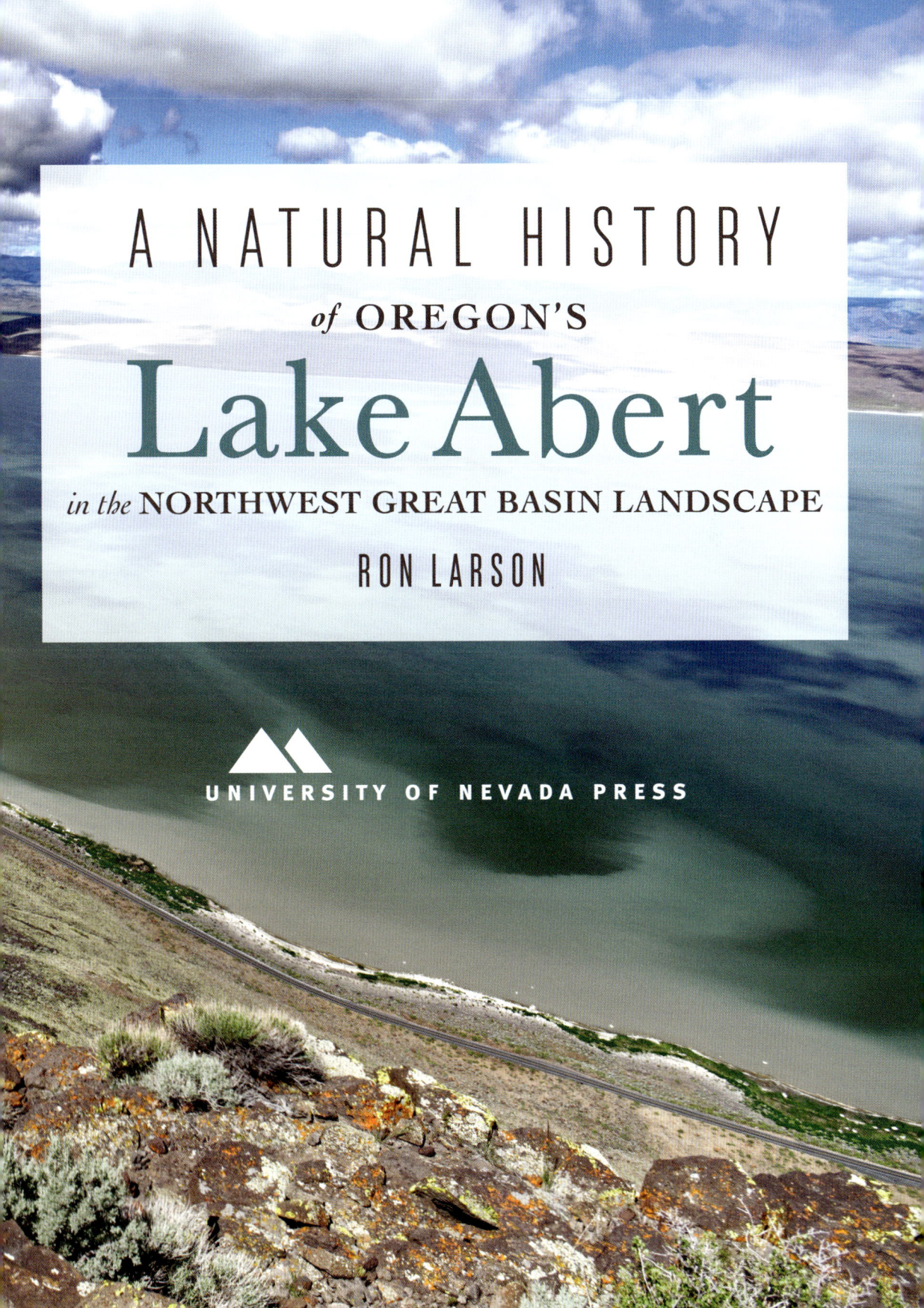

A NATURAL HISTORY

of OREGON'S

Lake Abert

in the NORTHWEST GREAT BASIN LANDSCAPE

RON LARSON

UNIVERSITY OF NEVADA PRESS

University of Nevada Press | Reno, Nevada 89557 USA
www.unpress.nevada.edu
Copyright © 2024 by University of Nevada Press
All rights reserved
Manufactured in the United States of America

FIRST PRINTING

Cover design by Diane McIntosh
Photographs provided by author unless otherwise noted.

LIBRARY OF CONGRESS CATALOGING-IN-PUBLICATION DATA
Names: Larson, Ronald J., author.
Title: A natural history of Oregon's Lake Abert in the Northwest Great Basin landscape /
    Ron Larson.
Description: Reno : University of Nevada Press, [2023] | Includes bibliographical
    references and index. | Summary: "*A Natural History of Oregon's Lake Abert in the
    Northwest Great Basin Landscape* focuses on the salt lake and includes descriptions
    and numerous photos of the region's geology, hydrology, and plants and animals—
    from lichens to pronghorn sheep—as well as its archaeology. Because birds are so
    conspicuous, both on the lake and in the uplands, there is an abundant amount of
    information included about them."—Provided by publisher
Identifiers: LCCN 2023002010 | ISBN 978-1-64779-088-2 (paperback) | ISBN 978-1-
    64779-089-9 (ebook)
Subjects: LCSH: Natural history—Oregon—Abert, Lake. | Salt lakes—Oregon—Abert,
    Lake. | Salt lake ecology—Oregon—Abert, Lake. | Abert, Lake (Or.)
Classification: LCC QH105.O7 L375 2023 | DDC 508.795/93—dc23/eng/20230825
LC record available at https://lccn.loc.gov/2023002010

The paper used in this book meets the requirements of American National Standard for
Information Sciences—Permanence of Paper for Printed Library Materials, ANSI/NISO
Z39.48-1992 (R2002).

Contents

Preface

But when I smell the sage,
When the long, marching landscape line
Melts into wreathing mountains,
And the dust cones dance,
Something in me that is of them will stir.

—MARY AUSTIN, *Going West,* 1922

Tucked away in the far, northwest corner of the Great Basin, in South Central Oregon's high-desert country, is Lake Abert, a place that I like to call "Avocet's Home." My special name for this area pays tribute to the elegant American Avocet, one of the most common and conspicuous waterbirds living there and in the wetlands throughout the Great Basin. For me, this beautiful bird exemplifies why the region comprising the lake and nearby Abert Rim is one of the West's outstanding natural wonders.

The rim is a nearly vertical, two-thousand-foot-high, fault scarp exposing hundreds of lava flows piled in layers like a rocky cake. There, deep-purple shadows linger long into the morning; sure-footed bighorn sheep easily make their way on the precipitous upper slopes; and Violet-green Swallows effortlessly glide on updrafts along the rim's rugged face. Situated at the base of the rim is the long sheet of shallow and salty water known as Lake Abert. In winter, the lake can be whipped up into whitecaps or partially covered by ice and snow. In summer, when the lake's water is low, what emerges is a brilliant-white, salt-encrusted ring, where dust devils swirl and dance across the dry, fissured surface.

Even under these harsh conditions, however, when there is enough water, the lake comes alive with tens of thousands of waterbirds. Foraging in the deeper areas, the long-legged avocets use their gracefully curved bills to scythe through the water as they search for invertebrate prey. Smaller shorebirds, including sandpipers, confine their foraging to the shallow, muddy shores, where they repeatedly probe the mud with their stout beaks in a sewing-machine-like action. Far out on the lake, almost mirage-like, myriad, tiny, animated objects pinwheel about. They are phalaropes, shorebirds adapted to floating on the open ocean or on salt lakes, and they swim in circles, creating a vortex to bring up prey to the water's surface, where they can catch it.

Near the lake on higher ground, drought-resistant bunchgrasses, and such common Great Basin shrubs as sagebrush, rabbitbrush, greasewood, and shadscale, sparsely cover the surrounding uplands. Hardy western junipers, surviving sometimes on less than one foot of annual precipitation, are the only trees growing on the lower slopes. Higher up on the north-facing slopes, where snowdrifts linger into spring, small groves of aspen, white fir, and ponderosa pine trees remain rooted. These trees require more water and are possibly relics of a wetter past.

In spring, along the few streams that tumble down from the rim, colorful wildflower gardens grow, decorated with red and yellow western columbine, pink geraniums, and purple horsemint. White milk vetch blooms on the mid-slopes in May and June, as do the yellow balsamroot—which has blossoms nearly as large as sunflowers—and the red paintbrush, which provides much-needed nectar to hungry hummingbirds. These plants support diverse insect pollinators, including butterflies, bees, and wasps. Nearby, multicolored, crust-forming lichens eke out a slow-motion existence on the hot and dry volcanic rocks. On the larger boulders that have fallen from the rim, six-inch-long, blue-bellied western fence lizards sun themselves. The lake and rim are alive with these varied, fascinating animals and plants, all part of the hardy fauna and flora that make their home in the sometimes-harsh landscape of the Great Basin.

This compelling terrain is also a cultural environment thousands of years old. Numerous archaeological sites, such as those revealing rock rings and petroglyphs, exist above the shoreline and show evidence of a centuries-long settlement of people who likely gathered fish and other resources from the lake. Archaeologists refer to them as the ancient Chewaucanians, a people who last lived in the Great Basin about three hundred years ago. In more recent history, although still more than a century ago, other settlers grazed sheep along the steep slopes of Abert Rim, but the only remaining artifacts of their lives are rock fences covered with orange lichen and the remains of a trail that the herders used to move their sheep onto the rim high above the lake.

Now, in contemporary times, the noise from semitrailers and motorcycles echoes off the rim and frequently drowns out the sounds of nature as traffic speeds along Highway 395, which courses above the eastern shore. A large, abandoned gravel pit and rock pile near the south end of the lake also mar the view and provide a habitat for invasive weeds that have spread throughout the area. Less obvious but much more damaging are the impacts from reduced inflows of water to the lake, caused by climate change, drought, and upstream irrigation.

The Great Basin is a place of startling contrasts, both for its scenery and for the feelings it evokes. Some people may think of it as the place where rivers go to die, because none of its surface water reaches the ocean. Others might characterize the terrain as one of endless sagebrush punctuated by blindingly bright salt flats and long, straight highways that shimmer in the summer heat. And still others may view the area as one of great beauty, featuring endless blue skies, star-filled nights, unmarred vistas, and twilight songs of coyotes, or they may remember moments of absolute silence there, marking it as a sublime place of wonder, solace, and rejuvenation.

Geologist and writer Frank DeCourten, in describing why he feels so strongly about the region, has asserted, "We desperately seek some external elixir to restore meaning and purpose to life, a search that I believe is doomed to failure without an emotional connection to the living world around us. We so seldom realize that the serenity we crave can be achieved simply through a deeper intimacy with the land and life that enfolds us."

For me, the evocative area encompassing Lake Abert and the Abert Rim is my place of solace and serenity. I decided to write about the natural history of this specific part of the Great Basin for several reasons. First, considering the vast size of the Great Basin—which covers nearly 200,000 square miles, including almost all of Nevada, a good portion of Oregon and Utah, and parts of California, Idaho, and Wyoming—there are surprisingly few books describing any of the region's natural history in detail or providing insight into why we should conserve it. Next, even the seventeen-thousand miles of the Great Basin in Oregon are too great to cover meaningfully in one book, so I decided to focus on a smaller geographic area that I know well. Finally, in 2014, Lake Abert nearly dried up, due to a drought and upstream water diversions, and its marvelous ecosystem was nearly lost because of high salinities. The urgency of these factors made me realize that the area's future is in jeopardy and that I need to draw attention to its plight.

Focusing on this particular location makes sense for many other reasons too. Although minor in terms of size, the Lake Abert/Abert Rim area is rich in factors that make the Great Basin special. For example, the Bureau of Land Management (BLM) has designated the lake as an Area of Critical Environmental Concern because of the large number of waterbirds that flock to it each summer to feed prior to making long-distance migrations south. Also, BLM has designated much of Abert Rim as a wilderness study area because it has outstanding wilderness characteristics and provides an environment for solitude and visual aesthetics. Furthermore, the National Park Service has designated the numerous, ancient cultural sites above the lake as the East Lake Abert Archaeological District.

Most biota in the lake and on the rim—from lichens and sagebrush to brine shrimp, leopard lizards, and pronghorns—are found elsewhere in the Great Basin, so what we learn about the ecology of this specific area can perhaps be more broadly applied. Similarly, we can compare geological processes occurring in this landscape to those happening elsewhere in the Great Basin, as well as to those taking place in the larger Basin and Range province. We may even be able to compare geological occurrences in this area to those in environments beyond Earth's boundaries. For example, scientists recently identified what appears to be a salt lake buried under ice at the Martian south pole, one possible intriguing comparison. With all these factors in mind, I decided that writing a book about this complex region is timely and of broad interest.

In her own book, *Refuge: An Unnatural History of Family and Place* (published in 1991), naturalist Terry Tempest Williams writes: "The landscapes we know and return to become places of solace. We are drawn to them because of the stories they tell, because of the memories they hold, or simply because of the sheer beauty that calls us back again and again." For me, Lake Abert and the massive Abert Rim make up such a place. When I view the lake, from the sagebrush-covered slopes, or even from the highway that skirts the eastern shore, I am awed by the fact that so much beauty and life can exist in that arid landscape. Spread out into the distance on a windless, summer day, the lake perfectly reflects a clear, blue sky. At times like this, it seems that the lake and the sky are one. Such days in this beautiful location make me think of the ancient Chewaucanians, who themselves looked out on the placid water and likely felt some of the same emotions, and that thought makes me feel as if we are somehow connected across that great chasm of time. This perceived relationship makes me feel an obligation to protect their home, to bring attention to the ecological damage that we have caused and which we are obligated to correct.

Besides the immediate, aesthetic response that many people feel when seeing the lake, I have also grown to appreciate it as a marvelous ecosystem comprising myriad species that inhabit an imposing environment: the lake's harsh waters can have a salt concentration up to eight times that of sea water, and an alkalinity so high that it can burn exposed skin. Furthermore, I value the solitude and connection to nature that I find when I hike up to the rim and away from the highway, or when I sit on the lakeshore watching the flocks of shorebirds as they fly by, flashing their alternating dark and light colors, and appreciating them all the more when they sometimes land nearby.

Although the jewel of this landscape is the lake, the complete story consists of countless plants and animals, from the very primitive, reddish, bacteria-like archaea that thrive only in the lake's high-salinity waters to the Golden Eagles and ravens that soar above the craggy slopes of Abert Rim. The myriad species in and around the lake are part of an ecosystem shaped by ageless processes including massive lava flows, repeated droughts, and blinding snowstorms. Thus, instead of being just a lake, it's a landscape and an ecosystem rich with biotic and physical interconnections that extend in many directions through time and space. In fact, its story goes back at least several million years.

One thing that I have learned about this region is that its biota faces incredible challenges, coping not only with the caustic water in the lake but also with the near-constant aridity in the surrounding uplands. Yet so many plants and animals persist there and even thrive when conditions are good. Whether or not these conditions remain favorable depends on humans. In reality, the area has long been impacted by people, from the ancient Chewaucanians, to the thousands of drivers who now speed along Highway 395, and to the few ranching families who raise cattle in the Chewaucan Basin upstream of the lake.

So, I have chosen to write this book about the natural history of the Lake Abert region, but what exactly is natural history? To me, writing about natural history is like composing a biography of nature: you tell a life story about a species, ecosystem, or natural landscape. In other words, you observe and describe the natural world. All of the natural sciences, including biology, ecology, and geology, are based on natural-history observations. And just as science starts with a hypothesis, natural history begins with a desire to understand the natural world. Natural history may not be a pure science, like physics, but it produces anecdotes that are the building blocks for increasing our understanding of the world in a way that's impossible to do in a laboratory, no matter how much high-tech equipment we use.

In a 2014 *Scientific American* blog titled "Natural History Is Dying and We Are All Losers," science writer Jennifer Frazer notes that fewer universities now offer courses in this field, a trend that has been occurring for some time. Yet, some college English and environmental studies teachers continue to help students learn about nature by guiding them in direct observation and in writing field notes and essays (Christensen and Crimmel 2008). Additionally, the study of nature appears to be flourishing in the general population, based on the popularity of birding, natural history tours, nature photography, and the countless books and websites devoted to the topic. Also, more parents seem to believe that their children should experience nature by being outdoors. Evidence for this belief in Oregon comes from voters who recently supported funding for elementary outdoor-science schools.

Indeed, the public's enthusiasm for studying nature has helped me to craft several themes for this book. First, Lake Abert is special, is threatened, and thus needs protection. In fact, the entire region is a kind of living natural-history museum that displays key parts of our natural legacy, and we should give it the care and thoughtful stewardship it deserves. Also, throughout the book I provide examples to show the interconnectedness of our world, helping readers to learn, for instance, that much of the water in the lake comes from snow falling on distant mountains; that nutrient-rich aerosols blown off the lake provide essential minerals to rock lichens and upland plants; that alkali (carbonate and bicarbonate salts with a high pH) from the playa (the desiccated portion of the lake) is picked up by dust devils and can travel hundreds of miles to affect other ecosystems; and that some waterbirds feeding at the lake in the summer do so after they finish nesting in Arctic Alaska, while others travel even farther to spend the winter in southern South America.

I also emphasize that some of the geological features seen around the lake exist beyond our planet. One such example is the basalt rock that forms Abert Rim and which is abundant on the moon and on Mars as well. Furthermore, I stress the theme of change. The Lake Abert environment has been in constant flux, encompassing the time when multiple lava flows covered the previous landscape millions of years ago; the period when the first people came into the region thousands of years ago to hunt mammoths; and the modern age, when climate change exerts considerable stress on people and ecosystems. Finally, and most crucially, my book underscores the importance of water in the West—a very scarce and valuable public resource that is underappreciated and, unfortunately, poorly managed. The Lake Abert ecosystem cannot exist without adequate inflows, and, therefore, we must do more to protect it so that it will be there for future generations and for all of the other species that need it too.

The direct impetus for this book came from my attending a 2011 mini-symposium held in Paisley, Oregon, where we focused on the status of the Lake Abert ecosystem and its declining water levels, a dire situation suggesting that it was at risk. During this meeting I realized that not all the participants shared a common sense of moral purpose about the lake's potential demise and about the value of working to correct this problem. I sensed that attendees lacked clarity about which factors most impact the lake and about why it is worth saving. More positively, I was impressed to meet a dedicated but small group of people who see value in a salty lake that few others seem to care about.

Although resource agencies were invited to the meeting, none stepped forward to show concern or to admit responsibility for the low lake levels, or even to say that Lake Abert is valuable. Therefore, I decided that I needed to write an informed synthesis about the lake ecosystem and to make the information available to anyone interested, rather than to just a few scientists reading a journal article. I got further confirmation about the need for this book in 2014, when the lake shrank to about 5 percent of its historical maximum size and its salinity reached lethal levels, problems caused largely by upstream water diversions and made worse by climate change. Then, in 2021, the lake hit bottom once again, making it clearer than ever that these stressful ecological events were likely going to be frequent in the future.

Unfortunately, the water shortage and increasing salinity affecting Lake Abert are also happening in many other parts of the world. Here in the West, the summer of 2021 was a bellwether. The Great Salt Lake in Utah recently hit its lowest recorded level, as have lakes all over the West, including Mono Lake and Owens Lake in California and Walker Lake in Nevada, all part of the Great Basin and the ones most publicized for experiencing this crisis. Because Lake Abert has already experienced a near-complete, but hopefully temporary, ecosystem collapse, owing to high salinities, perhaps its precarious situation can provide useful information that we can apply elsewhere, if people care to do so.

I am encouraged that Lake Abert is finally getting some much-needed attention, thanks to articles published by *The Oregonian* newspaper in 2022, and to hearings held in the Oregon House of Representatives. But I also know that there is much crucial work to be done to get enough water to the lake so that its valuable and beautiful ecosystem will persist.

Author's Note: As a natural historian, I must rely, at times, on secondary sources that come from earlier periods. The information I describe in this book about the lives of early indigenous people in the northern Great Basin was often obtained by archaeologists working in a very different era from that of contemporary researchers. Today, the descendants of indigenous people worldwide are themselves teaching us about the spiritual and cultural significance of the objects that their ancestors left behind, and they ask us not to disturb burial grounds or to study artifacts without proper consultation and careful consideration. I hope that my efforts to enlighten readers about the environment and lives of these ancient peoples honor them.

Throughout the book, I follow the standardized naming convention for birds now followed by scientists worldwide, capitalizing the first letter of each word (e.g., American Avocet and Violet-green Swallow). Peter Pyle and David F. DeSante of The Institute for Bird Populations have published a complete list of these names online at www.birdpop.org. For scientific nomenclature of plants and animals, I italicize the genus and species; capitalize the first letter of the genus; use the abbreviation *sp.* to refer to an unspecified or unknown species; and use the abbreviation *spp.* to refer to a group of species.

Acknowledgments

This book would have been impossible without the help, encouragement, and support of people who want Lake Abert to continue as a healthy and vibrant ecosystem. Foremost among these advocates are Keith and Lynn Kreuz, who opened up their Valley Falls cabin—"Shangri-La"—for me to stay in, and where Lynn fixed us wonderful, home-cooked meals. Trent Seager was a big influence who shared his love for the lake and its birds, as well as his extensive knowledge about the area. Steve Van Denburgh was an inspiration who modeled how to maintain a lifelong interest in science and who demonstrated how to apply this interest to environmental issues. Steve worked on the geochemistry of the lake in the 1960s and was still collecting water samples from the lake when he was in his mid-80s, having retired from the US Geological Survey.

I also appreciated discussions and advice from many other people who shared their knowledge of the lake and its biota and their concern over its future, and who greatly influenced me with their insistence about the importance of conserving Oregon's lakes. Chief among them are Joe Eilers of MaxDepth Aquatics Inc. in Bend, Oregon; Theo Dreher of the Oregon Lakes Association; and Lisa Brown from WaterWatch of Oregon. Other people who helped me include: Victor Camp, San Diego State University; Thomas Connolly, University of Oregon; Frank Conte, Sisters, Oregon; Donald Grayson, University of Washington; Susan Haig, Oregon State University; David Herbst, University of California, Irvine; Brian Mayer, Oregon Water Resources Department; Jonathan Muir, Oregon Department of Fish & Wildlife; Richard Pettigrew, Archaeological Legacy Institute; and Steve Sheehy, Klamath Falls, Oregon. Staff at the Klamath County Library were wonderful in helping me to borrow hard-to-find reference materials from other libraries. To all these people who assisted me with this project, including any others whom I might have forgotten, I am eternally grateful. I am also indebted to members of the East Cascades Audubon Society and to other volunteers who traveled long distances to survey waterbirds at the lake, and who made their data available. And to my wife, Kathy, who has helped sustain my interest in the lake, I want to express how truly thankful I am for her support and encouragement.

The book greatly benefited from valuable reviews provided by the following people: Michael Cummings, Portland State University (chapter 2); Joe Eilers, MaxDepth Aquatics Inc. (chapter 3); Keith

Kreuz, Portland, Oregon (chapter 4); Dennis Albert, Oregon State University (chapter 4, "Wetland and Shore Plants"); Steve Sheehy, Klamath Falls (chapter 5, "Lichens—Slow Life on Hot Rocks," and Steve identified other species, as well); Jherime Kellermann, Oregon Tech (chapter 6); Stan Senner, National Audubon Society (chapter 6); Thomas Connolly, University of Oregon, and Douglas Beauchamp, Eugene, Oregon (chapter 7); and Ryan Houston, Oregon Natural Desert Association (chapter 8). I am especially thankful, also, for helpful comments and suggested editorial changes provided by two anonymous reviewers.

Special thanks to University of Nevada Press Director JoAnne Banducci and her staff for believing that a book about an obscure Oregon salt lake was worth being published. I was also fortunate to have considerable help from Kathleen Chapman, whose careful editing greatly improved the manuscript. I couldn't have had a more capable copy editor!

Finally, I acknowledge the invaluable information developed by naturalists, explorers, scientists, natural resource staff, and others who have cared enough about the natural history of the Great Basin to share it with us, and I appreciate all the people who have helped protect public land everywhere so that we can enjoy it. In paying tribute to those who have written about the natural history of the Great Basin, I especially would like to point out the contributions made by Donald Grayson of the University of Washington. His exhaustive reviews of the prehistory of the region, published in 1993 and updated in 2011, provide a thoroughly interesting account of the many factors that have affected the flora, fauna, and people of the Great Basin, beginning with the Pleistocene period. His two books, *The Deserts Past: A Natural Prehistory of the Great Basin* and *The Great Basin: A Natural Prehistory,* need to be read and reread by everyone who is fascinated by this region. Other books about the Great Basin that I highly recommend are: Frank L. DeCourten's *The Broken Land: Adventures in Great Basin Geology;* Fred Ryser's *Birds of the Great Basin: A Natural History;* E. R. Jackman and R. A. Long's *The Oregon Desert;* and Stephen Trimble's *The Sagebrush Ocean: A Natural History of the Great Basin,* just to mention a few.

A Natural History of Oregon's Lake Abert in the Northwest Great Basin Landscape

CHAPTER 1

An Introduction to Lake Abert

A Great Basin Saline Lake Ecosystem

Abert Lake lies north of Goose Lake, covering only some sixty square miles. But of all the lakes in Oregon it is the most interesting…. The strange, wild beauty of the landscape here can hardly be described in words. Viewed from the south the deep blue-green water is seen stretching away in the distance; on the left side a rugged slope of rock, scantly overgrown with sage brush, rises from the shore; on the right huge boulders fallen from the cliffs above, lie in confused masses on the water's edge; above these tower the mighty cliffs, rising fully one thousand feet above the lake, black, silent, and majestic. Far into the distance stretch these awful heights, their colors mellowing and contours softening until they are lost in an indistinct mountain mass on the far horizon. We look in vain for a sign of life; a single sail upon the broad expanse of water; the smoke of a settler's cabin on the shore; all is silent and desolate; nature is alone in her grandeur.

—S. A. SHAVER et al., 1905

Lake Abert or Abert Lake, as it is sometimes called, lies in the far northwest corner of the Great Basin in South Central Oregon's Lake County, about eighteen miles east of Paisley and twenty-five miles north of Lakeview. It is also part of the massive Basin and Range province that extends from Mexico to Oregon, south to north, and to Utah, heading east. Lake Abert counts as the largest salt lake in the Pacific Northwest and as Oregon's sixth largest lake, recently measuring sixteen miles long and five miles wide. However, its size varies considerably from year to year, due to periodic drought and to the diversion of water upstream. Highway 395 snakes along the eastern shore of the lake, offering picturesque views to drivers traveling between Burns and Lakeview, who get the chance to witness an amazing wildlife spectacle each summer.

The lake has one main source, the fifty-four-mile-long Chewaucan River (pronounced *shee-wa-can*), which begins high on the

Leaf and three-petaled flower of the arrowhead plant (*Sagittaria latifolia*).

eastern flank of eight-thousand-foot Gearhart Mountain, a designated national wilderness area. *Chewaucan* is derived from the Klamath Native American word *cho-ä´*, which refers to an arrowhead or duck potato plant, and the suffix *-keni,* which denotes a place where indigenous people lived (Coville 1897). The arrowhead (*Sagittaria*) itself is a wetland plant found along the river, characterized by distinctive, arrow-shaped leaves and three-petaled white flowers. It has an edible, starchy root that was an important seasonal food for Native Americans in many parts of the West (Darby 1996) and apparently grew abundantly in the Chewaucan marshes. After explorer and politician Captain John Frémont and his men had passed through the basin in December 1843, he mentioned that the indigenous people, either Northern Paiutes or Klamaths, had dug up patches of ground while searching for the prized arrowhead roots.

The river's headwaters start at Dairy Creek, which was once a glacial cirque (a glacier formed in a bowl-shaped depression in the mountains) and later a small lake but which now consists of numerous cold springs, shallow, weed-filled ponds, and montane meadows (located on upland slopes) lush with colorful wildflowers in summer. I, myself, spent several summers in the 1990s documenting plants in the Gearhart Mountain Wilderness in Lake County (Larson 2007), not knowing that the verdant wetlands there fed a river and eventually a lake that I would someday come to know better.

After leaving the cirque, Dairy Creek tumbles down through lush, green, montane forests of white fir (*Abies concolor*), lodgepole pine (*Pinus contorta*), and quacking aspen (*Populus tremuloides*), which turn golden in autumn. Then at lower elevations, the creek becomes the Chewaucan River, which winds through a valley of drier ponderosa pine (*Pinus ponderosa*) and western juniper (*Juniperus occidentalis*) trees in Fremont-Winema National Forest and rushes over boulders that form rapids. (These rapids are home to Great Basin redband trout [*Oncorhynchus mykiss newberrii*], a variety of rainbow trout well known among fly fishermen and hardy survivors of severe droughts and violent volcanic eruptions.) Near the small town of Paisley, the river leaves its canyon and flows out onto the flat, seasonally marshy basin.

In the lower Chewaucan Basin, the river, modified by canals and drains, flows sluggishly for about twenty miles through an altered channel and two large, seasonal wetlands—the Upper and Lower Chewaucan Marshes—before entering River's End Reservoir. Leaving the reservoir, the river finally spills over a low-head dam and cascades over a rock reef before entering the lake.

Low-angle view of Lake Abert, looking north. Like a stretched-out and twisted piece of pie, Lake Abert lies nestled between the gently sloping hills of Coglan Buttes to the west and the precipitous cliffs of Abert Rim to the east. The Chewaucan River enters the south end of the lake when there is adequate flow. Google Earth, August 30, 2013.

Settlers began straightening and channeling the river in the 1880s to minimize flooding near Paisley. Unfortunately, the once-strong, unimpeded flow of the river was critical for sustaining the downstream wetlands, as well as the lake. Now, what remains of the river passes through an extensive network of nearly 175 miles of canals and drains, as it courses through fields of irrigated hay and alfalfa and through remnant wetlands.

The hay fields, pastures, and wetlands of the Chewaucan marshes provide seasonal habitat for a variety of birds, including Wilson's Snipe (*Gallinago delicata*), the Sandhill Crane (*Grus canadensis*), the Long-billed Curlew (*Numenius americanus*), the Red-winged Blackbird (*Agelaius phoeniceus*), the White-faced Ibis (*Plegadis chihi*), and several species of ducks. In May and June, the *hu-hu-hu* winnowing sounds of snipe and the loud bugle calls of cranes resonate across the fields between Paisley and Valley Falls. In years when sufficient precipitation creates flooded wetlands, you can hear lively choruses of breeding frogs and toads coming from scattered, small ponds.

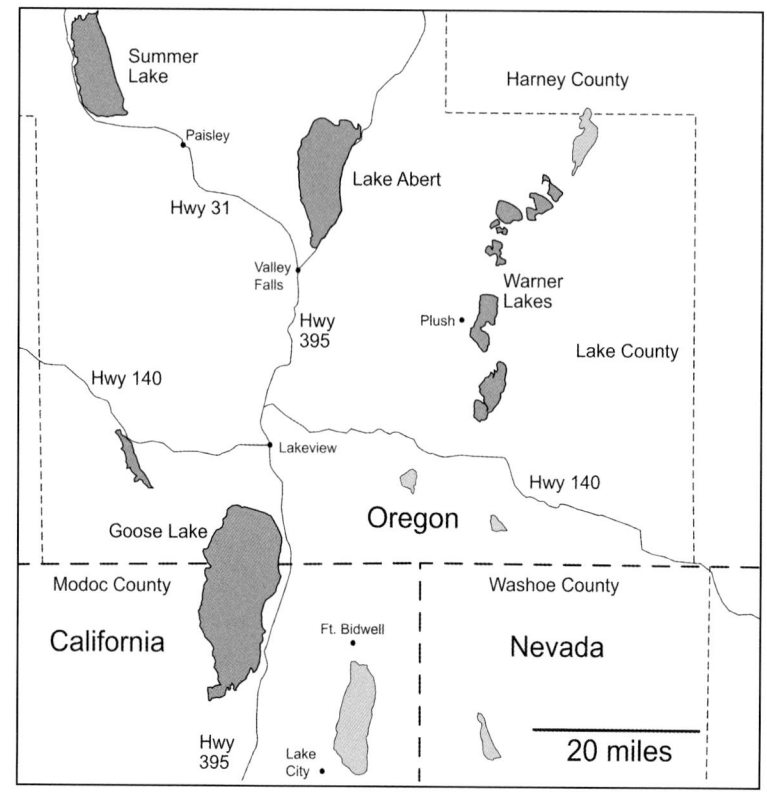

Lake Abert and adjacent areas of the northern Great Basin.

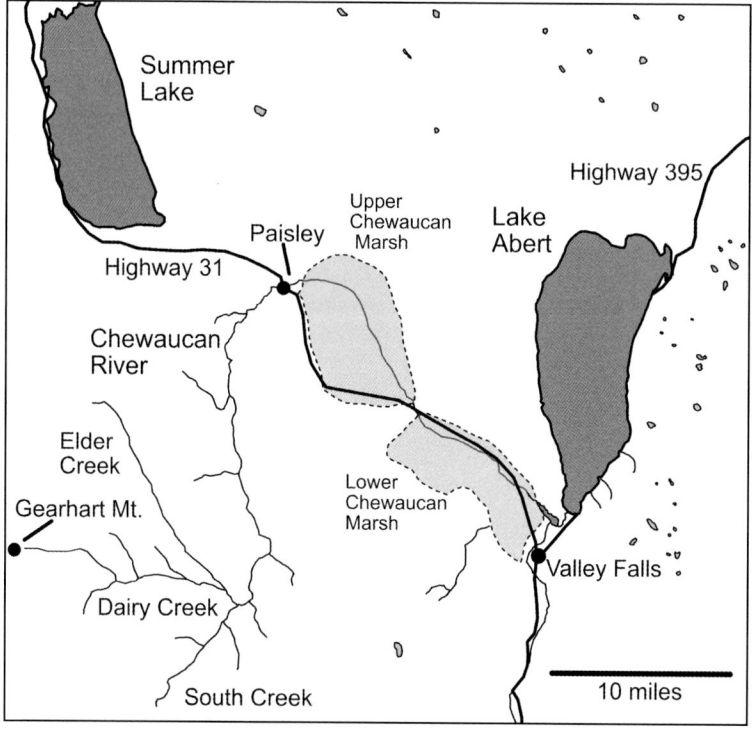

The landscape around Lake Abert, including smaller, unnamed lakes to its east and north.

Gearhart Mountain is the main source of the Chewaucan River, which flows into Lake Abert. The habitat around the mountain includes, *clockwise from top left*: wetlands in the Dairy Creek Cirque and Gearhart Wilderness (at a 7,700-foot elevation), which form the head-waters of Dairy Creek; and plants like the tall marsh marigold (*Caltha leptosepala*), the elephant's head (*Pedicularis groenlandica*), and the swamp onion (*Allium validum*).

Farmers in the Chewaucan Basin focus on producing grass hay, which they water using a flood irrigation system dependent on canals and drains. Flood irrigation requires no pumping, so it is inexpensive. Although providing some habitat benefits, flood irrigation results in uneven distribution of water and causes flooding in low areas, wasting water by allowing it to evaporate.

Before the river reaches the lake, it flows into River's End Reservoir, which developers supposedly built to provide a fish and wildlife habitat. Their unsupervised, harmful work, however, uncovered and disturbed Native American graves, and the reservoir is now a private hunting preserve. A low-head dam backs up water for the reservoir. When there is more water in the system than agriculture

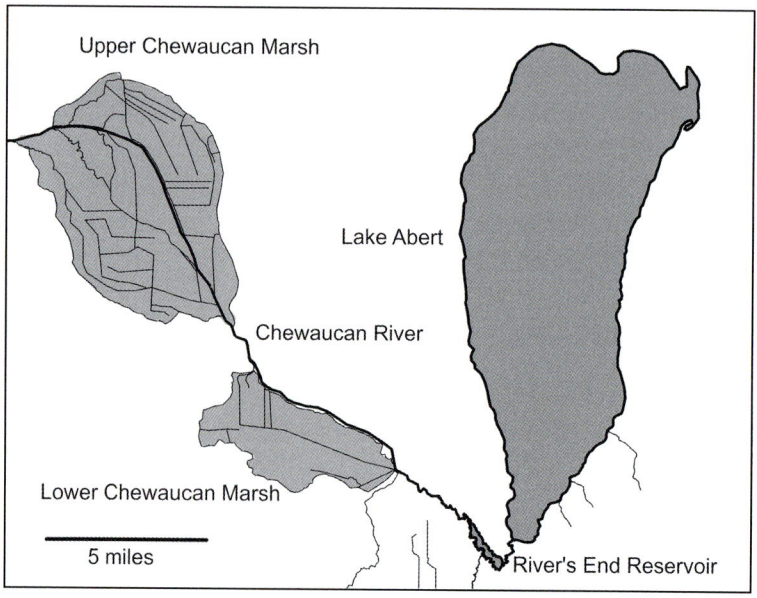

Rapids of the Chewaucan River several miles above Paisley, Oregon. Here the river is lined by aspens, ponderosa pines, white firs, yellow-flowered bitter-bushes (*Purshia tridentata*), willows, and other shrubs.

Upper Chewaucan Marsh

Lake Abert

Chewaucan River

Lower Chewaucan Marsh

5 miles

River's End Reservoir

Proximity of the marshes, reservoir, and river to Lake Abert (lines demarcate the larger irrigation canals and drains).

Chewaucan marsh birds: *left*—male Red-winged Blackbird harassing a Sandhill Crane that got too near its nest; *right*—Long-billed Curlew, a large grassland shorebird with an eight-inch-long bill.

5 miles

2 miles

Landsat images of Lake Abert and the two Chewaucan marshes: *top*— marshes on July 30, 2013, when they were nearly dry; *bottom*—Upper Chewaucan Marsh on April 19, 2014, when flood irrigation turned much of the marsh into impoundments and created several lakes in low-lying fields (the largest was 1,200 acres). USGS EarthExplorer.

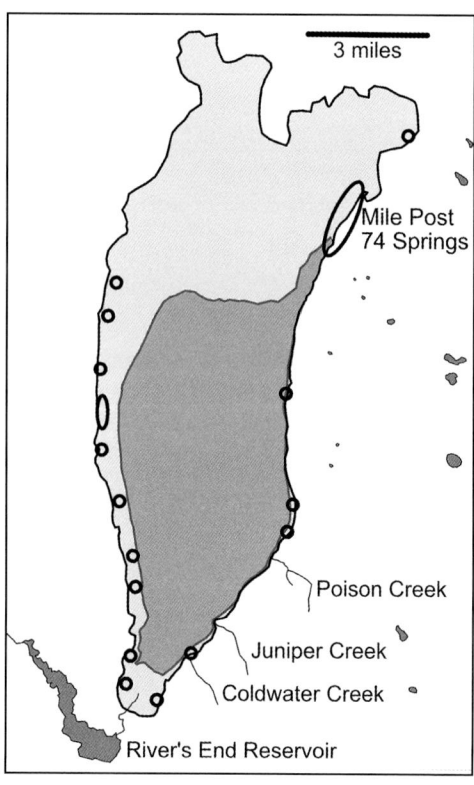

Many of Lake Abert's larger springs (designated by circles) were visible in August 2013 because of low water levels in the lake. Approximate extent of water is shown in darker gray. The lighter gray area represents playa exposed by low water levels. Drawing based on Google Earth image, August 30, 2013.

3 miles

Mile Post 74 Springs

Poison Creek

Juniper Creek

Coldwater Creek

River's End Reservoir

requires upstream and when the reservoir is full, the water flows into the lake at the southern end. However, by July in most years, little or no water reaches the lake. Although the primary source of water for Lake Abert is the Chewaucan River, three small streams sometimes also feed the lake for a short period in the spring. Only one of these, however—Poison Creek—frequently flows above ground and into the lake.

Numerous springs and seeps also drain into the lake. Although most of these have a small flow and only dampen the nearby mud, others, especially those along the northeast shore of the lake, have sufficient-enough flow that during low-lake levels, they form shallow creeks a few inches deep that drain into the lake. Although most springs near the lake are individually small, they provide critical fresh water to the lake ecosystem when the lake is low and salinities are high.

As the water from the springs flows toward the lake, it creates brackish-water marshes. Some of these marshes are lush with plants and provide scarce habitat for sundry invertebrates and birds, such as the Sora (*Porzana carolina*), a robin-sized, wading rail, noted for its ability to make itself so slim that it can slip through dense marsh vegetation (see chapter 6). The very shallow, northeast part of the lake, where the springs concentrate, is also where waterbirds often congregate in the summer and fall.

Historical Background

According to some historians, the first written account of Lake Abert comes from the 1832 journal of John Work, a fur trader working for the Hudson's Bay Company (Maloney 1943). In August 1832, Work traveled to California with a party of twenty-six men. Leaving Fort Vancouver, they proceeded up the Columbia River in three boats, and they then entered the Umatilla River in what is now Oregon, where they turned south. The weather, according to Work, started out "sultry and warm" but then turned unseasonably cold, and he made frequent, daily journal entries beginning with such phrases as "raw cold weather" and "stormy cold weather." Finding water to drink was also difficult, as he pointed out, writing, "Our situation is rather gloomy, the more so as a number of men have become quite discouraged and talk of turning back lest themselves and horses die of thirst."

Historians remain uncertain about the brigade's route after they left the Umatilla and headed overland, south through central Oregon, because Work did not include a map, coordinates, or place names in his journal. In 1942, historian Alice Maloney suggested that they may well have passed by Malheur Lake and by Alkali Lake, before finally reaching a salt lake on October 13 that she identified as Lake Abert. In her notes, however, she gave credence to an alternate theory, pointing out that L. A. McArthur of the Oregon Historical Society believed that the brigade had passed by the Warner Lakes and had not actually arrived at Lake Abert.

Some observations in Work's journal also make me think that he and his men were actually farther east of Lake Abert and that the salt lake he mentioned was one of the northernmost Warner Lakes, possibly Bluejoint, Turpin, or Stone Corral, which are mildly salinized and alkaline. In his October 13 journal entry, he wrote, "The valley where we are encamped…is a continuation of lakes close to each other as far as the eyes can reach tho we proceeded a considerable distance to the southward, and might with propriety be called the valley of lakes, the first below our camp is brackish but in those farther down the water is fresh and good." This "valley of lakes" stretching far into the distance is not an accurate depiction of Lake Abert. Instead, it is a good description of the Warner Lakes, which make up a fifty-mile-long chain consisting of ten large lakes, numerous smaller ones, and extensive marshes, oriented north to south (see chapter 3). Also, the northernmost lakes are, in fact, brackish because most of their water inflow is from Deep Creek near the south end of the basin, so they get progressively more brackish heading north. Therefore, Work should likely get credit for actually writing about the Warner Lakes.

In fact, the first white explorer to see Lake Abert was probably Captain John Frémont. A decade after Work and his brigade had traveled through Oregon's interior on their way to California, Frémont and his corps of explorers passed through South Central Oregon, heading south in December 1843. After making an arduous trek through deep snow, traveling from the west and traversing through the Klamath Basin, they finally reached Winter Ridge, a rim that sits three thousand feet above Summer Lake and the Chewaucan Basin. Describing their journey, Frémont wrote:

Engraving of Captain John Frémont from Henry Davenport Northrop's *Makers of the World History and Their Grand Achievements, 1903,* National Publishing Co., Philadelphia.

December 16th, 1843. We traveled this morning through snow about three feet deep, which, being crusted, very much cut the feet of our animals. The mountain still gradually rose; we crossed several spring heads covered with quaking aspen; otherwise, it was all pine forest. The air was dark with falling snow, which every where weighed down the trees. The depths of the forest were profoundly still; and below, we scarcely felt a breath of the wind which whirled the snow through their branches. I found that it required some exertion of constancy to adhere steadily to one course through the woods, when we were uncertain how far the forest extended, or what lay beyond; and, on account of our animals, it would be bad to spend another night on the mountain. Towards noon the forest looked clear ahead, appearing suddenly to terminate; and beyond a certain point we could see no trees. Riding rapidly ahead to this spot, we found ourselves on the verge of a vertical and rocky wall of the mountain. At our feet—more than a thousand feet below— we looked into a green prairie country, in which a beautiful lake, some twenty miles in length, was spread along the foot of the mountains, its shores bordered with green grass. Just then the sun broke out among the clouds, and illuminated the country below, while around us the storm raged fiercely. Not a particle of ice was to be seen on the lake, or snow on its borders, and all was like summer or spring. The glow of the sun in the valley below brightened up our hearts with sudden pleasure; and we made the woods ring with joyful shouts to those behind; and gradually, as each came up, he stopped to enjoy the unexpected scene. Shivering on snow three feet deep, and stiffening in a cold north wind, we exclaimed at once that the names of Summer Lake and Winter Ridge should be applied to these two proximate places of such sudden and violent contrast.

Thus, Frémont and his men had reached the Chewaucan Basin and were now in the Great Basin. They then proceeded farther east, where they spotted Lake Abert, and the narrative picks up again on December 20:

> Traveling for a few hours down the stream this morning, we turned a point of a hill on our left, and came suddenly in sight of another and much larger lake, which, along its eastern shore, was closely bordered by the high black ridge which walled it in by a precipitous face on this side. Throughout this region the face of the country is characterized by these precipices of black volcanic rock, generally enclosing the valleys of streams, and frequently terminating the hills. Often, in the course of our journey, we would be tempted to continue our road up the gentle ascent of a sloping hill, which, at the summit, would terminate abruptly in a black precipice. Spread out over a length of 20 miles, the lake, when we first came in view, presented a handsome sheet of water; and I gave to it the name of Lake Abert, in honor of the chief of the corps to which I belonged. The fresh-water stream we had followed emptied into the lake by a little fall; and I was doubtful for a moment whether to go on, or encamp at this place. The miry ground in the neighborhood of the lake did not allow us to examine the water conveniently, and, being now on the borders of a desert country, we were moving cautiously.

Charged with mapping the territory (and also set on seeking praise from his superior), Frémont named the lake for his corps's chief, Colonel John J. Abert of the US Bureau of Topographical Engineers, a branch of the Army Corps of Engineers.

Surveyors and Mapmakers

Although surveyors mapped Oregon's coast early in the nineteenth century to aid in ship navigation, they were less interested in mapping the vast region east of the Cascades because of its remoteness. As a result, maps of Oregon's interior from that era were crude and only showed the approximate shapes, sizes, and locations of major geographic features including rivers, lakes, and mountains.

A map from 1863 shows most of the larger mountains, lakes, and stream systems of South Central Oregon, but their shapes, sizes, tributaries, and locations are imprecise, especially those of the Warner Lakes and Warner Mountains, and Hart Mountain does not appear. Also, the location of Christmas Lake, which should be northwest of Lake Abert, is wrong. The shape of Lake Abert on the map is approximately correct, as is the general shape of the Chewaucan River, but the true headwaters of the Chewaucan River

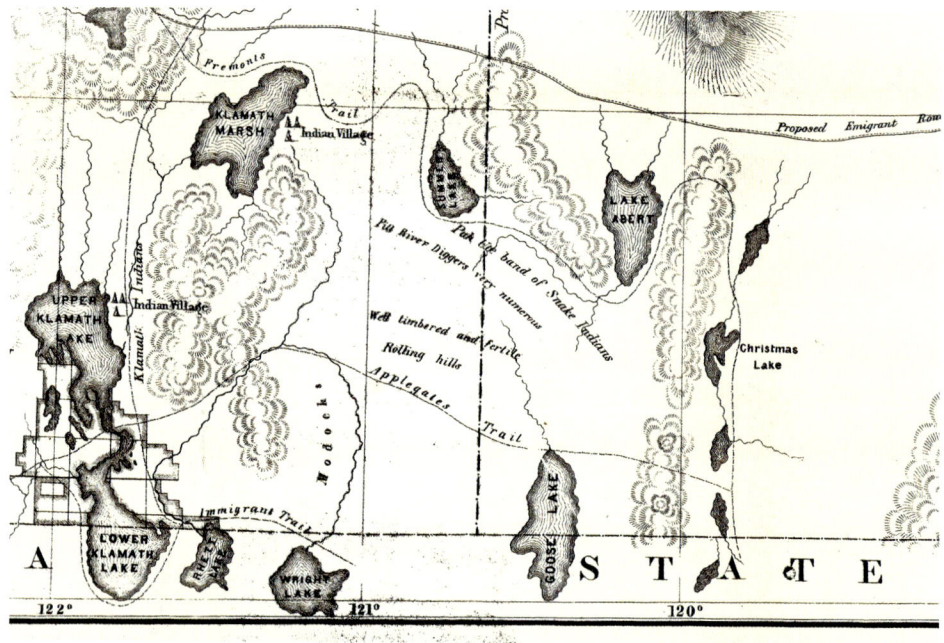

Part of the surveyor general's 1863 map of South Central Oregon. Four years after statehood, the region was still largely terra incognita, except to a few Euro-American explorers and to the remaining Native Americans who lived there.

lie farther to the southwest, and two tributaries flowing into Lake Abert from the north are fictitious.

In the early 1880s, the General Land Office (GLO) sent surveyors out to formally map public-domain lands. The GLO created the Public Land Survey System, a rectangular survey system encompassing townships, ranges, and sections that has been used throughout the West to show property ownership.

You can see the evolution in modern mapmaking techniques by comparing a map of Lake Abert, made by creating a composite from three GLO maps completed in 1881, 1901, and 1921, with a US Geological Survey (USGS) topographic map made more recently by using aerial photos. The GLO map composite is fairly accurate, especially when compared with even earlier maps, but it is noticeably inaccurate at the north and south ends of the lake, when compared with the more recent topographic map.

Although GLO maps were adequate for that period, they were not very accurate and did not include elevations, which are needed for transportation, mining, and other purposes. Consequently, in the 1880s, the USGS began producing topographic maps or "topo maps," also called quadrangle maps or "quad maps." Most of the quad maps were of the 7.5-minute series and were printed at a scale of 1:24,000, which means that each inch of the map equals 2,000 feet of land. The first 7.5-minute quad map of the lake was published

Surveyors' evidence: *left*—1912 GLO marker, located at the ¼ corner of section 36 near Lake Abert; *right*—1948 US Coast and Geodetic Survey reference marker from Abert Rim.

in 1966, with revised maps available in 2011 and 2014. Both historical and more recent USGS maps are available online (https://viewer.nationalmap.gov/viewer/).

The Developers

Interest in Lake Abert and the Chewaucan Basin extended far beyond mapmaking and surveying. In 1902 Congress passed the Federal Reclamation Act to fund development of irrigation projects in the West. Soon after the bill became law, the US Bureau of Reclamation, then called the Reclamation Service, sent out surveyors and engineers to identify reservoir sites in the West, including within the Chewaucan Basin. They selected two reservoir sites on the upper Chewaucan River above Paisley for this project. These reservoirs, had they been developed, would have stored an estimated one hundred thousand acre-feet of water when filled (Shaver et al. 1905). Neither reservoir was built, however, and although some Paisley residents have advocated for additional water storage in the Chewaucan Basin, the US Forest Service has determined that their construction on public land is incompatible with existing uses and values of the federally protected forest. Furthermore, upstream water storage would severely impact the Chewaucan River and Lake Abert in some seasons by reducing inflows, which are already critically low in many years because of drought and water diversions.

The first mention of a proposed development project at Lake Abert was made by USGS scientist Walter Van Winkle in a 1914 paper dealing with the quality of Oregon's surface waters. This proposed project aimed to recover salt, soda, and borax by damming the lake's north end to isolate that part and by allowing evaporation to concentrate the salt. A newspaper ad from that period, published

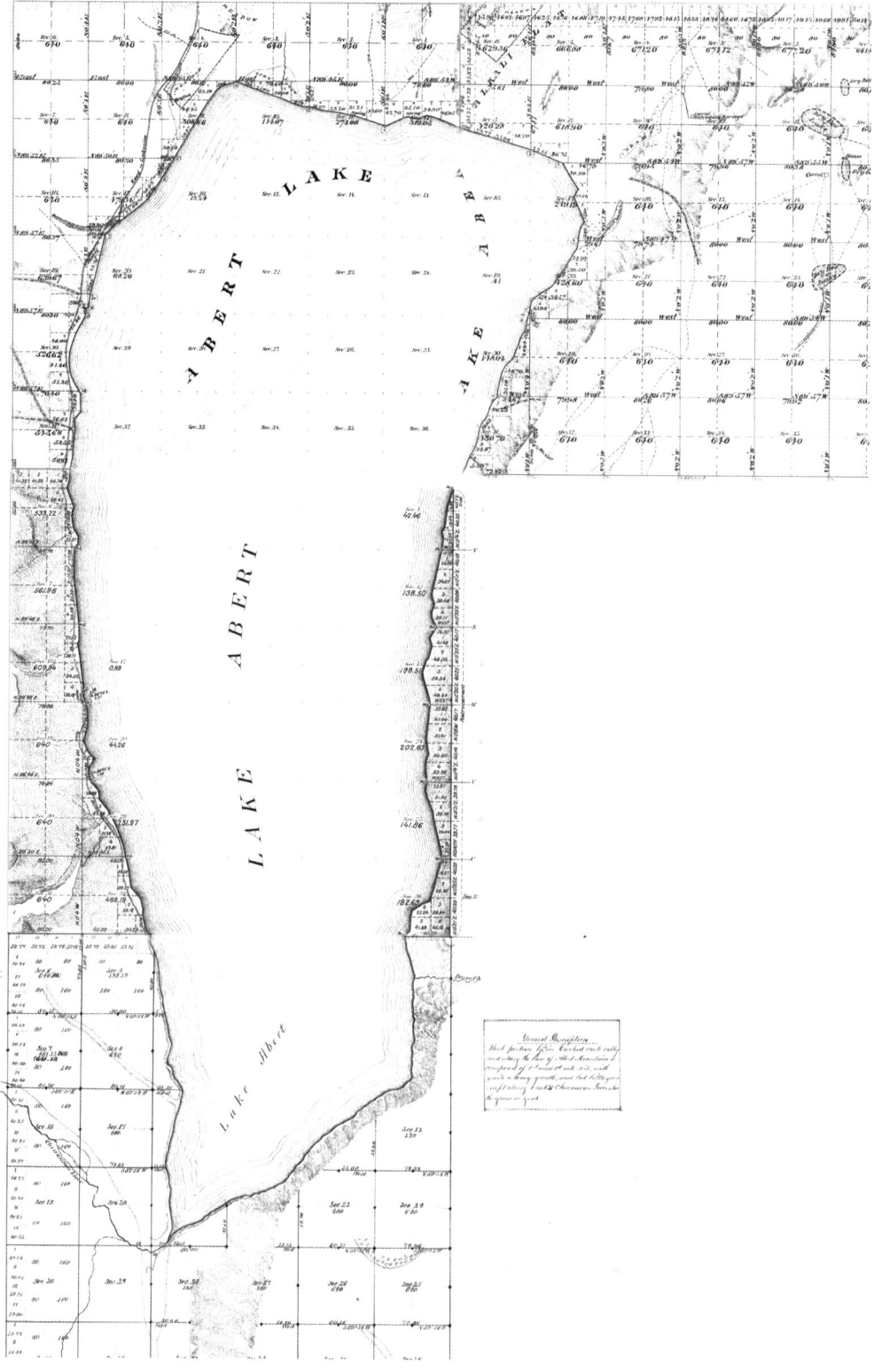

Two maps of Lake Abert illustrating how our view of the West evolved as cartography improved: *this page*—a composite of Lake Abert made from three GLO maps dated 1881, 1901, and 1921; *opposite page*—part of a 1974 USGS topographic map, 1:100,000 scale.

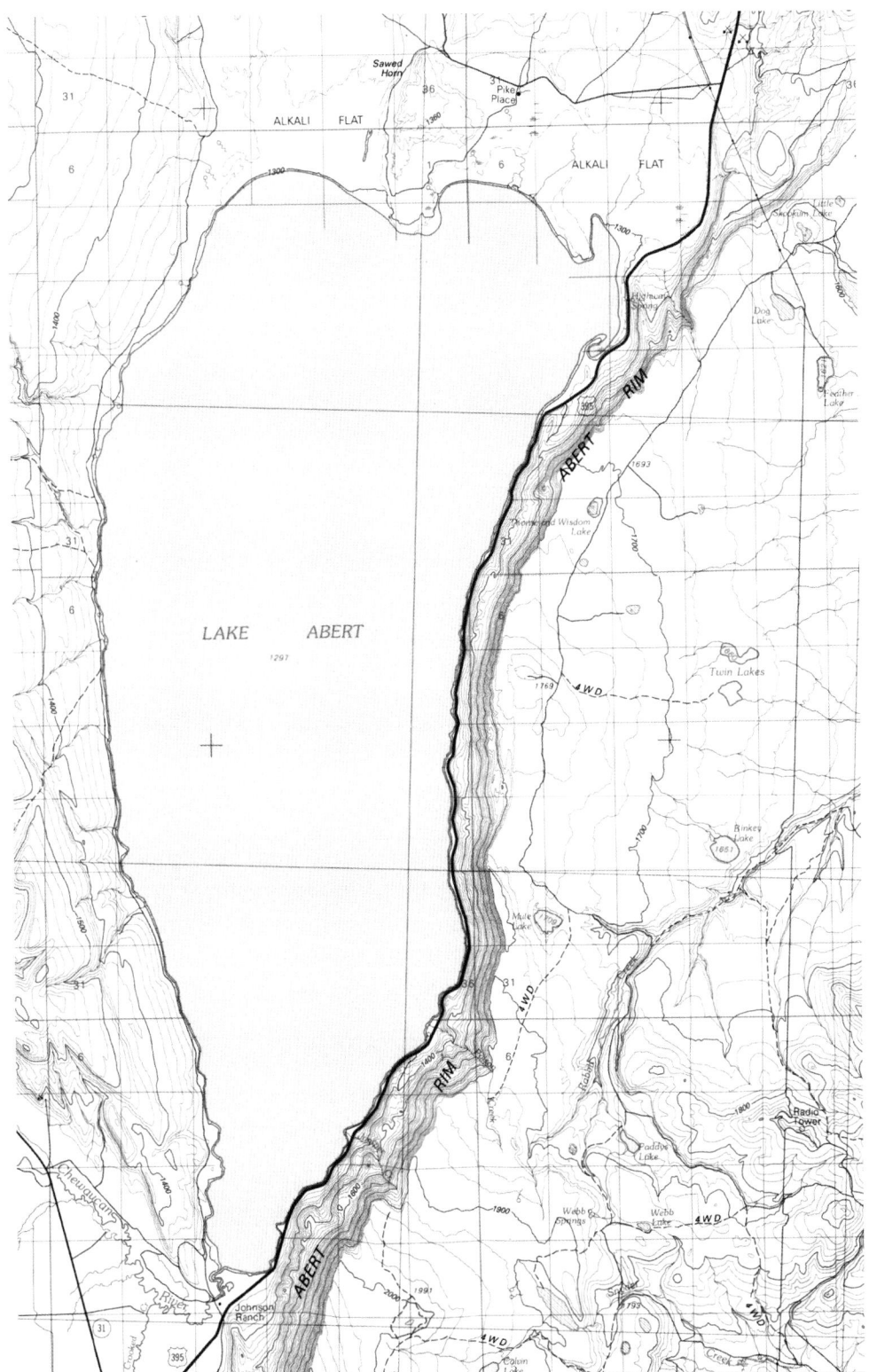

Sawed
Horn

Pike
Place

ALKALI FLAT

ALKALI FLAT

Skookum Little

1300

1300

Highway
Shops

Dog
Lake

ABERT RIM

Feather
Lake

1693

1700

ABERT

Thousand Wisdom
Lake

LAKE   ABERT

1297

1769

4WD

Twin Lakes

1700

Rinkes
Lake
1651

1500

Mule
Lake

1700

4WD

31

1800

RIM

6

4WD

Rabbit

Radio
Tower

1400

Paddys
Lake

1400

1900

6

Webb
Springs

Webb
Lake

4WD

1400

ABERT

2000

1991

4WD

Snow
1793

Chewaucan

River

Johnson
Ranch

31

1293

4WD

Creek

395

Colvin
Lake

Ad published in an early twentieth-century newspaper in Lakeview. I found no evidence that soda, potash, and salt were, in fact, recovered and sold. Jackman and Long, 1964, *The Oregon Desert,* with permission from The Caxton Printers Ltd.

in Lakeview, suggested that the American Soda & Potash Co. would produce 200,000 tons of soda, potash, and salt from Abert and Summer Lakes. However, no later references to this company appear to exist, so it apparently went bankrupt even before processing the minerals.

In 1988, the federal government recommended initiating what is arguably the most bizarre lake project ever envisioned: a plan to measure shock waves resulting from the blast of high explosives. The project was to be part of monitoring exercises called for by the 1963 Nuclear Test Ban Treaty signed by the United States and the Soviet Union. According to this proposal, four tons of explosives would be detonated in the lake, producing a blast one hundred feet high and creating a wave up to two feet high. Fortuitously, because of numerous environmental concerns, the blasting project did not get beyond the initial planning stage.

In the 1990s, a developer proposed the River's End Project for the terminus of the Chewaucan River. The project aimed to dam the outlet of the river to create a reservoir that was supposed to benefit a variety of species, including native redband trout and waterfowl. The Oregon Department of Fish and Wildlife (ODFW) initiated a study of the potential effects to the lake ecosystem from reduced inflows. As a result, in 1992, ODFW biologist George Keister wrote a report updating earlier work by USGS physical scientists Kenneth Phillips and Steve Van Denburgh, who conducted their own hydrologic analysis in 1971. Keister prescribed the optimum salinity and lake-level conditions necessary to sustain the lake ecosystem. His report also made recommendations to limit adverse ecological effects to the lake from the reservoir and put the State of Oregon on record for the first time in recognizing the importance of the lake for migratory waterbirds.

Keister's 1992 analysis was followed up by David Herbst in a 1994 report. Herbst is a University of California research scientist who, because of his studies of the lake's biota, could describe in greater detail the critical water levels and optimal salinities necessary to maintain the lake's productivity. By 1995, however, federal and state governments, largely ignoring the recommendations of both Keister and Herbst, issued permits for construction of the reservoir, and it was built with some public money. An even more serious failure was the unauthorized exposure of cultural remains, including human bones, during construction of the dam by the owner. Unbelievably, the agencies that authorized the project knew full well that there was a high density of indigenous cultural sites in the area.

Because the graves were disturbed, several Indian tribes filed a lawsuit against the landowner and the two federal agencies involved,

Satellite photo showing little or no flow from River's End Reservoir into the south end of Lake Abert. Google Earth, August 30, 2013.

the US Bureau of Land Management (BLM) and the US Fish and Wildlife Service. Federal District Judge Malcolm Marsh, in his ruling on the case, said that state and federal agencies who approved the project deserved a share of the blame, and the landowner was required to mitigate impacts to the cultural site. Jeff Mitchell, spokesperson for the Klamath Tribes, said, "It's not ended, as far as the families of the Yahooskin people who are still alive today [are concerned]. It will probably never end until they can have the opportunity to reinter the remains of their relatives." The Yahooskin are one of three Native American tribes comprising the Klamath people, and they are the descendants of the Northern Paiutes who lived in the Chewaucan Basin until about the mid-1800s, when they were forcefully relocated to the Klamath Indian Reservation.

In 2008, the Federal Energy Regulatory Commission (FERC), the agency responsible for the permitting of hydroelectric facilities, received a proposal for a pumped-storage hydroelectric plant that would use water from the lake to fill a reservoir located on the top of the rim. According to the proposal, lake water would be

North playa of Lake Abert, with scattered greasewood shrubs growing on alkali-encrusted soil.

pumped up to Mule Lake, some 1,300 feet above Lake Abert, via a 6,000-foot-long tunnel. To generate electrical power, the water would be allowed to pass back down through generators during peak-electrical demand periods, when the electricity is more valuable (Federal Register 2008). The proposal failed to mention that Lake Abert is rarely full; that the water is highly saline and caustic and therefore would readily corrode exposed metal parts; that there are numerous archaeological sites which could be destroyed by construction activities; that migratory birds and their habitats could be harmed; and other concerns. Fortunately, the project has not been permitted. In 2021, FERC received yet another proposal for a pumped-storage project that could adversely impact the lake. This project would be located south of Valley Falls and would use water from Crooked Creek and the Chewaucan River to fill and maintain its two reservoirs.

Efforts to Protect the Land

In 1995, the BLM designated Lake Abert as an Area of Critical Environmental Concern (ACEC). According to BLM *Manual 1613,* "ACEC designations highlight areas where special management attention is needed to protect, and prevent irreparable damage to important historical, cultural, and scenic values, fish, or wildlife resources or other natural systems or processes; or to protect human life and safety from natural hazards." Although the designation was an important recognition of this critical ecosystem, the BLM places no special requirements or restrictions on itself or on others who use BLM-managed lands. Other environmental laws, such as the Federal Clean Water Act, the Migratory Bird Treaty, and state laws protecting wetlands, may apply to Lake Abert, but these laws also do not prevent upstream water diversions that remove so much water from the river that it harms the lake ecosystem.

What future lies ahead for Lake Abert and other similar lakes in the Great Basin? In 2014 and 2015, and again during 2020–2022, nearly all terminal lakes (lakes without outlets) in South Central Oregon and adjacent parts of California and Nevada were dry or nearly so. If these conditions become the norm, or just become more frequent, the many species of waterbirds that depend on these lakes to replenish the energy reserves they deplete during migration will be in serious trouble. Where will they go?

CHAPTER 2

Geology

What the Rocks and Landscape Tell Us

The major part of the North American Continent is drained by
streams flowing to the ocean, but there are a few restricted
areas having no outward drainage. The largest of these was
called by Frémont, who first achieved an adequate conception
of its character and extent, the "Great Basin," and it is still univer-
sally known by that name. It is not, as the title might suggest, a
single cup-shaped depression gathering its waters at a common
center, but a broad area of varied surface, naturally divided into
a large number of independent drainage districts.

—G. KARL GILBERT, 1890

The history of Lake Abert and the other lakes in the northwestern Great Basin is part of the regional history that characterizes the larger hydrographic Great Basin and the even bigger
Basin and Range physiographic province. The Basin and Range is
exactly that—a vast region of the West, some 350,000 square miles,
or nearly 10 percent of the entire area of the United States, and it
extends from Oregon into Mexico and from California to Utah.
This region contains over four hundred short and nearly parallel
mountain ranges separated by broad, nearly flat basins, with an elevation range of up to seven thousand feet between the basins and
the mountain ranges, and with several peaks rising above twelve
thousand feet.

Within the Basin and Range is the Great Basin, an arid region
where almost every wet spot, no matter how insignificant, is important. Geologist G. Karl Gilbert, in his monumental 1890 report on
the origin of Lake Bonneville, attributed the name *Great Basin* to
Frémont, who decided on this appellation because of the region's
internal drainage, which prevents surface water from reaching the
sea. Ancient lakes like Bonneville, called paleolakes or pluvial lakes,
existed in the Great Basin up until nearly the end of the Pleistocene
epoch, about ten thousand years ago, after which a drying climate
caused most of them to desiccate and disappear.

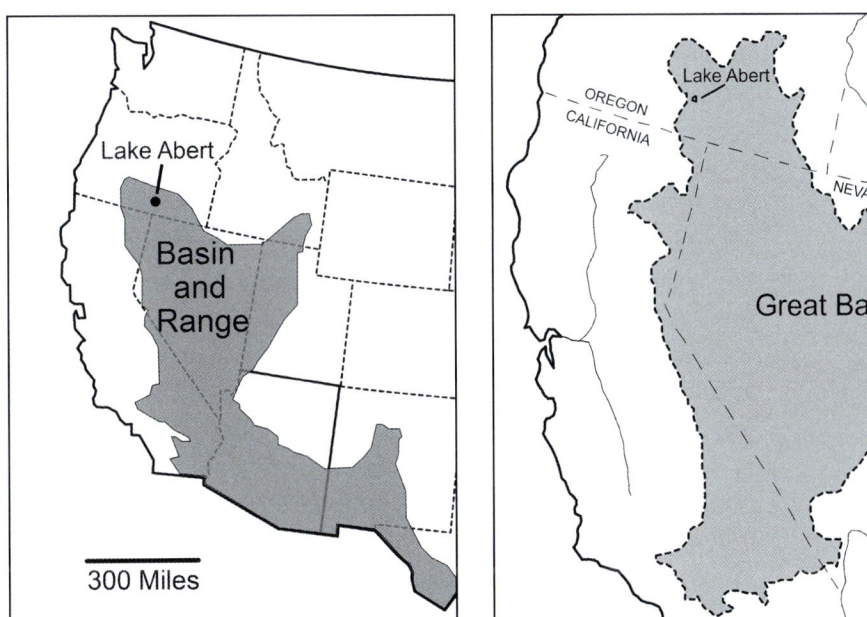

Exposed, cracked, and alkali-covered mud in 2014 at the south end of Lake Abert, with Abert Rim rising steeply to the basaltic palisade at its crest. Numerous lava flows created the Steens basalt that forms the rim. Landslides produced the low humps along the lower half of the slope.

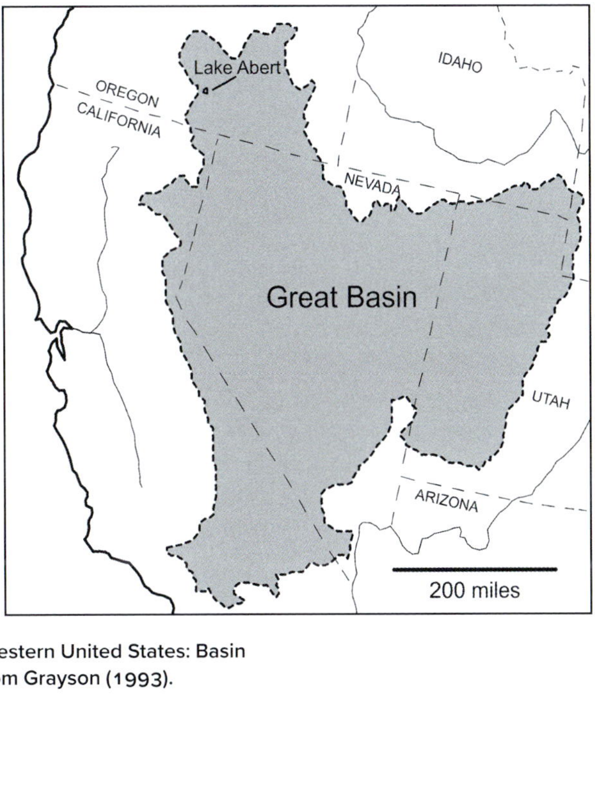

Boundaries of two physiographic provinces in the western United States: Basin and Range (*left*) and Great Basin (*right*). Redrawn from Grayson (1993).

USGS geologist Israel C. Russell: *left (seated at the far right)*—doing field work in Alaska in about 1890; *right*—serving as a professor at the University of Michigan, where he worked from 1892 until his death in 1906. Courtesy of the USGS Archives.

As Gilbert stated, the Great Basin actually comprises numerous, internally drained basins of widely varying sizes. It covers approximately 200,000 square miles and stretches southeast from Oregon to Utah and south to Southern California, and it includes most of Nevada. The Chewaucan Basin in which Lake Abert sits is one of the smaller basins, with an area of 650 square miles.

Historical Background

Geologists were among the first scientists to explore and report on the Great Basin, including in South Central Oregon. They discovered that the faulting which had created Lake Abert and Abert Rim had exposed rocks, and the rocks' exposure made them easier to study and also made the geological processes that had created them more apparent. The first geologist to write about the lake and rim was Israel Russell. In 1882, at the age of thirty, he traveled alone on horseback around South Central Oregon, interpreting the region's geology for the USGS. He did field work in the summer and spent the winter in Washington, DC, writing his reports. While in the District of Columbia, Russell helped to establish the National Geographic Society and became its first expedition leader.

A year earlier, Gilbert had first sent Russell to the western Great Basin to study its features. Gilbert directed the Great Basin Division of the USGS and was one of the first geologists to report that Basin and Range landscapes largely consisted of fault-block mountains and intervening basins that had contained paleolakes. He saw evidence of ancient shorelines around the Bonneville Basin and likely shared those observations with Russell. So, when the younger

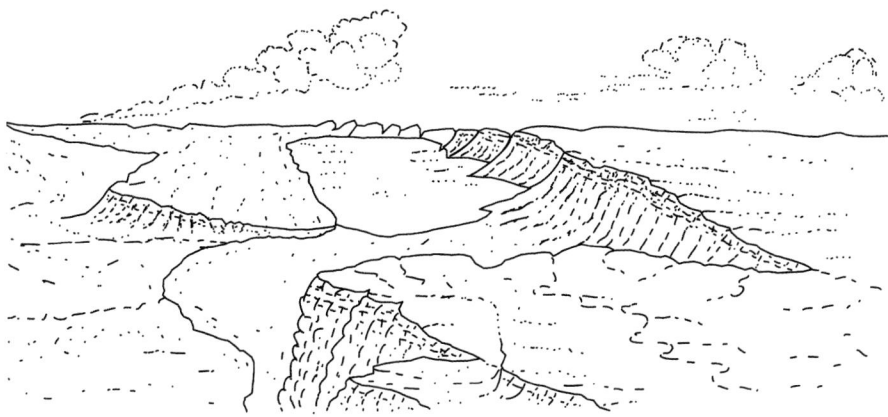

Sketch based on Israel Russell's 1884 drawing of Lake Abert, Coglan Buttes, the Chewaucan River, and Crooked Creek (the view is from the south along Abert Rim, east of the Valley Falls).

geologist first traveled to Oregon, he already had some knowledge of how the Great Basin landscape had formed.

Russell's accounts of South Central Oregon, which mix stiff, geological jargon with the colorful, scientific-travel narrative typical of that era, have a certain charm. From his writing, you sense how difficult travel was then, well before roads were built, but also how that same terrain greatly fascinated the young geologist. His observations constituted the first detailed description of what the landscape was like around Lake Abert. In his 1884 report titled *A Geological Reconnaissance in Southern Oregon,* he described Lake Abert and Abert Rim colorfully:

> The grand cliffs that present an impassable barrier along the eastern shore of Lake Abert expose the broken edges of strata on the heaved side of the fault; while the thrown side underlies the lake and forms the gently sloping western shore…. The palisade rising abruptly from the eastern shore of Lake Abert has a deep brownish tone, relieved by a growth of bright yellow lichens, and is grand in its barren ruggedness, especially towards evening, when its inherent richness of color is heightened with sunset tints.

Russell apparently climbed Abert Rim, not far from the village of Valley Falls, and made a sketch of the lake and rim that accurately captured the lay of the land just as it is today.

In his report, he described the region's geological landscape:

> From Silver Lake our explorations led us southward along the bold western shore of Summer Lake, and thence through the extremely rugged region embracing the Chewaucan Marsh, Abert Lake, Goose Lake, and Warner Lakes. The wild and all but impassable character of this remarkable lake region renders it difficult of exploration,

but at the same time of great geological interest. The ruggedness of its topography, and the abruptness of the precipices enclosing the valleys, is mainly due to orographic displacement. The numerous fault-scarps form precipitous palisades across the country, from a few hundred to 1,500 or 2,000 feet in height, trending approximately north and south, and sometimes unbroken by passes for a distance of 50 miles. The depressed blocks give rise to rock-basins, some of which are occupied by lakes that mirror on their placid bosoms the rugged grandeur of the surrounding walls. A few of the lakes are remnants left after long concentration by evaporation; while others, of even greater interest, are fresh, but occupy ancient basins from which the waters never overflowed.... This lake region is not only remarkable for the ruggedness of its scenery but furnishes a unique example of displacement on a grand scale, and at such a recent date that the fault-scarps still form beetling cliffs, that appear to have been upraised but yesterday.

Historian Patrick Sylvestre (2008) characterizes Russell as an "aesthetically conscious scientist," unlike most explorers of that age, who primarily concerned themselves with how to exploit nature. In 1885, Russell went on to write about paleolake Lahontan in Nevada, the second largest Pleistocene lake in the Great Basin. His report and Gilbert's 1890 report on paleolake Bonneville are considered to be the two most important nineteenth-century papers on Great Basin geology (DeCourten 2003).

Not long after Russell explored the northern Great Basin, another seminal geologist, Stanford University graduate Gerald A. Waring, made some major contributions. Hired by the USGS to assess the water resources of South Central Oregon, he also traveled alone, both on horseback and in a four-wheeled wagon called a buckboard. In 1906, he received a daily payment of $3, and he apparently had no expense account, so he covered all his own traveling costs with his meager salary. Waring started his reconnaissance by "buying a cheap horse, saddle and buckboard" (White 1971). His 1908 description of the area around the lake and rim was one of the first to refer to the region as a high desert:

Although the northern and eastern part of the area studied has a character of a broken plateau, one may travel in some directions for many miles in the level sandy lake valleys or over the approximately level rocky "high desert" without crossing more than an occasional depression. But the chief features that relieve the monotony of the region are the great scarps that have given it a broken character. These trend generally north and south, and the four principal lines border the principal lake valleys, or undrained basins…along the edge of Lake Abert it forms a very striking cliff that rises from the water's edge to a height of fully 2,000 feet, the upper 600 feet being nearly perpendicular.

Detailing the region's geological structure, Waring wrote:

> Probably in but few other places in the world is the geologic structure so well exhibited in the present land forms as it is in southeastern Oregon…. In most of the Great Basin region the typical Basin Range structure produced by faulting and tilting of long, narrow orographic blocks, is largely obscured by erosion or by earlier complex structures. But in Lake County erosion has acted very little on these great blocks, and little or no deformation proceeded the faulting, so that the typical structure is evident in the present conformation of the surface.

This dramatic exposure of huge, relatively young fault scarps, written about so strikingly by both Russell and Waring, continued to attract geologists to the region well into the twentieth century.

Lake Abert's Origin

The geologic history of Lake Abert likely goes back several million years. Like all of the larger lakes in the Great Basin, Abert was formed by tectonic movements of the earth's crust that caused the region to very slowly stretch, resulting in upward movements of the ranges and in down faulting of the basins. Lake Abert's basin tilts down toward the east and up toward the west, due to the block's clockwise rotation. The up-thrusted, or heaved, side of the crustal block, called a *horst,* forms Abert Rim, and the down-thrown block (or possibly blocks), called a *graben,* forms Cogland Buttes and the basin in which the lake sits. The block's arrangement under Lake Abert is called a half graben because only the west side is up-thrusted. By comparison, Goose, Summer, and Upper Klamath Lakes sit in full grabens.

Geologists believe that the Abert Rim fault became active approximately seven million to eight million years ago. No one knows when Lake Abert's basin began filling with water, but even if this process started several million years after the faulting began, this timing would still make the lake one of the oldest in North America. As a result of the rotation of the basin, several thousand feet of sediment have accumulated below Lake Abert on its eastern side (Scarberry 2007; Wagner 2007; Scarberry et al. 2010).

Lake Abert's origin intertwines with that of the rest of the Great Basin, and, in fact, with that of the entire Basin and Range.

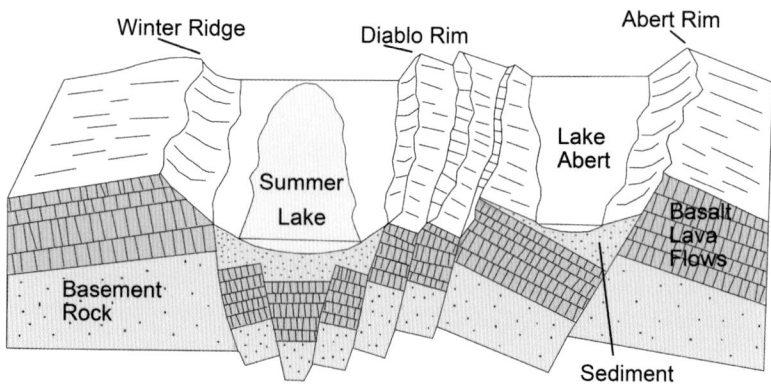

Possible arrangement of rotated, crustal blocks forming Abert Rim, Winter Ridge, and the basins containing Summer and Abert Lakes. Lake Abert sits in a basin tilting up to the west, while Summer Lake's basin is the result of the downward movement of multiple blocks forming a graben. At Abert Rim, the Steens Basalt lava flows, which happened between 16.8 million and 16.1 million years ago, covered an older basement of igneous or sedimentary rocks.

Geologist Frank DeCourten, in his 2003 book titled *The Broken Land: Adventures in Great Basin Geology,* refers to this region as "the broken land" because of the great geological upheavals that produced the basins and ranges. DeCourten describes the Great Basin as a "dome-shaped region, broken like a pile of bricks that have fallen from an ancient arch that extended across the region.… The dome cleaved into hundreds of blocks.… Some blocks foundered lower than others to become the foundations of the basins, while the high-standing blocks gave rise to the mountains." The collapse of the dome created a topography for the Great Basin in which all streams drain into isolated, smaller basins. DeCourten suggests that sometime in the far distant future, the stretching of the Basin and Range could break the continent apart, allowing an inland sea to form.

In their observations during the late-nineteenth and early-twentieth centuries, Russell and Waring noted that there are few places where the geology is as evident as it is at Lake Abert. The impressive, two-thousand-foot-high Abert-Rim fault scarp comes into view long before you can see the lake, no matter which direction you travel from. Prominent, near-vertical fault scarps are a characteristic feature of young fault-block mountains, including Abert Rim, and are the upturned edges of a crustal block forming an obvious cliff, as is visible along much of the rim. According to geologists Silvio Pezzopane and Ray Weldon (1993), Abert Rim is still elevating at an approximate rate of several inches per century.

Depending on how you define it, the fault scarp that includes Abert Rim is unbroken for nearly 30 miles, north to south, and can

Abert Rim, just south of Abert Lake, with its nearly vertical palisade, the result of numerous lava flows. Erosion has dissected the cliff into triangular buttresses that make it look like a row of pyramids.

Former historical marker that described the origin of Abert Rim and which once stood along Highway 395 above the lake.

OREGON GEOLOGY
ABERT RIM

DISCOVERED BY CAPTAIN JOHN C FRE-
MONT DECEMBER 20, 1843, AND NAMED
IN HONOR OF COLONEL J. J. ABERT

ABERT RIM, SOME 2500 FEET ABOVE
THE VALLEY FLOOR IS ONE OF THE
HIGHEST FAULT SCARPS IN THE
UNITED STATES. THIS BASALT FORMA-
TION, A BASIC LAVA COVERING MUCH
OF EASTERN OREGON, ISSUED FROM
GREAT FISSURES DURING THE MIOCENE
PERIOD OF GEOLOGIC HISTORY MANY
MILLIONS OF YEARS AGO. AFTER THIS
EXTENSIVE LAVA FLOODING, THE EARTH'S
CRUST WAS FRACTURED AT MANY POINTS
AND GREAT BLOCKS WERE TILTED. ABERT
RIM IS THE WESTERN EDGE OF ONE OF
THESE TILTED BLOCKS WHILE ABERT
LAKE LIES ON TOP OF ANOTHER.

be traced all the way to the western edge of the south Warner Mountains in California, 80 miles distant (Fuller and Waters 1929; Scarberry 2007). For comparison, the larger Steens Fault Zone, the fault scarp comprising Steens Mountain and the adjacent Pueblo Mountains, is nearly 125 miles in length, and Winter Ridge, or Winter Rim, which is located west of Abert Rim, consists of a 35-mile-long scarp. Hundreds of other faults are also present in the northwestern area of the Basin and Range (http://earthquakes.usgs.gov/hazards/qfaults).

Steen's Basalt

Abert Rim consists mostly of basalt, which is a dark, often black, fine-grained, igneous rock that is low in silica and comes to the surface as lava. Basalt is one of the most common rocks on Earth and constitutes much of the ocean floor's bedrock. It also occurs elsewhere in the solar system, on rocky planets including Mars and Venus and in dark areas of our moon called *mares*. Basalt has also been detected on the surface of an unusual asteroid named Vesta, which is the second largest object in the asteroid belt between Mars and Jupiter. Vesta has been called the "smallest terrestrial planet" because of its similarity to its two neighbors (Keil 2002).

Basaltic lava is the most fluid type of lava; however, its viscosity is still quite high, one hundred thousand times that of water. Consequently, basaltic lava flows slowly across level ground at less than one mile per hour, but it can still cover large distances before cooling. Nonexplosive, basaltic fissure eruptions can spread outward as *flood basalts,* and when they produce a large volume of basaltic lava, they can cover vast areas. A good example of this is the Columbia River Flood Basalt province, which covers over sixty thousand square miles of Oregon and Washington. Elsewhere in the world, even larger flood basalts are present, especially under the oceans.

Scientists have traced the source of the lava that formed most of Abert Rim to volcanic vents, or possibly to a shield volcano located near Steens Mountain, 80 miles east. From about 16.8 million to 16.1 million years ago, several hundred lava flows issued from these vents and eventually covered 20,000 square miles, depositing a total of more than 3,000 feet of basalt in some places (Jarboe et al. 2008; Camp et al. 2013; Mahood and Benson 2017). Apparently, most of the lava erupted over a short geological time period about 16.5 million years ago.

Geologists recognize that what they refer to as the Steens Flood Basalt province is part of the even larger Columbia River Flood Basalt province, created by lava flows from vents near the border of Idaho and Oregon and extending as far west as the Pacific Ocean. The Columbia River Flood Basalt flows were so extensive and thick

Approximate area covered by Steens and Columbia River flood basalts. Redrawn after Reidel et al. (2013). The western extent of Steens basalt in the vicinity of Lake Abert is unknown.

that some geologists jokingly complain that the lava covered all the interesting rocks, which remain buried below, according to Marli Miller in the second edition of the *Roadside Geology of Oregon* (published in 2014).

Looking at the cobble rocks on the Lake Abert shore, one can see that Steens basalt comes in a variety of colors, ranging from dark gray to russet or tawny. Although mostly massive in form, some of the rocks are filled with numerous small cavities formed by gas bubbles. Steens basalt is distinctive, often containing large crystals called *phenocrysts,* which consist of *plagioclase feldspar,* a silicate mineral that forms light-gray bundles or laths, up to two inches long, or snowflake-like clusters of radiating crystals. This mineral constitutes up to half the volume of a Steens basalt rock, and its characteristic appearance has led to its nickname, *turkey track basalt.* Plagioclase is also found on the moon and on the surface of Mars, where it is the most abundant mineral.

Other volcanic rocks, called *tuffs,* also helped form Abert Rim. Tuff is a sedimentary rock composed of volcanic ash and small pieces of *scoria*—cinders—that fuse together and which were discharged from nearby explosive volcanic eruptions. One can find

Turkey track basalt: an example of Steens basalt from the Lake Abert shore-line, consisting of numerous, light-colored laths of plagioclase feldspar forming turkey-scratch or snowflake-like patterns. The lens cap, shown for comparative size, is about two inches across.

Fenestrated outcrop of wind- and water-weathered, reddish, volcanic tuff, which is about twenty feet high and located on the lower slope of Abert Rim near Poison Creek.

exposed reddish tuffs, with a texture like course sandstone, part-way up the slope of Abert Rim near Poison Creek and at several other locations along the face of the rim, where they form promi-nent outcrops intricately fenestrated by wind and water into arches, cavities, and even small caves.

Sunstones

Several miles north of Lake Abert, other basalt flows produced a sometimes beautiful, precious gem called *Oregon sunstone,* which now sells for as much as one thousand dollars per carat for a large, colorful, high-quality specimen (Pay et al. 2013). Oregon sunstone, or *labradorite,* as the mineral is called, is similar in chemical com-position to the plagioclase feldspar that makes Steens basalt so dis-tinctive, and both may have originated in the same volcanic magma chamber, before coming to the surface in a lava flow.

Sunstones are translucent and come in a variety of hues, includ-ing nearly colorless, red, green, and multicolored, but most are straw-colored and resemble glass shards or nuggets—up to 4 inches across—rather than flat crystals. Labradorite crystals form as the lava cools, and larger crystals form from thicker lava flows that cool slowly, sometimes over many decades when the flow is very thick. The best Oregon sunstones come from mines located north of Lake Abert, but the BLM manages the public Oregon Sunstone Collecting Site, located 15 miles east of Lake Abert on BLM Road #6115. Most of the sunstones at this site are small, about 0.4–0.5 inches (10–12 mm) long, range from transparent and colorless to pale yellow or orange, and resemble broken glass fragments.

Sunstones resembling glass shards, from the BLM's Sunstone Collecting Site near Lake Abert (scale is twenty-five mm per inch).

Magnetic Reversals Recorded in Basalt

Another interesting aspect of basalt is its ability to record the polarity of Earth's magnetic field at the time that the lava solidifies. Iron oxide particles suspended in fluid lava orient themselves, like microscopic bar magnets, with the Earth's magnetic field. Later, geologists can measure that orientation with sensitive instruments, after the lava has cooled and formed rock. The capacity of lava to record the direction of Earth's magnetic field has been one of the key sources of evidence for plate tectonics and has also provided proof of magnetic reversals, where the north and south magnetic poles have switched positions. The thick basalt flows forming Steens Mountain were produced over a 1,000-year period and recorded one magnetic reversal that happened approximately 16.6 million to 16.7 million years ago, called the Steens Mountain Reversal (Jarboe et al. 2008).

Gravity Eventually Prevails: Landslides and Rock Falls

Boulders of Steens basalt, some as large as small vacation trailers, are conspicuous at the base of Abert Rim and even out in the lake. Some of these boulders, which tumbled down from the rim, are likely evidence of the seismic activity that occurred when the crustal plates moved, creating rock falls and large landslides. In fact, much of the lower slope of Abert Rim comprises materials from landslides, especially at the south end. Visible evidence of these slides also appears on the rim as the notches where Coldwater, Juniper, and Poison Creeks flow. The largest was the Poison Creek landslide.

Proof of an even larger landslide exists south of Valley Falls and east of Highway 395, where there is a 1.5-mile-wide, crescent-shaped notch in Abert Rim. Over 4 miles west of the notch are two similarly shaped, conical, twin hills that rise to several hundred feet. Both the presence of the hills and the concave notch in the rim are confirmation of a huge landslide that sent rock and other debris traveling several miles across the valley floor. Such landslides can go surprisingly long distances because they ride over a cushion of air, a process that reduces friction (Badger and Watters 2004).

Landslides also created seven large scallops, visible on USGS topographic maps, located west of Abert Rim at nearby Winter Ridge above Summer Lake. Several of these slides traveled over a mile out onto the basin floor (Badger and Watters 2004). One of the largest of these landslides produced a steep scarp, appropriately named Slide Mountain. Earthquakes likely caused these slides, while simultaneously causing the landslides at Abert Rim. Evidence

Poison
Creek

Juniper
Creek

Coldwater
Creek

500 feet

Evidence of landslides and rockfalls from Abert Rim: *top*—view south across the lake's eastern shore showing prominent hills—formed by landslide debris from the Coldwater, Juniper, and Poison Creek slides—below the rim, as well as large boulders that rolled down (a staff gauge used to measure lake levels has been attached to one of these boulders); *bottom*—view north along Abert Rim showing how the three major drainages—Coldwater, Juniper, and Poison Creeks—were formed by landslides near the south end of the lake, and also how the landslide debris that slid down the slope piled up near the lake's edge. Google Earth.

also shows that landslides from Winter Ridge entered the Summer Lake basin, and if this happened when water levels were high, they would have triggered small tsunamis that could have moved anything mobile in their path, including sediment and plants.

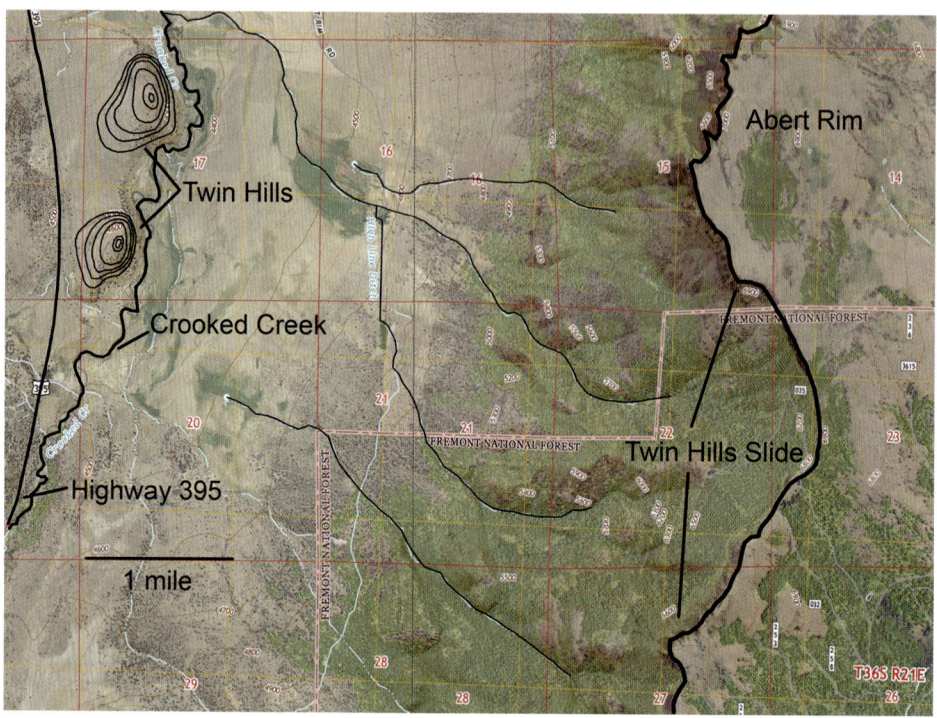

The Twin Hills, located south of Valley Falls near Highway 395. These conical hills were likely formed by a large landslide that carried debris west from Abert Rim for over four miles.

Desert Varnish

Rocks in the Great Basin and in other arid regions of the world often have a very thin, black or brown, glossy coating called *desert varnish* or *rock varnish* (Perry and Sephton 2008; Parsons and Abrahams 2009). Throughout history, humans have scratched petroglyphs onto boulders, which are visible because the carvers cut through microscopic layers of varnish that coated the rocks' surfaces. For decades, geologists have debated about how desert varnish forms, and they continue to acknowledge some uncertainty (Parsons and Abrahams 2009). The discovery of possible rock varnish on Mars has generated further interest in this subject (Perry and Sephton 2008).

Desert varnish contains manganese and iron oxides, and sometimes silica, and there are often associated microorganisms and organic compounds present. The recent discovery of bacteria that can precipitate manganese provides support for a microbial role in varnish formation. Nevertheless, further work is needed to determine whether this biological process is widespread (Northup et al. 2010). Water also seems to be involved in the development of desert varnish because rocks in ephemeral stream channels and around the edges of playa lakes often have so much varnish that they appear black. What we know for sure is that this substance forms slowly

Desert varnish on basaltic rocks: *left*—varnish with a brownish, patina-like finish on a three-foot-high boulder located above Lake Abert, with circular petroglyphs scratched into it; *right*—black varnish on rocks and boulders that are up to two feet across and which lie in an ephemeral stream channel near Egli Rim, Lake County.

over thousands of years, a fact that becomes evident when you observe the petroglyphs at the lake, which still have little visible varnish on their etched surfaces. Some research suggests that geologists can accurately assess the age of desert varnishes by using a relatively new technique called *varnish microlamination dating,* which involves studying the buildup of the varnish by shaving thin layers and studying their chemical makeup under a microscope. This process has great potential for determining not only the age of petroglyphs but also the time frame for broader geological events (Liu and Broecker 2013; Whitley 2013).

Desert Pavements, Patterned Ground, and Moving Rocks

Cold deserts, including those in South Central Oregon and in adjacent areas of the Great Basin, provide an environment conducive to the formation of some unusual soil features. Two such features are particularly prevalent on Abert Rim and at the north end of the lake, where visitors will find large areas of clay-rich soils strewn with an abundance of rock fragments.

The first of these unusual features is called *desert pavement* and consists of soil covered by randomly scattered gravel. Geologists generally believe that frost heaves create desert pavement. Frost heaving forces the rock fragments upward in the soil, and the sorting action of the wind gradually removes finer soil particles. Some geologists suggest an alternate theory, believing that moisture

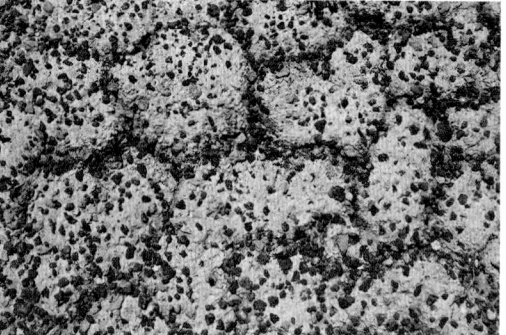

Desert-soil phenomena: *left*—desert pavement almost completely covered by small, angular rocks; *right*—patterned ground where pebbles are organized into hexagon-like networks. Both images show two-to-three-foot areas.

changes in the Great Basin's clay soil induce shrinking and swelling that forces rock to the surface. Because we know that the Lake Abert region is windy and that clay soils predominate, it seems likely that both processes help create desert pavement.

The second prevalent soil feature is called *patterned ground,* and it is characterized by pebbles sorted into polygonal networks, rather than by the randomly scattered gravel found in desert pavement (Kessler and Werner 2003). Geologists mostly find patterned ground in the Arctic or in high-elevation, cold environments, where a variety of stone stripes, circles, and polygons are present, ranging in size from a few inches to dozens of feet across. Scientists have long known that sorted, patterned ground results from repeated soil freezes and frost heaves, but a general model to explain all forms of this soil feature has been elusive.

However, in 2003, geologists Mark Kessler and Brad Werner developed a model in which polygonal networks grow in a two-step process. First, ice forms in the soil, causing it to expand upward toward soil-rich, central areas and forcing the rocks toward the rock-rich peripheral zone. Second, as the freezing soil expands, it squeezes the rocks and confines them into narrow, marginal areas. The result is a central area composed mostly of soil and bordered by a polygon of rocks. When this occurs over an area of several square feet, a network forms. Apparently, patterned ground is also present on other rocky planets: the National Aeronautics and Space Administration (NASA) has identified it in photos taken of some high-latitude areas on Mars (NASA 2008).

A third unusual, geomorphic soil feature, called *ice rafting,* is also present on the Lake Abert playa (a part of the basin that is generally dry but sometimes refills) but is much more visible at Summer Lake. Ice rafting is characterized by the presence of small

Ice-rafted boulders on the Summer Lake playa. The largest is approximately three feet long and weighs several hundred pounds, and it has a different texture from that of the other boulders, which suggests that it did not come from the same source.

boulders weighing hundreds of pounds and located far from shore on playa-lake sediments.

Out on the Summer Lake playa, numerous baseball-sized rocks are visible far from shore. They are called *drop stones* because ice moved them there, dropping them into the deep water while melting. More unusual are small boulders that weigh several hundred pounds and which are located too far—a mile or more from the nearest slope—to have rolled downhill and out onto the mud. The best explanation for the location of these boulders on the playa is that they were ice rafted to their present location.

Ice rafting can occur under at least two sets of circumstances. First, it can happen when rocks roll downslope onto the ice, which then transports the rocks to a new location. Second, the rafting can occur when ice forms around rocks in shallow water and lifts them. The wind then moves the ice to another site, the rocks become grounded, and when the ice melts, the boulders are left behind. Rafting, which moves rocks weighing several hundred pounds from shallow water, requires ice that is sufficiently thick and water that is deep enough so that the friction of the rock on sediment is overcome (Lorenz et al. 2011). However, a light wind can easily move the rocks, once they are lifted clear of the lake bottom.

The best examples of ice-rafted boulders in Oregon are the very large, glacial *erratics* (large, displaced rocks) found in the

Willamette Valley, and in particular, a forty-ton boulder at Erratic Rock State Natural Site near McMinnville. Geologists believe that the Missoula floods—catastrophic ice-dam outbursts that surged across western Montana, northern Idaho, eastern Washington, and western Oregon—ice rafted this shale boulder hundreds of miles downstream in the Columbia River system from Montana or Alberta, Canada, fifteen thousand years ago (Miller 2014). The floods spread across the land when the reoccurring ice dams that formed Lake Missoula (in what is now western Montana) broke, which happened about once every fifty years over a two-thousand-year period.

Another possible explanation for the presence of rocks far out on the playa may be that ice has shoved the stones there. This unusual process, called *sliding rocks,* has gotten a lot of attention at Racetrack Playa in Death Valley. Observers had been aware for a long time that Racetrack Playa rocks had been moving because the stones had left conspicuous trails in the sediment, but no one had actually seen the rocks travel or had understood what mechanism causes this phenomenon.

However, in the winter of 2013–2014, automated cameras and GPS units attached to the rocks captured evidence of boulders that weighed up to eighty pounds moving hundreds of yards across the wet playa (Norris et al. 2014). We now know that the process involves thin, broken-up, ice sheets that "bulldoze" the rocks in relatively light winds of only ten miles per hour across the slick, water-saturated mud. Thus, in the sliding rocks phenomenon, the ice is thin and does not lift the stones, but instead, the wind pushes them.

I have not seen evidence of sliding rocks occurring in Oregon. However, ice rafting may occur at Summer Lake because the conditions for it to happen, that is, rocks, water, ice, and wind, are all present in the winter. If so, this may be a relatively rare event, especially for boulders weighing several hundred pounds, because ice thickness and adequate water depth are not guaranteed.

Clearly, flood basalts, basin and range faulting, landslides, and moving rocks have all had a profound effect on the landscape of southeastern Oregon and on the rest of the Great Basin. Throughout history, such geological events have completely altered previous stream-drainage patterns and have created new ones, including Lake Abert and its now-threatened environment.

CHAPTER 3

Lake Abert Limnology

The Water Story

> In the remote high desert of western North America there are
> ancient saline lakes from which no water flows. Once these
> were vast inland oceans, rolling between the rimrock cliffs of
> the Pleistocene Great Basin. Lakebed desert floor is now all that
> remains of many of these pluvial lakes. Some however have per-
> sisted. These closed basins, like the seas, retain the sediments
> and solutes of the streams and springs they receive. Millennia
> of desert heat and dry winds have evaporated the waters, leav-
> ing salt-enriched brine lakes, or a shoreless expanse of salt flats.
>
> —DAVID B. HERBST, 1986

I can't remember a time when water didn't play a large role in my life. Some of my first childhood memories are of splashing in mud puddles, scooping up tadpoles, or peeking into tide pools. Fortunately, I have been able to live near and to study a wide variety of aquatic environments, including the deep sea and the polar oceans, but I never expected to become so fascinated with the Great Basin lakes.

Water is so basic to life that it's nearly impossible to talk about the Great Basin without mentioning it, or the lack of it. Perhaps the first explorer to grasp this reality, and to write about it, was John Frémont, who said, when he and his men had entered the "Great Interior Basin" of South Central Oregon at a latitude of 42° 57' 2", a few days before Christmas 1845, "We were now in a country where the scarcity of water and grass makes traveling dangerous, and great caution was necessary." He was correct, of course, because the Great Basin is a region of scant rainfall and high evaporation rates, and consequently, it has very few perennial rivers and lakes, compared with the well-watered eastern United States.

The more we learn about lakes, the more we realize that they are more than just blue spots on a map. We can't easily define them because they are part of a continuum of water bodies that encompasses everything from puddles and ponds to inland seas.

We usually know a lake when we see one, but when we encounter most Great Basin lakes, we quickly realize that they don't fit our usual assumptions.

I know my own concept of what a lake is has evolved because of the time I have spent in the Great Basin, where these bodies of water seasonally disappear and reappear and where they come in all sizes. As geologist Israel Cook Russell aptly stated in his 1895 report, *Present and Extinct Lakes of Nevada,* in which he colorfully described the Great Basin lakes:

> These ephemeral water bodies frequently come and go almost as erratically as the shadows of clouds cast on their tawny surfaces. In some instances, waterbodies of the same type appear during the winter months and remain until the heat of summer reaches a maximum; they then give place to smooth plains of mud.

Few people realize that the Great Basin features more lakes than rivers. In fact, archaeologist and mammalogist Donald Grayson has listed forty-five substantial lakes for the region but only twenty-three rivers, and the only lakes he includes are the large ones found in the bigger basins (Grayson 2011). And in some parts of the Great Basin, lakes are actually far more abundant than his summary suggests. For example, some sections of the Great Basin located in Oregon and in adjacent areas of California and Nevada contain several thousand minor, wet-season lakes or playas, similar to those that geologist Israel Russell once described, but that same area has only a handful of perennial streams. The region has more lakes than rivers because lakes in arid landscapes need just enough water to flood part of their basin seasonally. In contrast, the area's rivers, because of the dry climate and high evaporation rate, require a large watershed that is able to supply enough surface water and groundwater to keep them flowing over long distances.

Furthermore, nearly all Great Basin rivers originate at higher elevations, where there is more precipitation, whereas lakes exist at all elevations. Although the Great Basin does have perennial, or all-season lakes, they are either fed by rivers or occur at higher elevations with more precipitation and less evaporation, and they are mostly spread around its northern, western, and eastern edges, which receive more precipitation than its central and southern regions (Grayson 2011). It is in these latter, drier regions where we find the truly seasonal lakes, which disappear and reappear.

Lakes, in their broadest sense, are still or weakly flowing water bodies, and thus they are distinct from more-strongly flowing streams and rivers. They are arguably the simplest of hydrologic systems because they basically consist of just two things—a basin and

Cumulus clouds gather around Lake Abert on a warm July day. A rarely seen plume of muddy water, caused by a recent rainstorm, extends along the eastern shore.

the water that fills it. They occur in low areas of land, where more water accumulates and remains than overflows, leaks, or evaporates. They form when landslides, lava flows, sand dunes, or other events dam rivers or when glaciers, volcanic calderas, tectonic activities, or wind leaves a depression that fills with water. Lakes take on different shapes depending on their origin and on the geomorphic processes occurring in them and around their shores. For example, a tectonic lake like Abert takes on an elongated shape because its basin is associated with a linear fault.

Lakes are generally short-lived, at least in geological time, because they gradually fill with sediment, no matter how deep they are, or they eventually drain because a discharging stream at their outlet *down cuts,* or erodes, their shores. However, lakes like Abert, which formed in tectonically active areas like the Basin and Range, can persist over many millennia, due to continued downward displacements of the earth's crust. Tectonic lakes are therefore the earth's oldest ones. Geologists don't know exactly how old Lake Abert is, but because they believe that Abert Rim started forming about seven million years ago, they think that the lake is at least several million years old.

Because a tectonically formed lake is generally quite old, it can contain relict species—remnants of once more-widespread

Lake Abert at sunset.

species—that arrived many thousands or even millions of years ago via a river system that no longer connects and drains into the lake. Consequently, relict species can provide a record of the changes that occurred in drainage patterns over time. One such example is an aquatic snail called the robust springsnail (also known as the Jackson Lake springsnail, *Pyrgulopsis robusta*), which, despite its name, is only ⅛ of an inch long. The snail (see chapter 4) lives in XL Spring at the north end of Lake Abert but nowhere else in the Chewaucan Basin. However, robust springsnails still survive in the Columbia-Snake River system as far east as Idaho and Wyoming (Hershler and Sada 2002; NatureServe Explorer). The fact that this snail species only lives in one part of the Chewaucan Basin suggests that several million years ago, the Snake River flowed into the basin, but the course of the river changed over time, leaving the springsnail behind in XL Spring.

South Central Oregon Lakes

A lake is the landscape's most beautiful and expressive feature. It's earth's eye.... Nothing so fair, so pure, and at the same time so large, as a lake, perchance, lies on the surface of the earth.
—HENRY DAVID THOREAU, 1854

Oregon's only well-known, named lakes are Crater and Wallowa, but in fact, the state boasts 1,400, according to the *Atlas of Oregon Lakes* (Johnson et al. 1985), and they encompass about a half million acres, or about 1 percent of the state's area. Additionally, many more unnamed lakes, perhaps as many as five thousand, exist.

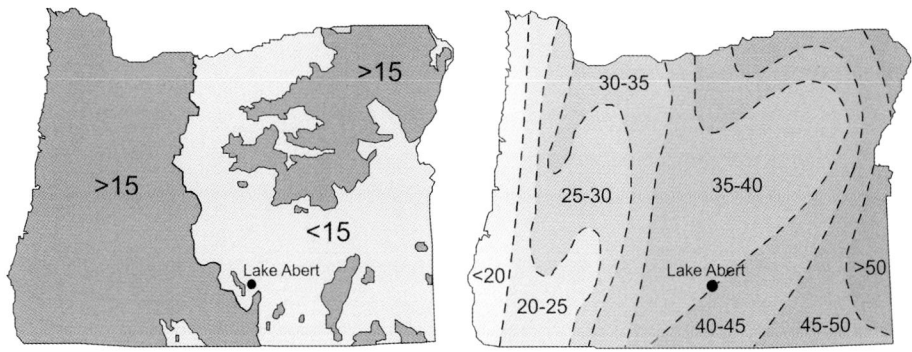

Measuring Oregon's climate: *left*—average annual precipitation, in inches; *right*—open water evaporation, also in inches. Drawings based on various sources.

Although some states have many more lakes, Oregon likely has the greatest lake-basin diversity, due to the varied landforms—including coastal dunes, volcanic calderas, glacial valleys, and basin and range faulting—that created these bodies of water.

Lakes are not necessarily more numerous in areas with high precipitation, as South Central Oregon's Great Basin country shows. Basins in this region mostly get less than one foot of annual precipitation, in the form of snow and rain in the winter and spring, and yearly evaporation is several times that amount. Despite the limited precipitation, this region has many large lakes, including Abert, Goose, Summer, Upper Klamath, and Warner. Additionally, there are several thousand seasonal lakes, the smaller vernal pools and larger playa lakes.

So, why are there so many lakes? The presence of lakes in a region characterized as a high desert results primarily from three factors: (1) basin and range faulting has created mountains that catch moisture and basins that hold water, thus forming lakes and wetlands; (2) the basins have water-retaining clay soils that prevent leakage; and (3) most of the precipitation falls in the winter and spring before evaporation rates are high. Thus, many of these basins are at least seasonally flooded. These lakes also lack an outlet, and so they are called *terminal* or *closed-basin lakes.* As a result, they capture all the water that comes into them, holding onto it until it evaporates, or, as Mark Twain said of the Nevada terminal lakes, "Water is always flowing into them; none is ever seen to flow out."

Because most lakes in the Great Basin are terminal and lie in areas with scant precipitation and high evaporation rates, they fluctuate greatly in depth and surface area, depending on the balance between inflows and evaporation losses. Seasonal and longer-term changes in surface areas of some of these lakes would be even greater, were it not for the groundwater that springs provide. This is especially true for Summer Lake, which is fed by the large springs forming the Ana River. Great Basin lakes are also shallow because sediment has filled their basins and has created nearly flat lake bottoms, which accentuate changes in the lakes' surface areas as water levels rise and fall due to changes in precipitation and evaporation. All these factors make the Great Basin lakes very different from those in the Cascade Range and Sierra Nevada, and from those in areas west of the mountains, because mountain and westside lakes are generally perennial, have a stream at their outlet, and are often deep. Additionally, Great Basin lakes are usually salty, and, in fact, several, including Lake Abert and the Great Salt Lake, have a salt concentration higher than that of the ocean.

What Is a Salt Lake?

All lakes in the Great Basin contain some salt, but Lake Abert in South Central Oregon is among the saltiest. To understand why, we first need to appreciate that all water bodies contain some salts, or dissolved minerals, that are derived from airborne dust, the weathering of rocks, agricultural runoff, and other sources. However, the concentration of salt in most lakes is quite low, usually consisting of less than 0.1 percent (or 1 gram of salt per 1,000 grams of water), because their outflow usually removes most of these dissolved minerals. However, Lake Abert is typical of Great Basin lakes in that it is old and lacks an outlet, and consequently, it has accumulated salt over a very long time. Lake Abert's water and upper lake-bed layers contain 40 million to 50 million tons of salt, and even more is present as brine deeper in the sediment (Van Denburgh 1975). Nearby Summer Lake is not as saline as Lake Abert because the wind transports salt away from its basin, thereby reducing its salinity.

Just as Lake Abert's water level and surface area vary, depending on the amount of fresh water that enters the lake and on the extent of the evaporation that occurs, so too does its salt concentration. The lake is most dilute in years with high inflows, particularly in the spring, when inflows are generally highest; and it is most saline in years with low inflows, especially in the late summer and fall. After a series of wet years, when the lake's volume has increased substantially, its salinity can be as low as 2 percent, and during such times, it is more dilute than the ocean, which has a salinity of about 3.5 percent. In contrast, during periods of low inflows, such as those that occurred between 2013 and 2016, and more recently between 2020 and 2022, the lake's salinity reaches an estimated 25 percent or more (Larson et al. 2016).

Salt actually formed a crust over the lake bed beginning in 2013, covered most of the lake bed in 2014, and remained, to a lesser degree, into 2016. Then in the winter and spring of 2016 and 2017, increased inflows brought the salinity back down to 3 percent. This cycle happened again in 2021 when the lake nearly dried up, and as it desiccated, pools of concentrated brine with a salinity of about 25 percent remained behind. These changes show how variable salinities can be in the lake, and such inconsistency can be stressful or even lethal to the lake's biota.

Classifying Salt Lakes

We classify salt lakes based on their salt content, categorizing those with very little as *subsaline,* or *brackish,* and those with high concentrations as *hypersaline.* Usually, we don't classify a body of water as a salt lake unless it has a salt concentration above 0.3%, the equivalent of 3 grams of salt per 1,000 grams of water (Hammer 1986). Lake Abert is Oregon's only hypersaline lake, with salinities normally exceeding 5 percent. In addition to their salinity classifications, Oregon's salt lakes also have high alkaline levels because they contain carbonate salts, meaning that they have a pH above 7, with Lake Abert having a pH of 10, and therefore we also refer to them as *soda lakes.*

Classification of Oregon's Saline Lakes

SALINE LAKE TYPE	SALT CONCENTRATION		PERMANENCE AND FREQUENCY OF FILLING	SOUTH CENTRAL OREGON LAKES
Subsaline	0.05–0.3%	0.5–3 g/L	Intermittent and Episodic	Crump, Goose, and Hart
Hyposaline	0.3–2%	3–20 g/L	Seasonal/ Intermittent and Episodic	Bluejoint
Mesosaline	2–5%	20–50 g/L	Semi-seasonal/ Intermittent and Episodic	Summer and Harney
Hypersaline	Greater than 5%	>50 g/L	Semipermanent and Episodic	Abert

Sources: Based on data from Phillips and Van Denburgh (1971), using Hammer's classifications (1986).

Salt lakes occur worldwide in arid regions, even in Antarctica, and at widely varied elevations, ranging from below sea level to over twelve thousand feet (Hammer 1986; Williams 2002). They are most numerous on the east side of the southern Andes; in central Asia (especially in China); and in Australia. China has nearly

Worldwide distribution of salt lakes. Redrawn after Williams (2002) and Jellison et al. (2008).

one thousand salt lakes, as does Australia. In China, they cover an estimated twenty thousand square miles (Qinghai Institute of Salt Lakes, http://english.isl.cas.cn).

Collectively, inland salt lakes hold about 50 percent of the volume of all lakes, but because most salt lakes are very shallow, the area they cover is likely much greater than that of all combined freshwater lakes. Saline water bodies also vary tremendously in size, from small lakes like Don Juan Pond in the McMurdo Dry Valley of Antarctica, which covers only 7 acres, to the largest such body—the Pacific Ocean—which covers nearly one-third of the globe. The Caspian Sea is the greatest inland saline-water body and contains two-thirds of the total volume of all inland salt lakes. Lake Turkana in Kenya and Ethiopia is not only the world's fifth largest salt lake, but it is also the largest soda lake.

The world's saltiest body of water is Don Juan Pond, which can't freeze, even at −58°F, because its 44 percent salinity is so high (NASA 2014). Lakes with salinities above 35 percent, however, are very rare because at those salinities, nearly all salts will have reached their solubility limit: they can't remain dissolved at such high levels and therefore settle to the bottom. With the exception of Don Juan Pond, most high-salinity lakes are usually present in hot climates because the solubility of most salts increases with rising temperatures. Salt

World's Largest Salt Lakes

LAKE	COUNTRY	APPROXIMATE SURFACE AREA (SQUARE MILES)	SALINITY (%)	REFERENCES
Caspian Sea	Kazakhstan, Russia, Turkmenistan, Azerbaijan, and Iran	143,000	1.2	Koriche et al. 2021
Balkhash	Kazakhstan	6,700	0.35–0.6	Hammer 1986
Eyre	Australia	3,700	0.5–33	Deocampo and Jones 2014
Turkana	Kenya and Ethiopia	2,700	1.7–2.7	Avery 2013; Spigel and Coulter 1996
Aral Sea	Kazakhstan and Uzbekistan	2,600	~10	https://kazaral.org/en/aral-sea/general-information/; nasa 2016a
Issyk-Kul	Kyrgyzstan	2,400	0.6	Savvaitova and Petr 1992
Urmia	Iran	2,300	12–30	Asem et al. 2014; nasa 2016b
Mar Chiquita	Argentina	1,900	2.7–23	Bucher and Curto 2009
Qinghai	China	1,750	0.12	Zhang et al. 2011
Great Salt	United States	1,700	5–27	https://ffsl.utah.gov/wp-content/-uploads/GSLSAC_SalinityInfluencesRangesTM_Final_July2021.pdf
Van	Turkey	1,450	2.1	Kaden et al. 2010
Uvs Nuur	Mongolia and Russia	1,300	1.9	https://whc.unesco.org/en/list/769/
Poopó	Bolivia	1,000	Up to 12	Zolá and Bengtsson 2006
Tuz	Turkey	650	32	Oyewusi et al. 2021
Dead Sea	Israel and Jordan	360	34	Bodaker et al. 2009

Note: Lake size and salinity vary over time, so these data might not represent current or future conditions.

ponds, lakes, seas, and oceans represent a continuum of saline water bodies, are an essential part of the global water cycle, and comprise diverse and important, but often unappreciated, ecosystems (Williams 2002; Jellison et al. 2008; Wurtsbaugh et al. 2017).

The Great Salt Lake in Utah, covering about 1,700 square miles, is the largest saline lake in North America and is one of the ten largest salt lakes in the world. Other large salt lakes in the western United States include the Salton Sea (350 square miles) and Mono Lake (75 square miles) in California; and Pyramid Lake (195 square miles) and Walker Lake (50 square miles) in Nevada.

Now, there is even evidence that salt lakes might exist elsewhere in our solar system, including possibly on Mars (Orosei et al. 2018). A satellite equipped with ground-penetrating radar has detected what is believed to be a Lake-Abert-size brine pool on the planet, approximately one mile below its southern polar ice cap. Scientists believe that the brine consists of sodium and magnesium perchlorate. The latter chemical can reduce water's freezing point, so that it remains liquid, even in the extremely cold environment of Mars.

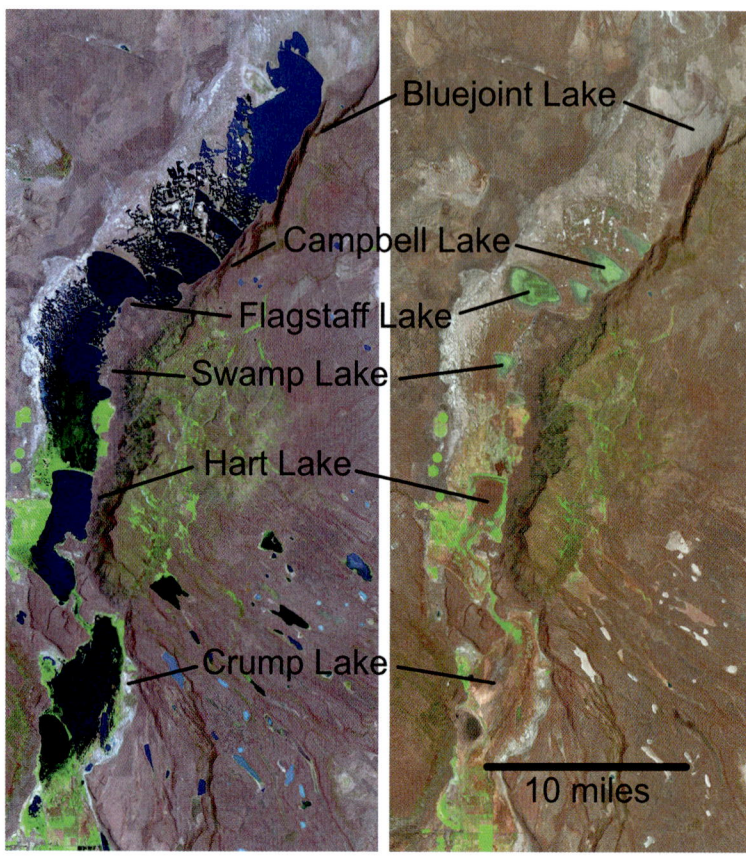

Bluejoint Lake

Campbell Lake

Flagstaff Lake

Swamp Lake

Hart Lake

Crump Lake

10 miles

Landsat satellite images of the Warner Lakes under wet conditions in 1999 and dry conditions in 2014. All Warner Lakes were dry by late August 2014 and again in 2021 and most were dry by August 2022. We know of only one other time when the lakes dried up, in 1992, but we can assume that they probably desiccated during the Dust Bowl era too. USGS EarthExplorer.

Another way that scientists classify lakes is by how much of the year they hold onto their water. Those that don't dry up, or which rarely desiccate, are called *perennial* or *semi-perennial.* Those that contain water sporadically are called *intermittent,* and ones that dry up consistently in some seasons are called *seasonal.* Most lakes in South Central Oregon and adjacent areas of the Great Basin are intermittent and seasonal, meaning that they rarely fill up, they contain water mostly in the spring, and they are usually dry by summer or fall. We classify Lake Abert as semi-perennial and categorize Summer, Goose, Harney, and the Warner Lakes as intermittent. All these lakes are also episodic, meaning that they fill infrequently and irregularly, and then they experience declining water levels. For example, during the past one hundred years of its history, Lake Abert reached its highest level in 1958; was nearly dry for several years during the Dust Bowl era, in the 1920s and 1930s; and nearly dried up again in 2014–2015 and 2021–2022.

The Prevalence of Playas

Most lakes in the Great Basin reached their highest water levels and greatest areas during the Pleistocene epoch, which began 2.6 million years ago and ended about 11,700 years ago. We also refer to the Pleistocene as the Ice Age because 20 or more glacial cycles occurred then. Today, what remains of most of these paleolakes is a playa, which in addition to being seasonally dry, typically has a nearly level lake bed consisting primarily of clay. Playas are common in the Great Basin. More than 80 percent of the nearly 200 major Great Basin drainages contain them (Madsen et al. 2002), a fact that reveals just how plentiful lakes were in the Great Basin during the Pleistocene epoch and also how many of them are now mostly dry.

Playas result from the natural evolution of terminal lakes. Such lakes start out as freshwater bodies and then, over hundreds of thousands of years, accumulate dissolved minerals and sediment. The amount of salt they accrue depends on several factors, including: the amount of runoff and groundwater that they receive; the geology of their watershed; and how frequently they dry up. At some point in their history, these lakes store so much salt that they become subsaline. Then, as more minerals reach them, their salinity gradually increases, and they can cycle through all the stages that lead to becoming hypersaline. This process speeds up during droughts, when the salts become concentrated, due to reduced lake volume.

If the climate becomes wetter, then a lake can return to an earlier stage of salinity, or, if the climate gets dry, then it can desiccate

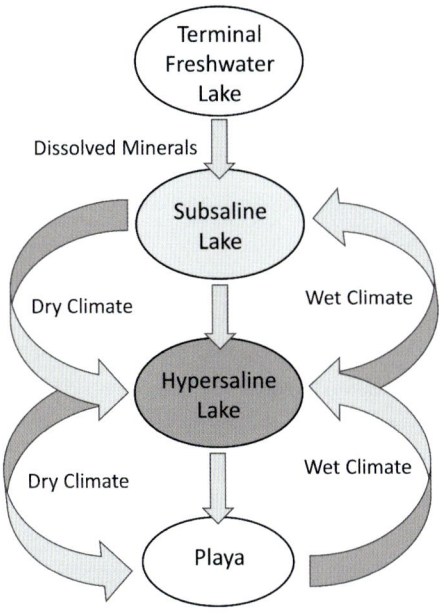

Hypothetical evolution of a playa from a terminal freshwater lake. The arrows show that climate can partially reverse the process.

and become a playa, either entirely or in part. This cycle can repeat itself indefinitely, so long as a lake basin is present. It is worth noting that this evolution from playa to salt lake and back to playa can happen in a single year. Also, during each dry period, capillary action (in which fluids move up through the soil, carrying dissolved salts to the lake's surface) and wind can then blow the salt away. This removal of salt by the wind is occurring at Summer Lake and also happened dramatically at California's Owens Lake when its inflows were diverted to Los Angeles, starting in 1913.

South Central Oregon, as well as the eastern part of the state and adjacent areas of California and Nevada, have numerous playas. In fact, based on an inspection of more than 200 USGS quadrangle maps, I estimate that there are more than 2,500 of them just in Oregon. Alvord Desert, located on the eastern side of Steens Mountain, is the largest playa in Oregon, at eighty square miles. Other big Oregon playas include: Coyote and Catlow in Harney County; and Alkali, Christmas, Coleman, Fossil, Greaser, Guano, Silver, and Lower Chewaucan in Lake County. Summer and Harney Lakes both have extensive playas with areas of approximately forty square miles. Additional large playas lie in Surprise Valley in northeastern California, not far from Nevada's and Oregon's borders. Many are found in Nevada too, with some—such as the Black Rock Desert—covering nearly one thousand square miles.

Because some playas, especially the larger ones, are what remains of a lake or are the parts of a lake that passed through a saline stage (the latter being the case for both Lake Abert and Summer Lake), they can have high concentrations of alkaline salts, which create a brilliant-white, efflorescent crust. The crust forms when the soil's capillary action draws alkali-rich groundwater to the surface, and then the water evaporates, leaving the salt behind. This crust, along with light-colored clay sediments, make playas conspicuous on satellite images. Bolivia's Salar de Uyuni playa, in particular, regularly attracts the attention of astronauts in the International Space Station because it is so visible (NASA 2022). That playa is the world's largest—covering four thousand square miles—and is located on the high-elevation Altiplano, the extensive Andean plateau of southwestern Bolivia.

Playas are amazingly flat because of thickly deposited, fine sediment that blankets any irregular features of the lake basins. In fact, they are the largest natural, level, terrestrial surfaces. We also refer to them as alkali flats, salt flats, or salt pans. Bonneville Salt Flats, part of the Great Salt Lake Desert in Utah and covering 3,800 square miles, is a famous example of a playa and was formerly part of Paleolake Bonneville. Another well-known Great Basin playa is the

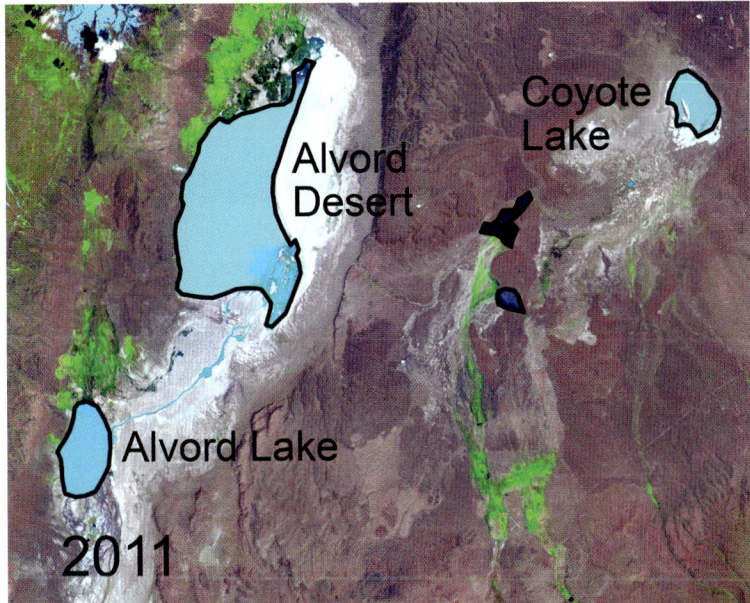

2011

2014

Landsat images of several Great Basin playas in eastern Oregon: *top*—lake and desert playas during a wet year (image captured in July 2011, when the playas contained shallow water); *bottom*—lake and desert playas during a dry year, when little or no new surface water flowed in (image captured in July 2014). The approximate water areas in July 2011 are outlined and were as follows: Alvord Desert—30.5 square miles; Alvord Lake—6 square miles; and Coyote Lake—3 square miles. USGS EarthExplorer.

Black Rock Desert in northwestern Nevada, once part of Paleolake Lahontan and now home to the annual Burning Man event.

Though generally flat, large Great Basin playas often feature dunes formed of wind-blown, fine-grained sediment, which consists of volcanic ash and aggregated silt, clay, and evaporites, all blown from their lake beds when exposed. The dunes at Lake Abert's north end are not especially well developed or numerous, but those at Oregon's Fossil Lake, northwest of Abert, are both large and extensive, with some reaching sixty feet high. Dunes also

Alkali Lake dunes. These dunes, located in the Alkali Lake Basin north of Lake Abert, were created from wind-blown sediment coming from the lakebed of Paleolake Alkali. The dunes have scattered clumps of basin wild rye and greasewood.

dominate the landscape north of Lake Abert at Alkali Lake, where Paleolake Alkali once lay. These dunes regularly drift onto Highway 395 and require removal by the highway department. Because they are dry and unstable, they have virtually no plants other than scattered basin wild rye grasses (*Leymus cinereus*) and greasewood (*Sarcobatus vermiculatus*). However, the dunes are home to an unusual beetle, which I have affectionately nicknamed the "golden-bristled dune beetle" (*Eusattus muricatus*), for the stout, golden bristles near the base of its thorax and elsewhere on its body. This beetle usually makes its home in dunes and lives in several other western states, including Texas and Utah. You can also find it in Oregon's Alvord Desert dunes east of Steens Mountain.

Salt lakes and playas make up a significant part of the global water cycle and an integral component of inland aquatic ecosystems. They provide aesthetic, cultural, economic, recreational, scientific, and ecological value to all of us (Williams 2002; Wurtsbaugh et al. 2017). Worldwide, salt lakes are finally receiving long-overdue attention because of their unique ecosystems and also due to growing concerns that climate change, upstream water diversions, and other factors already impact them adversely.

"Golden-bristled dune beetle" (my nickname for *Eusattus muricatus*), which lives in the Alkali Lake dunes and is about 0.4 inches long.

Storm Surges—Wind Tides

A lake's elevation depends on a variety of factors, particularly on climate and wind, and the largest lakes actually experience small tides caused by the gravitational pull of the moon and of the sun. Lake Abert is too small to experience a noticeable lunar/solar tide, but its water levels do change quickly when the wind pushes water downwind and up onto the opposite shore. Over a day's time, strong, southerly winds can push a thin layer of water far up onto Abert's north shore, creating a small surge, due to the lake's shallowness and its north/south orientation. Surges appear most visibly at the north end of the lake because of its nearly flat bed, which very gradually slopes downward to the south at less than two feet per mile. Some observers in the early twentieth century reported that south winds forced water two miles onto the playa at the north end of the lake (Waring 1908). Recent satellite images reveal that water has surged repeatedly onto the playa for several miles during strong autumn windstorms.

Ice Cover

During unusually cold periods, ice can form on Lake Abert: very cold temperatures are actually necessary for ice formation, due to the lake's salinity. According to one observer, two to three inches of ice covered the entire lake in January 1985 (Kreuz pers. com. 2016). Landsat satellite images of the lake have revealed partial ice cover during the winters of 1989, 1993, and 1997 too. Ice has likely formed on the lake at other times as well, but we lack a record because wintertime cloud coverage has interfered with satellite capture of additional images.

Lake Abert's Geochemistry

Beginning in the 1960s, USGS scientists initiated a series of geochemical studies at Lake Abert, focusing on the origin and chemistry of the clay minerals that constitute most of the muddy lake bed. They also investigated the presence of brine in Lake Abert's

Evidence of a wind-caused surge: *left*—Landsat image taken on October 15, 2016; *right*—Landsat image taken on November 1, 2016. These two satellite images of Lake Abert, taken approximately two weeks apart, show how strong winds can push a thin layer of water far up onto the lake's shore. Strong southwest winds occurred after the first image was taken. Note the difference in the expanse of the water. USGS EarthExplorer.

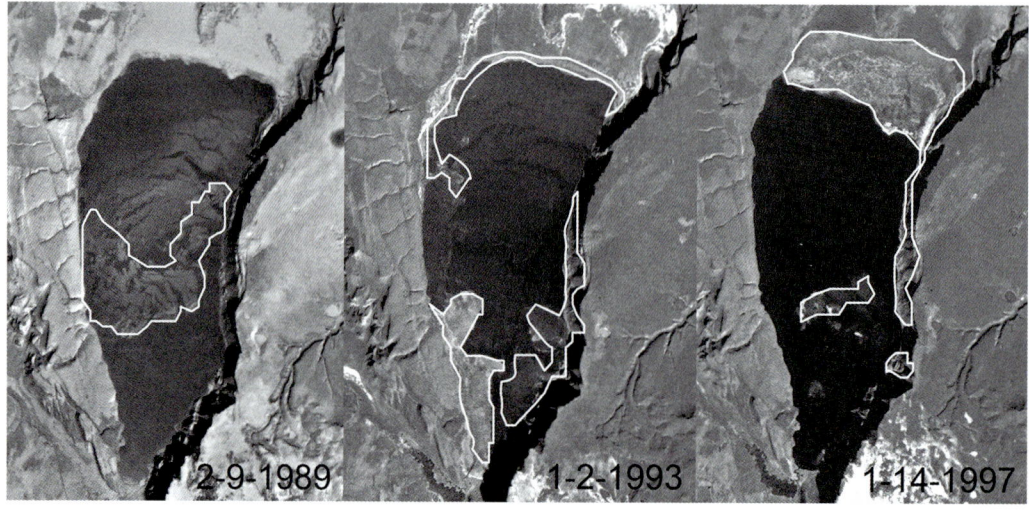

Landsat images showing ice on the surface of Lake Abert. Areas of ice coverage are outlined with white borders. USGS EarthExplorer.

playa sediments to determine how it affects the lake's geochemistry (Jones et al. 1969). The 1960s also saw the initiation of hydrological and water chemistry studies at Lake Abert and at other South Central Oregon terminal lakes. USGS physical scientists Kenneth Phillips and Steven Van Denburgh, who conducted these latter studies together, provided reports in the 1970s that are still considered to be the best sources of information on hydrology and water chemistry for these lakes. (I had the pleasure of meeting Steve Van Denburgh in 2013, and since then, we have become friends and have met at the lake several times. Together, we measured the extremely low lake elevation that occurred at Abert in June 2014).

To understand the geochemistry of Lake Abert, we must first appreciate that its source, the Chewaucan River, carries a variety of dissolved and suspended materials downstream, which flow into the lake. Because the lake acts as a sump—or a kind of natural reservoir—most of these materials remain there, building up as sediment, although some materials do leave from the lake as aerosols or from its playa as wind-blown dust. During its several-million-year history, the lake has accumulated a lot of this sediment, perhaps as much as one thousand feet in depth, and a wide variety of dissolved, inorganic chemicals. These chemicals, or ions, consist of atoms or molecules in the water that have either a positive or negative charge. The major ions in Abert's water consist of sodium, chloride, carbonate, and bicarbonate. The presence of carbonate and bicarbonate ions causes the water to have a high pH of ten (Phillips and Van Denburgh 1971), which makes the water alkaline and caustic. Pure, neutral water has a pH of about seven. Many Great Basin salt lakes are alkaline because of the presence of carbonates and bicarbonates, and this is why we often also call them soda lakes.

As Lake Abert's water levels recede during droughts, the ionic concentrations of dissolved minerals increase until they reach a point called the "solubility limit," where the individual ions can no

Trying to find water: Steve Van Denburgh, USGS scientist emeritus, measuring lake levels at Lake Abert in June 2014, when the lake was extremely low.

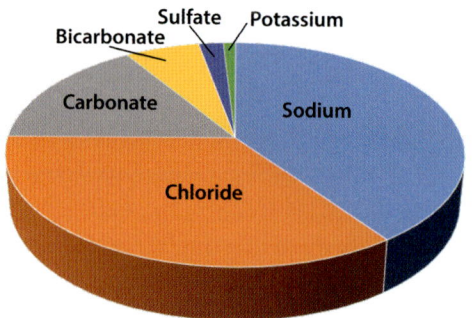

Relative proportions of major ions in Lake Abert's water. Phillips and Van Denburgh (1971).

longer remain dissolved in water but precipitate out as salts. As the salinity of the lake reaches this solubility limit, waves splash water onto the shore that then evaporates, leaving the salt behind. Additionally, salt forms on the lake's surface on warm, windless days when evaporation creates a thin, salty sheen. High evaporation rates in August 2013 led to noticeable salt deposits on the shore and to visible salt on the lake's surface, when Abert's depth was lower and its water saltier than is typical.

Salts precipitate from water in a sequence based on their solubility—how easily they dissolve in the water—with the least soluble salts forming first and the most soluble ones forming last. Formed

Alkali salts covering much of the exposed Lake Abert lake bed during the 2021 desiccation event (photo taken on May 17, 2021). A shallow, brown-tinted channel snakes its way south from near the Mile Post 74 Springs. Smaller springs are visible as dark areas emerging from near the shore. Highway 395 serves as a reference scale.

by evaporation, all such salts are called *evaporites,* or *evaporate minerals.* At Lake Abert, the first salt that precipitates in this sequence is called *trona,* a hydrated sodium carbonate-bicarbonate salt. Sodium chloride, or table salt, precipitates last during this sequence because of its high solubility.

Sodium chloride has a distinct crystal structure that looks like tiny cubes, while trona's looks like sharp teeth. In August and September 2013, trona salt appeared in shallow pools on Abert's lakeshore as ¼- to ½-inch-long, translucent to opaque, shark's-toothlike crystals. Other carbonate salt crystals materializing that year were shaped like the radiating petals of the chicory flower. The high salinity at that time was above 20 precent.

Glaringly white, salt crusts up to an inch thick on the playas at Abert and Summer Lakes consist mostly of trona (Peters et al. 1994), and this valuable mineral is mined in some areas of the western United States, including at Owens and Searles Lakes in eastern California. One of the world's largest trona deposits lies in southwest Wyoming, within the fifty-million-year-old Green River Formation, famous also for its trove of beautifully preserved fossilized fish (one of which sits on my desk). We use trona to make baking soda, various cleaning compounds, and products that control the pH of water.

Salt crystals from Lake Abert, collected in August 2013, *clockwise from top left*: translucent trona crystals, each about one half inch long; smaller, quarter-inch-long, pointed, opaque crystals covering a four-inch-long rock in the nearshore splash zone; cubic halite crystals about one eighth inch across; and radiating, rectangular crystals, each about a quarter inch long.

Mineral Composition of Lake Abert Salt Crusts

SALT MINERALS	PERCENT OF SALT CRUST BY WEIGHT
Trona ($Na_2CO_3 \cdot NaHCO_3 \cdot 2H_2O$)	79
Table Salt (NaCl)	16
Potassium Sulfate (K_2SO_4)	1.4
Sodium Sulfate (Na_2SO_4)	1.1
Sodium Bicarbonate ($NaHCO_3$)	0.6

Source: Data from Allison 1982.

Another salt, called *natron,* forms in cooling, alkali-rich water. Ancient Egyptians used natron extensively for cleaning and in the mummification process, because it absorbs water and its high pH retards bacterial growth.

Perhaps the best-known evaporate mineral mining site in Oregon is at Borax Lake in the Alvord Desert. As its name suggests, the water in this small, alkali lake is rich in borax (sodium borate), which comes into the water from geothermal springs. This unique lake sits on top of a thirty-foot-high mound created through the precipitation of minerals from the water; the rising mound gradually lifted the lake over an extended period of time. In the late 1800s, the Rose Valley Borax Company mined the mineral and hauled it one hundred miles in wagons pulled by teams of mules to the railhead in Winnemucca, Nevada. Mining there lasted for ten years and ended in about 1902.

Perhaps the most surprising fact about Borax Lake is that it contains a relict fish, the Borax Lake chub (*Siphateles boraxobius*), whose ancestors lived in Paleolake Alvord. The chub, a two-inch-long minnow, is only found in this ten-acre lake. The federal government once listed this chub as endangered because its only habitat was (and is) Borax Lake, its population size was small and variable, and it was considered vulnerable to adverse impacts from uncontrolled recreation and from the withdrawal of groundwater from its aquifer. However, due to successful conservation efforts, the species was considered recovered in 2019 (Federal Register 2020).

Dangerous Dust

> [I] gazed down upon the Owens Lake and thought about all of the freshwater flowing into it and going to waste.
>
> —FRED EATON, Mayor of Los Angeles, 1898–1900

Although we have eagerly made use of the trove of minerals provided by salt lakes, our work to mine and process these evaporates, especially by diverting water from salt lakes to expose the salts, has had a distinct down side.

Trona, for example, has proved to be not only a valuable mineral: it can be hazardous to the environment and to our health. The high pH level of its dust irritates respiratory tissues and exposed skin (Rom et al. 1983). At Owens Lake in eastern California, dust containing trona and a variety of other substances, including heavy metals, became a critical health issue after Los Angeles began diverting water from the Owens River, causing the lake to go dry in 1926 (Robinson 2018). In 1878, the lake was fifty feet deep and covered one hundred square miles. However, agricultural water diversions in the late 1800s, coupled with drought, brought the lake level down dramatically.

But it was the completion of an aqueduct diverting even more water to Los Angeles starting in 1913 that finally desiccated the lake. As a result, frightening dust storms stirred up seven thousand tons of alkali particulates from the playa, sending them high into the sky and significantly reducing visibility.

Californians euphemistically called the storms "Keeler Fog," referring to the small town at the edge of the playa that was most impacted. Dust coming from the Owens Lake playa was once the largest source of small-sized airborne particulates in the United States. Wind carried the dust hundreds of miles from the lake, and the particulates were recognized as a human health hazard that likely caused many millions of dollars in economic impacts to nearby communities (Reheis et al. 2022). Controlling the dust has been extremely costly for Los Angeles residents, who have already spent over one billion dollars and who will likely pay millions more, far into the future. Additionally, the state has had to redivert some extremely valuable water back to the dry lake bed because, not surprisingly, reflooding has proved to be the easiest means of mitigating the dust (Robinson 2018).

Oregon's Summer Lake playa frequently produces its own alkali dust, a "Summer Lake Fog." In fact, this phenomenon made the national news in February 2015, when "milky rain" fell in parts of the Pacific Northwest, as far away as five hundred miles from Summer Lake, which was eventually traced as the source by a university meteorologist. Areas of Washington State all the way up to Spokane were affected (Washington State University 2015). The Weather Channel listed this phenomenon as one of the strangest weather events of early 2015. More recently, a similar occurrence of milky rain, or "mud-rain" as it was also called, happened in February 2022 in Boise, Idaho, and was associated with cold, high winds

Summer Lake viewed from the northwest at Fremont Point on Winter Ridge. White plumes of alkali dust, rising hundreds of feet into the air, are evident along the southeastern shore.

(KTVB 2022). Unfortunately, because Lake Abert's elevation has recently become quite low, and is likely to experience low water levels more frequently in the future, it likely is also a source of alkali dust.

Almost any afternoon in the summer, as well as in the spring and fall, white columns or clouds of dust continue to rise hundreds of feet from the Summer Lake playa. Dust devils from the dry shoreline of Lake Abert also transport alkali dust, but so far are insignificant compared to the large storms rising from its nearby counterpart.

Water diversions have also produced dangerous levels of dust from the lake beds at Mono Lake, Salton Sea, and the Great Salt Lake, as well as from salt lakes around the world, raising concerns globally about adverse health and environmental effects (Blank et al. 1999; Abuduwaili et al. 2010; Wurtsbaugh et al. 2017). Owens Lake has served as a cautionary tale, teaching us that dust abatement at desiccated, alkali lake beds can be extremely expensive and that it is better to keep them flooded.

Desiccation Polygons—Patterns in Mud

Playa sediments have a high clay content, due to the microscopic clay particles that flow downstream and settle into a lake. Clay, which actually consists of microscopic plates, has a great ability to attract and hold water: it can increase its volume by up to 50 percent and substantially swell. As the mud then dries, the loss of water

Desiccation polygons in a variety of shapes and sizes, covering the dry lake bed at the north end of Lake Abert in June 2016. Most of the polygons were three to four feet across.

causes the clay to shrink and form cracks, which merge into what we call *desiccation polygons* (El-Maarry 2014). These developing cracks tend to meet at angles of 90 to 120 degrees and can assume a variety of complex and interesting geometric shapes with varying numbers of sides.

Desiccation polygons vary considerably in size, from a few inches to many feet across, with the largest forming in thick mud. The cracks are V-shaped in cross section because the top layer of mud dries and shrinks faster than the deeper layers. The length of the individual cracks relates to the moisture content of the mud and is longest in mud containing the most water. As the mud dries, the cracks become more numerous and shorter. In strongly layered mud, curls form on the uppermost strata and are susceptible to being blown away by the wind. Various environmental factors can impact these phenomena, including the chemical and physical nature of the mud and the air's temperature, and so mud's desiccation features on the same playa can vary considerably in form and size and can change over time. You can easily see this at Lake Abert, where mud polygons are diverse in shape and size, even over short distances. Freezing and thawing cycles also cause polygons to form in the uplands around the lake.

Observers have found exceptionally large desiccation cracks at Arizona playas, ones that are three feet wide, one thousand feet long, and up to fifty feet deep and which form polygons that are

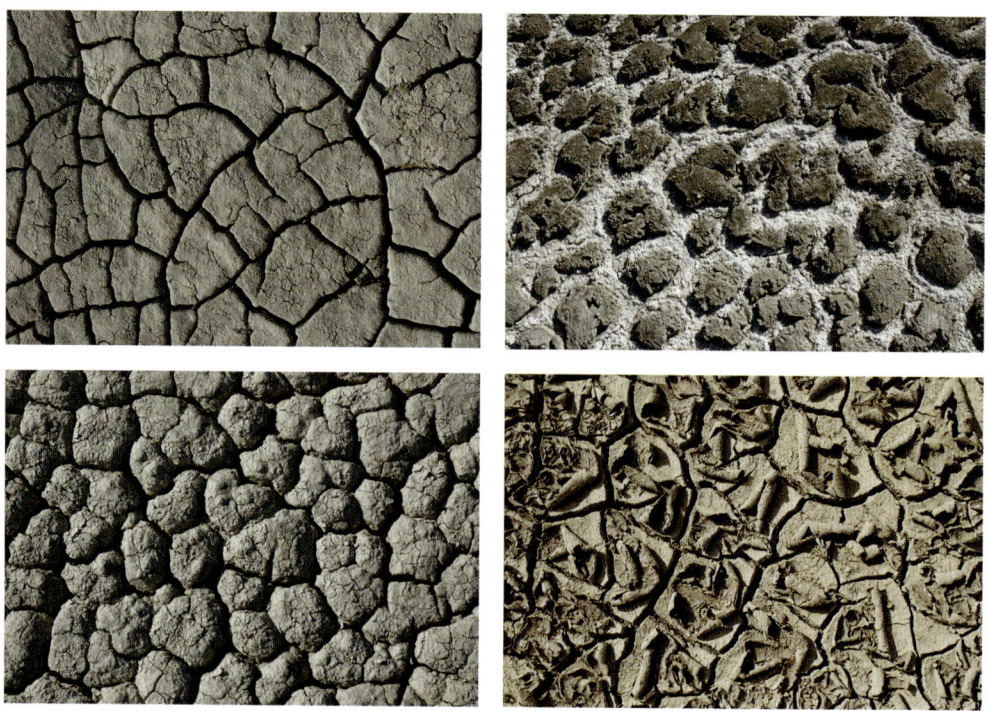

Desiccation polygons from the Lake Abert playa. Each photo represents about one square foot in area.

hundreds of feet across (Harris 2004). Far beyond our planet, spacecrafts have taken images of giant polygons at the bottom of impact craters on Mars. These polygons are similar in shape and size to desiccation polygons present on Earth, and they serve as evidence that liquid water once existed on the red planet (El-Maarry et al. 2010). In 2017, the Mars rover Curiosity captured images of what appear to be mud cracks and provided conclusive evidence that a paleolake was present in Gale Crater several billion years ago (Stein et al. 2018).

Tufa—Rocks from Water

Lake Abert's shore holds additional geological curiosities, including rocks with a conspicuous, light-gray, mineral covering called *tufa,* primarily made of *calcite,* a calcium carbonate mineral. Abert's tufa ranges in thickness from a thin veneer to a ½-inch thick, rind-like covering on rocks and boulders near the lake, but it is also present on boulders exposed in road cuts along Highway 395 above the lake. You can also find much larger tufa formations, muffin-shaped mounds that are up to twenty-five feet in circumference and eight feet high (Bartruff 2013; Hudson et al. 2017). They tower over the XL Ranch Road (Lake County Road #3–09), north of the lake. Twelve of them, arranged along a low ridge, spread out in an area about one acre in size.

These mounds formed from the discharge of calcium-rich spring water into shallow areas of the lake during the Pleistocene epoch (Hudson et al. 2017). Interestingly, Native Americans used one of these formations, which is located on the west side of the lake, as a shelter, now called Rattlesnake Cave (see chapter 7). According to a recent study, some of the Lake Abert tufa mounds are up to twenty-five thousand years old (Hudson et al. 2017). The mounds and deposits left on rocks around Lake Abert at various elevations are evidence of how lake levels have varied in the past.

The tufa formations at Lake Abert are actually quite small compared to those at Nevada's Pyramid Lake, which are unique, both in terms of their size and form, with some rising up to over forty feet (Benson 2004). During the Pleistocene, Pyramid Lake was part of Paleolake Lahontan, which covered 8,500 square miles when the lake reached its maximum size about thirteen thousand years ago. Pyramid Lake tufas are remarkably diverse, resembling giant mushrooms, broccoli, spheres, or barrels. Captain John Frémont

Tufa deposits near Lake Abert, *clockwise from top left*: one-eighth-inch-thick, rind-like tufa partially covers a boulder (one foot in diameter) on the shoreline; a massive tufa formation, twenty feet across and six feet high, dominates the landscape more than one mile north of the lake along the XL Ranch Road, near what is known as Pikes Place; and tufa-coated cobblestones lie above laminated sediments that were deposited in the lake when it was much deeper, near the end of the Pleistocene (they were exposed by a road cut along Highway 395 near the south end of the lake).

wrote the first-known descriptions of these formations, and scientific dating indicates that most of them materialized between 13,000 and 26,000 years ago. Supersaturated calcium carbonate coming from springs in the lake bed provided the minerals required to form these tufas (Benson 2004).

At Nevada's nearby Winnemucca Lake, geologists have used radiocarbon dating to assess the age for tufa-covered petroglyphs there, which they have determined to be approximately ten-thousand years old and the oldest-documented petroglyphs in North America (Benson et al. 2013). Diverse tufa formations also exist at Mono and Searles Lakes in California, and smaller ones are present at both Summer Lake and the Surprise Valley Lakes in northeastern California.

Tufa formation in ancient lakes interests an array of scientists, including: geochemists, who study its chemical makeup and its usefulness in carbon dating; biologists, who are fascinated by the microorganisms involved in its formation; and paleolimnologists, who pursue evidence of how lake levels varied in the past.

The Great Basin's Paleolakes

The Pleistocene epoch, lasting from 2.6 million to 11,700 years ago, was a period of fluctuating climate and numerous ice ages, when glaciers advanced during cool and wet periods and then retreated during warmer, interglacial periods. Those same climatic events that affected glaciers also had dramatic effects on Great Basin lakes: near the end of the Pleistocene era, 15,000–20,000 years ago, large parts of the basin in South Central Oregon, eastern California, northwestern Nevada, southeastern Idaho, and western Utah lay submerged under lakes, some of which were quite large.

In total, as many as 120 paleolakes inundated more than 40,000 square miles, which means that over 20 percent of the Great Basin flooded (Morrison 1991; Minkley et al. 2007; Grayson 2016). In comparison, current Great Basin lakes cover only about a few thousand square miles, or only about 10 percent of the Pleistocene's total. By convention, scientists give paleolakes different names from those of the modern lakes that occupy parts of the same basins, for example, referring to Paleolake Chewaucan, or Lake Chewaucan, which covered the subbasins where Lake Abert and Summer Lake— its remnants—now lie.

The two largest Pleistocene lakes in the Great Basin were Paleolake Bonneville in western Utah and Paleolake Lahontan in northwestern Nevada. The Great Salt Lake is a remnant of Bonneville, which existed until about thirteen thousand years ago and which appeared and disappeared—refilled and dried up—numerous

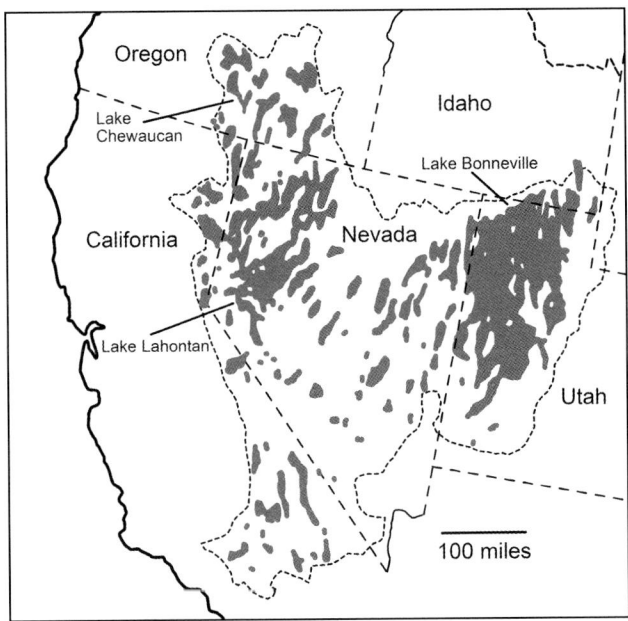

Locations of Great Basin paleolakes, late in the Pleistocene. The two largest were Lake Bonneville and Lake Lahontan. The outline of short dashes is the approximate boundary of the Great Basin. Redrawn, based on a sketch originally from Grayson (1993).

times as the climate changed. It was a truly massive paleolake, covering an estimated twenty thousand square miles, nearly as large as Lake Michigan (at twenty-two thousand square miles), and it was almost one thousand feet deep. At its highest level about eighteen thousand years ago, Paleolake Bonneville even overflowed into the Snake River drainage.

At its peak thirteen thousand years ago, Paleolake Lahontan, which covered over eight thousand square miles and had a depth of nine hundred feet, was smaller than Bonneville but was still substantial, and its many lobes extended outward, some spreading over one hundred miles. Interestingly, one arm of Lahontan apparently even overflowed north into eastern Oregon's Alvord Desert, located east of Steens Mountain and the nearby Pueblo Mountains (Hemphill-Haley 1987). Lahontan cutthroat trout, popular in recreational fishing in the West for the past thirty years, apparently arrived with this overflow. Paleolake Lahontan had mostly disappeared nine thousand years ago, but Nevada's Pyramid, Walker, and Winnemucca Lakes are its remnants.

When geologist Israel Russell studied Winnemucca Lake in 1895, it was thirty feet deep and covered about one hundred square miles. However, by the 1930s, it had dried up because of the construction of both Derby Dam, upstream on the Truckee River, and a road that blocked the remaining inflows from reaching the lake (Eilers and

Oregon's Largest Pleistocene Paleolakes

PALEOLAKE	BASIN	AREA (SQUARE MILES)
Malheur	Harney	920
Fort Rock	Fort Rock, Christmas, and Silver Lake	750
Warner	Warner	500
Alvord	Alvord	490
Chewaucan	Abert and Summer Lake	480
Goose	Goose Lake	370
Catlow	Catlow	350
Alkali	Alkali	230
Coyote	Coyote	175
Guano	Guano	Unknown, but likely near 30

Sources: Data from Orme (2008) and Grayson (2011).

Walker 2014). In an effort to save Winnemucca, President Franklin Roosevelt designated it as a national wildlife refuge in 1936; nevertheless, no water reached the lake, and, consequently, in 1962 it became the first national refuge to lose its designation (Pratt 1997; Eilers and Walker 2014). In wet years, however, local runoff creates some short-lived inundation of the Winnemucca Lake playa, so we don't yet consider it dead.

Oregon's segment of the Great Basin encompassed only ten large Pleistocene lakes. They covered over four thousand square miles, compared with the one-thousand-square-mile coverage of Oregon's modern lakes, which shows how much wetter this region was during the Pleistocene. Paleolake Malheur was the biggest of the ten lakes, covering an estimated nine hundred square miles, and it was the third largest Pleistocene lake in all of the Great Basin. Paleolake Chewaucan spread over an area of nearly five hundred square miles and was approximately three hundred feet deep. It was the eleventh largest Pleistocene lake in the Great Basin (Allison 1982; Licciardi 2001; Pezzopane 2001; Grayson 2011).

At the same time, there were also several thousand minor paleolakes in the region that now includes South Central Oregon and the southeastern part of the state, and each of these smaller water bodies covered anywhere from a few acres to one square mile. In modern times, these paleolakes have become playas and seasonal lakes. Today, an area of southeastern Lake County, located about thirty miles east of Lakeview and called High Lakes Plateau, includes over one hundred playas, a fact that reveals just how numerous the minor paleolakes were.

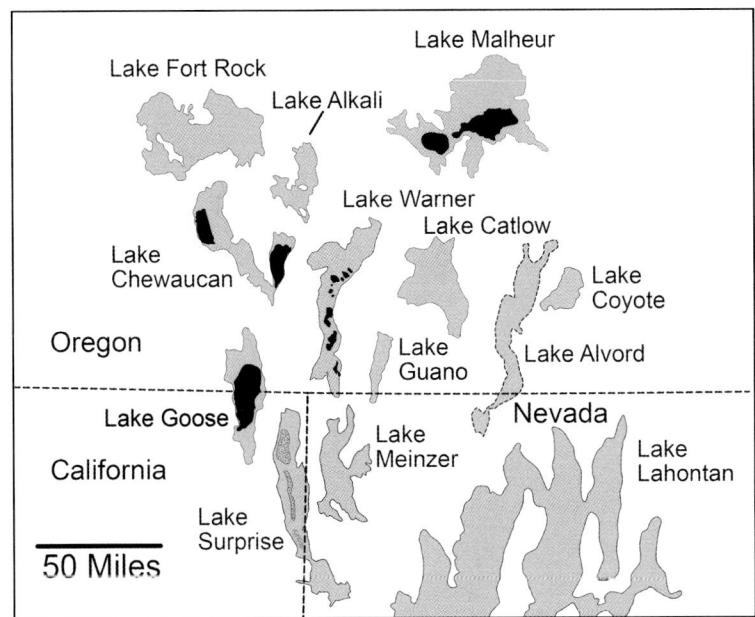

Approximate locations and relative sizes of the largest late Pleistocene lakes in the northwestern Great Basin. The larger, existing perennial or intermittent lakes are shown in black. Lake Lahontan in Nevada is only partially shown because of its large size. Drawing based on data from Dugas (1998), Reheis (1999), Negrini et al. (2001), and Reheis et al. (2014).

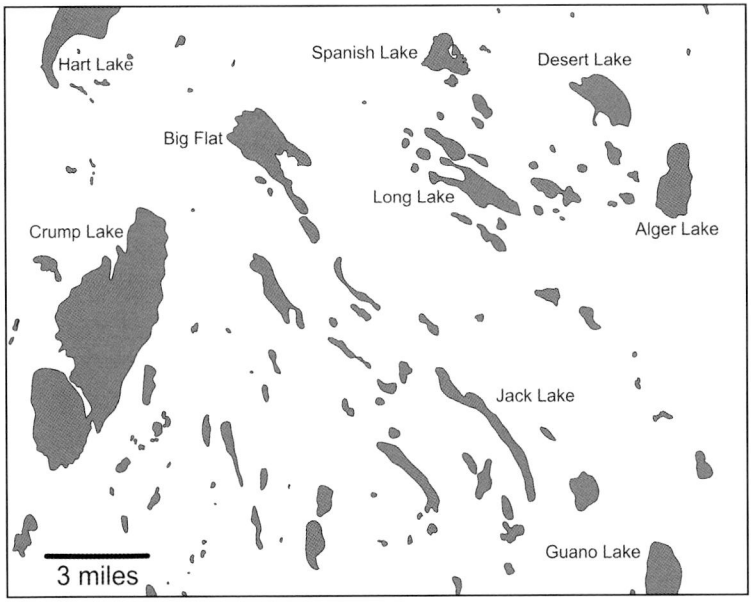

Playa lakes located east of the Warner Lakes in South Central Oregon in the High Lakes Plateau. Approximately 120 playa lakes of varying sizes lie in this 300-square-mile region.

Ira S. Allison (1895–1990), professor of geology at Oregon State University. Allison pioneered studies on Oregon's paleolakes. Courtesy of Oregon Digital, Item P046:244, Oregon State University Special Collections and Archives Research Center, Corvallis.

Paleolimnology—The Study of Ancient Lakes

In 1939, Ira Allison, a young geology professor at Oregon State University, began a study of South Central Oregon's paleolakes. By then scientists had recognized that a wide variety of data—including the chronology of volcanic ash deposits, climatic interpretations of pollen preserved in lake sediments, and the identification of animal fossils—could provide clues to the age of the lakes and evidence about how they had changed over time, objectives of the modern field known as *paleolimnology.* Allison published reports on Paleolake Fort Rock—whose 750-square-mile basin covered much of the modern-day Fort Rock, Christmas Valley, and Silver Lake subbasins—in 1979 and on Paleolake Chewaucan in 1982 (Allison 1979; Orme 2008).

Scientists became increasingly interested in following up on Allison's pioneering work, and on that of geologists Gilbert and Russell, and so these new paleolimnologists began to develop their own research tools to study Oregon's ancient lakes. They quickly realized that lake sediments are deposited one by one, on top of each other—like a layer cake, with the oldest layer at the bottom—providing a rich archive of past conditions. Understanding what is written in this record, however, is not easy and requires considerable scientific detective work.

Early paleolimnological studies in Oregon focused on just the most visible geological evidence of past lake levels, such as old shorelines. A breakthrough came with the discovery of fossilized bones of fish, birds, and mammals, such as those in the rich trove at Fossil Lake, Oregon (Allison 1966; Hargrave 2009), which then led to an exciting realization: the marriage of geology and paleontology could help scientists date and explain the environmental conditions under which the lakes had formed. As this science of paleolimnology advanced, its practitioners created a variety of new tools to help them determine the age of the sediments, as well as the environmental changes that had occurred in the lakes' water depth, water chemistry, temperature, and productivity (the capacity to serve as a rich ecosystem for a variety of interconnected plants and animals).

Paleolimnologists learned how to find clues about these ecological changes by analyzing the chemical composition of fossilized remains, including those containing ancient mollusk shells and ostracod valves. Ostracods—tiny, aquatic, two-shelled crustaceans—are often preserved in lake sediments, and researchers can recognize each species by observing the minute details of its shells. Throughout the history of the lakes, each ostracod species has lived under a limited and specific range of ecological conditions.

Fossilized ostracod valves collected from sediments exposed by a road cut along Highway 395 at the north end of Lake Abert. Most of the specimens are from *Candona* sp., which is the largest. The valves are mostly 1–1.5 mm long.

Therefore, knowing which species were present at any given time provides clues about the conditions of the lake in that era.

Materials deposited in the lakes can also provide evidence of conditions on the nearby land. For example, some paleolimnologists study the pollen preserved in oxygen-poor lake sediments, and, by doing so, they can identify plants that once lived in the surrounding area and can determine what the environment and climate were like.

Paleolake Chewaucan

In South Central Oregon, Paleolake Chewaucan has become a key site for regional paleolimnological studies (which began with Ira Allison's 1982 report) because its basin has had low rates of erosion, owing to scant precipitation, and thus, some ancient landscape features have remained intact. In addition, the downcutting action of the Ana River has exposed old sediment deposits there. These deposits contain *tephra,* volcanic ash and larger particles that erupted in the Cascades and elsewhere over tens of thousands of years and which fell into the paleolake. Scientists have been able to discretely sample and analyze the tephra because it appears in recognizable strata. In fact, nearly one hundred layers of tephra, exposed along the steep Ana riverbank, were deposited long ago from volcanic eruptions hundreds of miles away, with the most recent ash falling from nearby Mount Mazama about seven thousand years ago. The near-ideal paleolimnological conditions present at Summer Lake have also led to a considerable number of published studies.

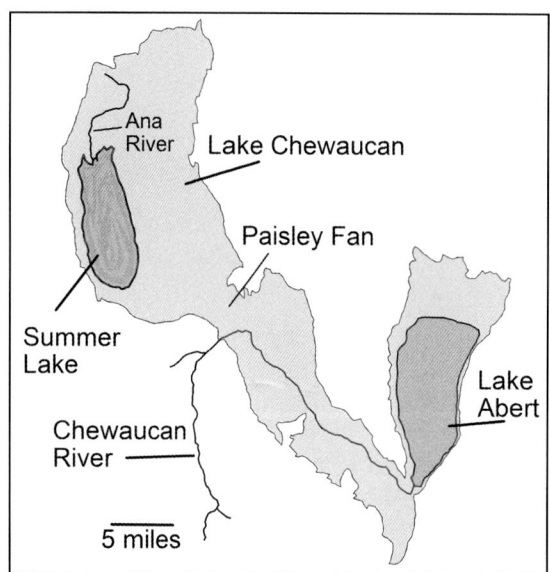
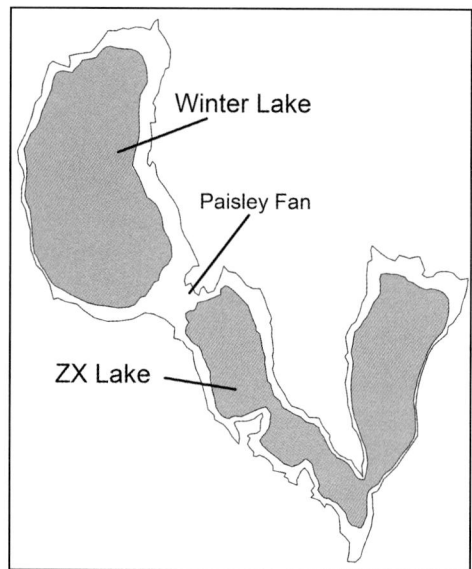

Approximate expanse of Paleolake Chewaucan in different eras: *left*—during the late Pleistocene; *right*— early in the Holocene, during the development of Winter and ZX Lakes. The darker gray areas in the map on the left approximate the current areas of Abert and Summer Lakes. Redrawn, based on original sketches from Grayson (1993).

Although Paleolake Chewaucan—whose maximum area was 480 square miles and maximum depth was over 300 feet—was small compared to Paleolakes Bonneville and Lahontan, it has attracted numerous paleolimnologists. They have found varied evidence for dating the lake's shifting levels, including: distinct shoreline features, fossils of freshwater organisms, nearly one hundred recognized layers of volcanic tephra, and lake sediments that have preserved evidence of reversals in the earth's magnetic field (Cohen et al. 2000; Negrini et al. 2000; Licciardi 2001; Negrini 2002). These multiple sources of evidence have helped researchers to understand not only when the lake's elevation changes occurred but also what the climate was like then.

Although there are numerous indicators of Paleolake Chewaucan's water-level changes throughout its history, the most conspicuous ones are shoreline terraces formed by the action of waves cutting bench-like shapes into the hillside. Drivers on Highway 31 can see two of these terraces at Tucker Hill, also called Shoreline Hill, located between Valley Falls and Paisley. Here, on the west side of the highway, you can see two long benches near the base of the hill. If you look carefully, you can also find less-distinct marks, evidence of ancient shorelines, on the lower slope of Abert Rim, just south of Poison Creek, and at many other places around the basin.

In the 1980s, Allison identified at least six Paleolake Chewaucan shoreline features that formed during *standstills,* or periods when the lake remained at one level long enough for wave action to cut into the hillside (Allison 1982). Paleolimnologists, however, are somewhat limited in their ability to fully document the chronology of lake levels based solely on such shoreline features, because the ones formed during earlier standstills at lower lake elevations could be erased by erosion during more recent standstills at higher water elevations. Thus, using landscape features alone to assess old lake levels can be misleading.

Fortunately, other evidence can help scientists assess the history of a lake's levels. One such source of additional evidence for Paleolake Chewaucan's water-level changes is the presence of tufa on the basin's rocks and boulders. Although researchers have found most of this tufa no more than fifty feet above Lake Abert's current level, they have also found some on rocks one hundred feet or more above the lake.

Additional tools are required to date the changes in Paleolake Chewaucan's elevation levels. One accurate way to do this is by measuring the amount of radiocarbon in organic materials like fossils. Both carbon-12 and carbon-14 isotopes (similar chemical elements) are present in the atmosphere, and scientists can measure their ratio, which varies over time. This radiocarbon decays at a known rate, called its half-life, for the time required for half of it to break down. During the lifetimes of ancient snails and ostracod crustaceans, their shells acquired these two carbon isotopes in the same ratio that was present in the atmosphere at that time. Scientists can estimate the time in history when the lake creatures' shells acquired the carbon isotopes by measuring the isotopes' ratios in the current samples and knowing the rate that the isotopes change over time.

Two distinct Paleolake Chewaucan shorelines revealing themselves at Tucker Hill, at elevations of 4,370 and 4,455 feet and visible along the west side of Highway 31 between Paisley and Valley Falls.

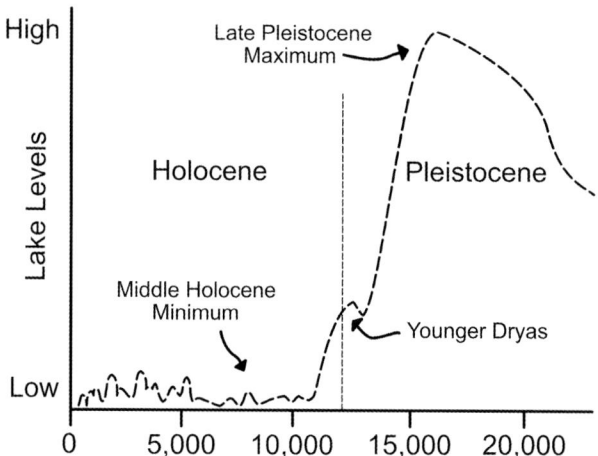

Approximate lake levels during the past 20,000 years in the northwestern Great Basin. Because the water levels are based on only a small number of carbon-14 dates, and thus are incompletely known, the water level line is dashed. The dashed vertical line near Younger Dryas represents the Pleistocene-Holocene boundary. Redrawn, based on data from Negrini (2002), Ibarra et al. (2014), and Wirston and Smith (2017).

Based on paleolimnologists' extensive studies in the northwest Great Basin, the region's paleolakes reached their highest recent elevations between thirteen thousand and seventeen thousand years ago, several thousand years after the last maximum glacier coverage in the southern Cascade Mountains, which occurred between eighteen thousand and nineteenth thousand years ago (Rosenbaum and Reynolds 2004). Like lakes, glaciers advance when the climate is cool and wet and retreat when the climate is warm and dry. Thus, the high lake levels present near the end of the Pleistocene epoch and during the early Holocene period were not due to the glaciers melting, because by that time, the glaciers were too small to account for the large rise in the levels.

High elevations of the northwestern Great Basin paleolakes apparently did not occur simultaneously during this late Pleistocene period. Lake Chewaucan's water level probably peaked more recently than Lake Warner's or Lake Surprise's and much more recently than that of Fort Rock Lake. Many factors likely contribute to these apparent discrepancies, some related to geological factors and some related to scientific approach, including local topography, regional tectonic activity, different data sets employed, and the kinds of evidence used (Wirston and Smith 2017).

After northwestern Great Basin paleolakes peaked, they may have temporarily risen again during the *Younger Dryas,* a brief period of global cooling and increased precipitation. The Younger Dryas caused a temporary rise in lake levels and an expansion of wetlands in the Chewaucan Basin. This finding is based on a study by a researcher who found pollen from fir trees, pine trees, and cattails in Paisley Cave, a cave located north of Paisley that was formed by Lake Chewaucan's waves (Saban 2015). This discovery

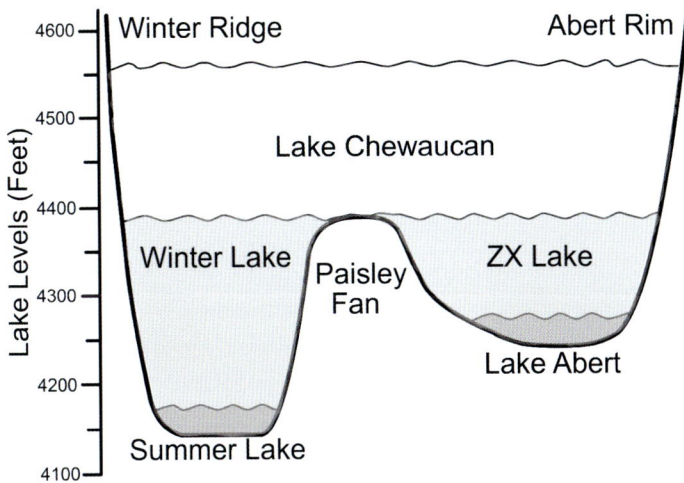

Diagrammatic cross section of the Chewaucan Basin, from Winter Ridge to Abert Rim, showing three representative water levels from sometime in the late Pleistocene or early Holocene. The upper water line of approximately 4,560 feet represents the last period of maximum high water in Paleolake Chewaucan, which is thought to have occurred around 26,000 years ago (Egger et al. 2018). The middle line represents the lake near the end of the Pleistocene, around 14,000 years ago, when water levels declined to Paisley Fan's elevation of approximately 4,400 feet, creating Winter and ZX Lakes (Egger et al. 2018). After that, much drier conditions prevailed and likely caused Winter and ZX Lakes to recede further, and by about 9,000 years ago, water had receded so much that it created Abert and Summer Lakes.

provides strong support for the likelihood of relatively cooler and wetter conditions, because pine and fir trees no longer grow in this arid environment.

After the Younger Dryas, Lake Chewaucan once again receded, decreasing to a level below 4,390 feet, which is the elevation of the Paisley Fan, a high point of land separating Summer Lake's basin from that of Lake Abert. When the water level dropped that low, Lake Chewaucan split into two lakes, referred to as Winter Lake and ZX Lake by geologist Allison (1982). As these two lakes gradually dried up during the Holocene period, only the lowest areas in the Chewaucan Basin contained water, areas that eventually became Abert and Summer Lakes. Summer and Abert Lakes are now in separate subbasins, and the former is one hundred feet lower.

More-Recent Geological History of the Great Basin Lakes

During our present Holocene period, Great Basin lakes have never reached the high elevations that occurred during the Pleistocene epoch and have varied a great deal in size and depth, depending on the amount of precipitation and evaporation that have occurred (Santi et al. 2019). However, it seems clear that the northwestern Great Basin was especially dry during the middle Holocene period between six thousand and nine thousand years ago. When this middle period ended, effective moisture increased, and lakes and wetlands briefly expanded. For example, in an area called Paleolake Beasley in the Fort Rock Basin, wetlands expanded between four thousand and six thousand years ago and again between three thousand and four thousand years ago (Jenkins et al. 2004). Summer Lake's levels may have elevated during that same period (Egger et al. 2018). While climate has been highly variable during the latter part of the Holocene period, the lowest water levels we have seen may be similar to those that occurred during the arid conditions present in the middle Holocene.

Paleolimnologists have also found evidence of variable water levels in Lake Abert that occurred during the past several thousand years in lakeshore archaeological sites with dates ranging from 500 to 3,500 years ago and located over 30 feet above the current lake level (Pettigrew 1980, 1985). However, by about 500 years ago, humans apparently had abandoned their villages at Lake Abert and along the lower Chewaucan River, which suggests that the lake levels had gotten so low and the salinity so high that critical foods in the lake and marshes had disappeared. Based on tree-ring analyses, drought conditions prevailed at Goose Lake during the 1420s, the 1430s, and in the mid-1630s (Nebert 1985). In addition, central Oregon experienced a severe, decade-long drought in about 1480 (Pohl et al. 2002). Thus, we have evidence of one or more severe droughts in South Central Oregon at about the time of the last documented human occupation of the Lake Abert shoreline. Clearly, despite periods of expanded lakes and wetlands, the middle and more recent Holocene eras have never had lakes and wetlands as expansive as those during the late Pleistocene period.

Lake Abert water levels have also varied in the past several hundred years. Researchers have estimated that gravel terraces visible above the Lake Abert shoreline are a few hundred years old (Phillips and Van Denburgh 1971). These are present ten to fifteen feet above any recently observed water levels and nearly twenty feet above recent averages.

Deeper Isn't Always Better

Although we know that Paleolake Chewaucan experienced widely varying lake levels in the late Pleistocene epoch, we don't know what its ecosystem was like when water filled much of its basin. However, because the climate was cool—perhaps as much as 12°F cooler on

Gravel beachlines at Lake Abert providing evidence of higher water levels in the past. Studies of tree rings show that a wet period occurred in the early 1800s and suggest that these beachlines are about two hundred years old.

average than it is now (Hudson et al. 2017)—we can assume that the lake was undoubtedly cold, deep, and relatively unproductive. Under such conditions, and because of extended periods of cloud cover that reduced light levels, the lake's nutrients would have been concentrated in deeper water. Therefore, the growth of algae—the basis for the food chain—would have been minimal. Thus, lake conditions during the period were likely similar to those of the low and ultra-low productivity lakes now found in the Cascade Mountains (Johnson et al. 1985).

During the period when Paleolake Chewaucan was deep, marshes would have been poorly developed, due to steep shorelines, low temperatures, ice damage, and scarce nutrients. However, once the Holocene period began, temperatures increased, lake levels declined substantially, and conditions for marsh development improved. In the Warner Basin, marsh peats started accumulating early in the Holocene, approximately ten thousand years ago (Wirston and Smith 2017). More marshland meant a more productive lake, and vice versa, in a feedback loop. Under those conditions, the new lake and its wetland environment would perhaps have been similar to Upper Klamath Lake today, which is highly productive. Marshes and other shallow-water habitats would have provided ideal conditions for waterbirds, as well as for fish like tui chubs. An abundance of food, as well as such useful marsh vegetation as bulrush and cattails, would also have attracted people to settle nearby.

Historical Lake Abert Hydrology

> We measure the humidity by the amount of sand in the air. When it rains, we keep our hired man in—we want all of the water on the land.
>
> —R. A. LONG, 1964

Hydrology—the study of water quantities in lakes, streams, and rivers—is hugely important in the arid West, where there is scant precipitation and high evaporation rates, and where the water that remains is so critical for humans and ecosystems. Understanding stream flows and lake levels is not only vital now: it will become even more essential in the years ahead, as the climate gets drier and more variable.

In the United States, we usually measure water quantities in two ways: as flow—cubic feet per second (CFS)—for rivers and streams; and as volume—acre-feet (AF)—for lakes and reservoirs. The two units are related, so that one CFS equals approximately two AF per day. An acre-foot of water is enough to cover one acre to a depth of one foot. It is often convenient to measure large water volumes in units consisting of one thousand AF (one TAF, or one thousand acre-feet). We can also refer to annual stream flow, or discharge, as an average in CFS.

Hydrologists in the West usually report annual water quantities on a "water-year" (WY) basis, which lasts from October 1 to September 30 and which they date based on the next calendar year. Thus, on October 1, 2022, the 2023 WY began. Use of the WY makes sense in the West because snowfall in the mountains starts accumulating in November and may not melt until the following spring, and so it actually covers parts of two calendar years.

Lake Abert's water levels have varied considerably over time, both because of changes in climate and, more recently, because of flow modifications by humans. Based on a variety of pioneering hydrologic studies, we know that the lake's annual water budget, or volume, depends on four main factors: (1) inflows from the Chewaucan River; (2) inflows from springs and creeks around the lake; (3) rain and snow falling onto the lake; and (4) evaporative losses.

Because Lake Abert is a terminal lake and has no outlet, and because its seepage losses through the clay-rich sediment are insignificant, its volume at the end of the water year equals the amount of water present at the beginning of the water year—measured as the lake's level—plus the sum of inflows and precipitation, minus evaporation. Thus, if inflows plus precipitation exceed evaporation,

Lake Abert's water budget, which is created by inflows from the river, inflows from springs and creeks around the lake, rain and snow falling onto the lake, and evaporation losses.

water levels will rise, and if evaporation is greater than the sum of inflows plus precipitation, lake levels will decline.

Although this model is simple in theory, its application proves difficult. This is because hydrologists can only accurately measure the level of the lake. The model's other parameters vary both seasonally and annually, so assessing them is tricky. Also, the state's gauge for measuring inflows is located far upstream above Paisley, but there are numerous ungauged diversions downstream, so the amount of water that reaches the lake must be calculated based on estimates of the other parameters.

Chewaucan River Flows

It is the proper destiny of every considerable stream in the West to become an irrigation ditch. It would seem the streams are willing. They go as far as they can, or dare, towards the tillable lands in their own boulder fenced gullies—but how much farther in the man-made waterways. It is difficult to come into intimate relations with appropriated waters: like very busy people they have no time to reveal themselves. One needs to have known an irrigation ditch when it was a brook, and to have lived by it, to mark the morning and evening tone of its crooning, rising and falling to the excess of snow water; to have watched far across the valley, south to the Eclipse and north to the Twisted Dyke, the shining wall of the village water gate; to see still blue herons stalking the little glinting weirs across the field.

—MARY AUSTIN, 1903

In *The Land of Little Rain,* Mary Austin mourns the demise of many western streams, especially those in the Intermountain West, which have dramatically changed, due to the development of irrigation. The Chewaucan River shares this fate.

Lake Abert gets almost all its water from the river, which enters the lake at the south end when there is a sufficient flow. Hydrologists have calculated the discharge of the river, using the flow-measurement gauge located upstream of the town of Paisley, for nearly a century. Each day, the Oregon Water Resources Department (OWRD) posts flow measurements and historical flow data on its website (http://apps.wrd.state.or.us/apps/sw/hydro_near_real_time/display_hydro_graph.aspx?station_nbr=10384000).

Historical records show that the highest Chewaucan River discharge occurred in December 1964 at 6,500 CFS. That was during the Christmas flood of 1964, which caused loss of life and was very destructive in many parts of the West. At the other extreme, the river has stopped flowing for brief periods, during severely cold winters, due to freezing. Above Paisley, river flow is mostly a function of discharge from tributary streams like Dairy Creek, which drains the east side of the 8,000-foot-high Gearhart Mountain. Below Paisley, the river system has been highly altered by channelization and by the construction of a complex network of weirs (or dams) and ditches that divert water in what was once the Upper and Lower Chewaucan Marshes (see chapter 1).

Early maps and notes from General Land Office surveyors suggest that no distinct channel used to run through the Upper Chewaucan Marsh. Instead, during high spring flows, the river flooded out into the marsh, and as upstream flows receded, the water in the marshes drained back into the river and flowed downstream to the lake. Just how much of the marshes' overflow actually reached the lake is unknown. However, we can assume that substantially more river water reached the lake in the past than does so today.

Prior to modern-day water diversions, evaporation loss from the marshes was small because peak flows occurred in April and May, before the marsh vegetation had developed and before the temperatures were warm enough to greatly increase evaporation. Today, however, the river system was altered specifically to help farmers apply water to as many acres as possible. Thus, agriculture removes more water from the river, especially under low-flow conditions, than naturally evaporated in the past. Furthermore, flood irrigation causes water to pool in low areas, forming temporary lakes (see chapter 1), which increases the loss of water through evaporation.

The OWRD, which regulates water use within the state, reported in 1989 that farmers had irrigated 45,000 acres in the Chewaucan Basin with surface water. The *duty,* or the amount of water that farmers with water rights can legally apply to their land, averages 4.6 AF per acre, meaning that those with water rights can apply a total of over 4.5 feet of water per acre each year. If farmers fully accessed all available water rights, they would use nearly twice the median annual flow of the river's water, which would be higher than the total annual flow of the river 96 percent of the time. While it is unlikely that farmers would remove all of the legally available water from the river every year, it is clear that their reliance on irrigation has the potential to take away nearly every drop.

Additionally, the quantity of diversions permitted by OWRD has exceeded the historical seasonal river flow for every month except May. This condition—in which permitted water rights exceed available flow—is called *overallocation.* This means that in most years, especially in dry ones, there is little or no water remaining to support the river's environment and downstream lake ecosystems.

Western water law is based on the Prior Appropriation Doctrine, which allows a person to receive individual water rights when they are the first to use the water for an apparently beneficial purpose. The doctrine was developed as a means of ending quarrels among miners, who would remove all the water in a stream to operate their equipment, leaving downstream users dry. Oregon enacted such water rights in 1909 to settle disputes among these water users. The state developed these laws at a time when much of Oregon was sparsely settled and political corruption was rampant, so the needs of future generations and the health of the environment were largely ignored. Unfortunately, water policies have not kept pace with the needs and priorities of contemporary Oregonians (Bastasch 2006), nor do the laws and the way that OWRD regulates them reflect the reality that climate change is quickly reducing supplies. Consequently, the very agency whose job it is to manage Oregon's valuable water supplies for all the state's citizens may leave the river and lake ecosystems high and dry.

Historic ZX Ranch barn, located near Paisley.

According to OWRD's 2016 Water Right Point of Diversion Summary Report, the state has awarded permits for more than 450 water-diversion projects in the Chewaucan River Basin. The Chewaucan Land and Cattle Company, also known as the ZX Ranch, has received the water rights for over 60 percent of these diversions. A rancher founded ZX in the 1880s with land that he likely obtained under the 1870–1895 Swamp Land Act of Oregon. Although the Act was designed to help small farmers acquire land east of the Cascades, the mismanaged and corrupt program mostly benefited a few wealthy land speculators and cattle barons who intended to monopolize both the land and the water (Pintarich 1980). Under the Swamp Land Act, these speculators and barons acquired many acres of public land at no cost at all or for as little as 20 cents per acre. Eventually, just a few cattle barons acquired much of the land in Oregon east of the Cascades, and that created tensions with other settlers.

The original ZX Ranch owner eventually sold the land to the Kern County Land Company in California. Since then, it has been resold six times, most recently to J. R. Simplot, a multibillionaire who was once listed among the one hundred richest Americans and who purchased the ranch in 1994 (Welch 2017). The multinational company that he founded, J. R. Simplot Inc., is one of the largest privately held agribusinesses in the world, running cattle on the seventy-three-thousand-acre ranch and leasing over one million acres of public land for raising and grazing thirty thousand cow/calf pairs. In addition to ZX, the company owns other large ranches in

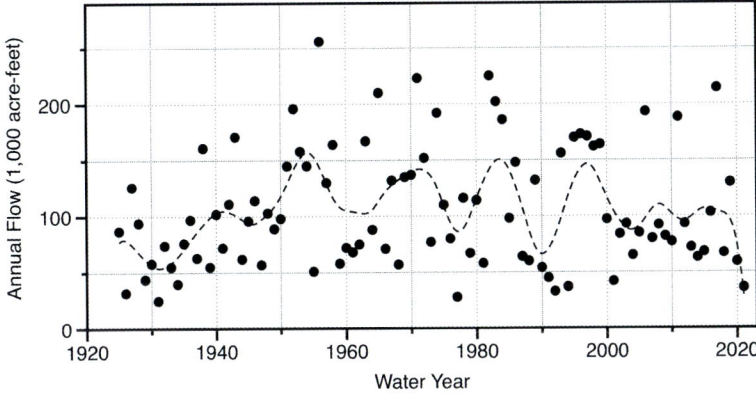

Total annual Chewaucan River flow, WY 1925–2021, measured above Paisley at the OWRD gauge. The dashed line is a smoothed curve that shows longer-term variations in discharge.

California, Idaho, Nevada, and Utah, as well as a huge cattle feed-lot in Idaho. Before his death in 2007, Simplot said that ZX Ranch was "not just a ranch. It's an empire!" (Hadley 1998).

The spread of such massive agribusinesses has made tracking the Chewaucan River's flows more crucial than ever. Hydrologists now have a broad basis for making their comparisons, having calculated the average daily flows of the river for more than ninety years. From 1925–2021, these average daily flows have varied considerably, from a low of twenty-five CFS in 1931 to a high of 350 CFS in 1956. Average annual flow over this extensive period of record-keeping has equaled approximately 146 CFS, and median flow has equaled 128 CFS.

Closely examining this lengthy record reveals a twenty-five-year period of low flows extending from about 1925 to 1950, followed by a period of higher, but variable, flows extending to 1985. Since 1985, flows have continued to vary and to be somewhat lower. Over fifty years, from about 1950 to 2000, there have been peaks and valleys at approximately fifteen-year intervals. A recent series of mostly declining flows that began in the late 1990s is likely an indicator of climate change.

Lake Abert Inflows

Besides the Chewaucan River, Lake Abert has three small creeks—Coldwater, Poison, and Juniper—that contribute seasonal surface or subsurface flow to the lake in wet years. Additionally, there are numerous springs and seeps located around the lake (see chapter 1). As a result of low water levels in 2014, observers could actually see these numerous springs and seeps as the water flowed across the alkali-covered sediment. Many of these springs were too small to

October 2016 aerial view of the braided "river," which is located along the eastern shore of Lake Abert and which flows south from the Mile Post 74 Springs during periods when the lake is very low. Highway 395, located on the right side of the image, provides a distance scale. Image courtesy of Joe Eilers, MaxDepth Aquatics, Inc.

produce enough flow to reach the lake during the recent desiccation event, but some larger ones contributed water to the lake.

The biggest of these was the Mile-Post 74 Spring complex located near Highway 395's milepost marker. The presence of emergent marsh vegetation, easily seen in Google Earth images, shows the several-mile-long extent of this spring complex. Its significance became apparent in the summers and falls of 2014–2016, and again in 2021–2022, when it created a sluggish "river" over one hundred yards wide and several inches deep, which flowed south to what was left of the lake. In 2014 and 2021, water from the Mile-Post 74 Spring complex likely accounted for most of the increase in lake area during the summer, because little or no Chewaucan River flow reached the lake.

The Ups and Downs of a Great Basin Playa Lake

When Frémont and his men reached Lake Abert in December 1843, he described the presence of "a white efflorescence that lined the shore like a bank of snow." This appearance of salt on the shore suggests that the lake's water was relatively low then. Geologist Israel Russell created the first known illustration of the entire lake in the

Water-level staff gauge at Lake Abert, which is attached to a large boulder on the eastern shore about midway along the lake. When the lake is low in a dry year, as shown here in a photo from May 15, 2014, the water level cannot be directly measured at the gauge. A hydrologist, instead, needs to measure the water level by using a surveying instrument and the gauge as an elevation reference.

early 1880s (see chapter 2). His depiction appears to show a high lake level, which seems likely because Goose Lake had reportedly overflowed in 1881. Within recorded history, Lake Abert reached a maximum elevation of approximately 4,260.5 feet above sea level in 1958, and at that time, the water drowned juniper trees growing along the shore that were over 50 years old (Phillips and Van Denburgh 1971). At that elevation, the lake covered 64 square miles, contained an estimated volume of 500,000 acre-feet, and had a maximum estimated depth of 15 feet.

In 1951, hydrologists installed three permanent staff gauges in the lake. After that, staff from the Soil Conservation Service and from the Oregon State Engineers office in Lakeview measured Lake Abert's levels at varying intervals. Recently, OWRD installed a new lower-elevation gauge because the previous one was destroyed. Oregon state officials very rarely read it, so volunteers from as far away as Portland carry on this work.

The infamous Dust Bowl drought of the 1920s and 1930s caused extreme hardships in the High Plains, devastation captured on film by the noted photographer Dorothea Lange during the Great Depression and, more recently, by documentarian Ken Burns in 2012. Few

Oregonians know that the Dust Bowl also severely impacted South Central Oregon. Some natural scientists describe that period in Oregon as a once-in-500-years event (Pohl et al. 2002), meaning that, based upon historical records and measurements made of tree rings going back nearly a millennium, such conditions should occur, on average, only twice every 1,000 years. During the 1920s and 1930s, Chewaucan River flows were only 25 percent of their long-term average, and as a result, Lake Abert was nearly dry six times and reached its lowest documented elevation of approximately 4,244 feet above sea level, just about a half foot above the deepest reported part of the lake (Phillips and Van Denburgh 1971). At that elevation, Lake Abert covered approximately 10 square miles and had an estimated volume of only three thousand acre-feet. The lake again decreased to that low elevation in 2021–2022.

Beginning in about 1950, and extending into the 1960s, above-average precipitation brought the lake up gradually to a record high level of 4,260.5 feet in 1958, a level it nearly achieved again in the early 1980s. Then, due to very low precipitation and to removal of water by upstream diversions in 1992 and 1994, the lake steadily declined, reaching a low level of less than 4,250 feet in 1994. The lake's level cycled up and then back down over a number of years, eventually sinking to an elevation of less than 4,248 feet above sea level by 2009. Consequently, a 3-mile-wide mudflat emerged, extending across the north end of the lake. The lake had not been that low since about 1950.

Despite high inflows in 2011, Lake Abert continued to experience a steady loss of surface area, which was most pronounced

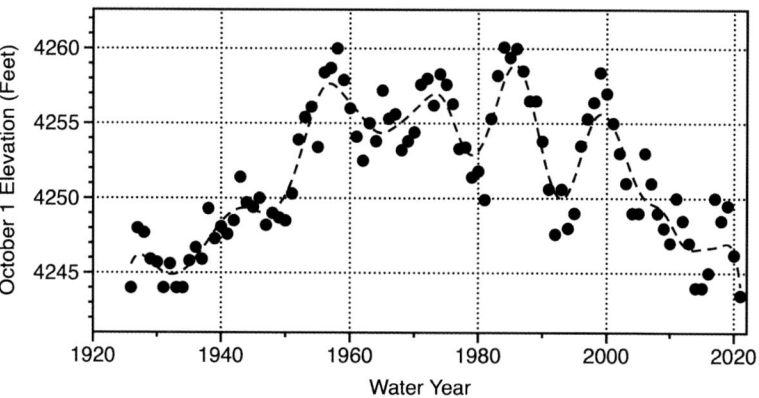

Estimated Lake Abert elevations on October 1, from 1926 through 2021. The dashed line is a smoothed curve showing trends in water levels. Data are from Phillips and Van Denburgh (1971), Keister (1992), Larson et al. (2016), and unpublished sources.

between 2013 and 2014, when it shrank from 35 square miles to only 10 square miles. In 2014, the lake reached the lowest elevation level and smallest surface area that it had experienced since the end of the Dust Bowl, falling to a level of 4,245 feet or possibly less: I could not directly measure the elevation then because the lake was several miles away from the gauge. By 2021, the lake had become nearly completely desiccated again and had shrunk to an elevation of approximately 4,244 feet, which is approximately the lowest elevation ever, and in 2022 it was also desiccated except for a pool along the east side that was created by springs.

This extensive record shows us that Lake Abert's hydrology includes episodic periods of high water, which occur infrequently in high-precipitation years, and longer periods of declining elevations, when evaporation exceeds the total amount of water from inflows and precipitation. Because yearly evaporation from the lake amounts to more than three feet, and precipitation is often less than one foot, it takes only a few years of low inflows to dramatically shrink the shallow lake.

Lake Abert is not the only Oregon lake to experience multiple desiccation events in recent years. Landsat satellite images taken in July and August 2014, and again in 2021 and 2022, show that Summer, Goose, Harney, and the Warner Lakes had dried up and that nearly all the other lakes in the region had lost all or most of their water.

We don't totally understand the causes of this recent desiccation, but we realize that multiple factors are probably involved. For Lake Abert, upstream water diversions were likely key to 2014's

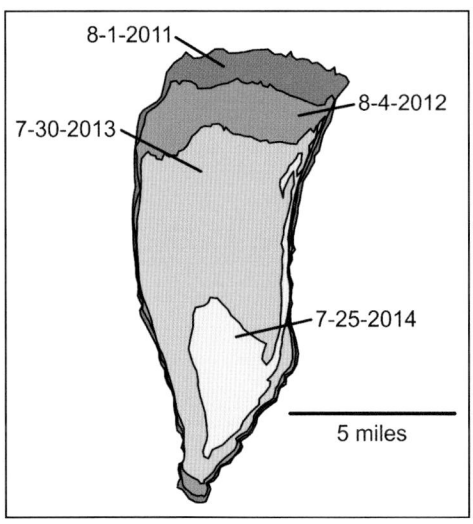

Changes in Lake Abert's surface area, occurring from August 2011 to July 2014, based on four mid-summer Landsat 8 images. The gray areas represent the approximate water-surface areas on the four dates: fifty-two square miles on August 1, 2011; forty-five square miles on August 4, 2012; thirty-five square miles on July 30, 2013; and ten square miles on July 25, 2014.

Landsat images showing changes in the sizes of Goose, Harney, and Malheur Lakes during wet and dry periods: *top left*—Goose Lake during a wet period in 1999; *top right*—Goose Lake during a dry period in 2014; *middle*—Harney and Malheur Lakes during a wet period in 1999; and *bottom*—a similar image of Harney and Malheur Lakes taken during a dry period in 2016. USGS EarthExplorer.

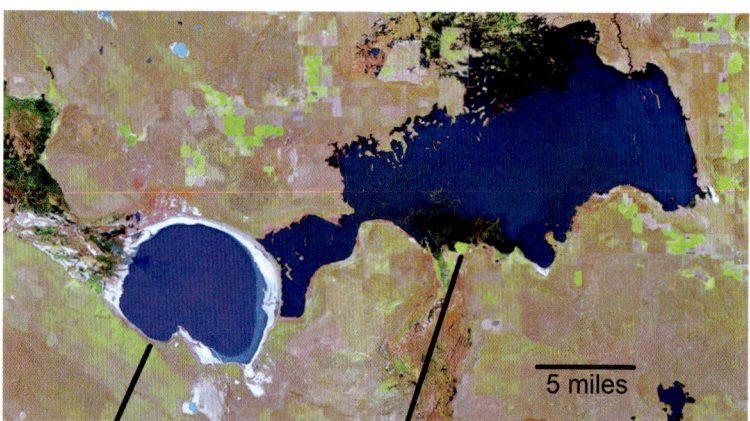

Harney Lake   Malheur Lake

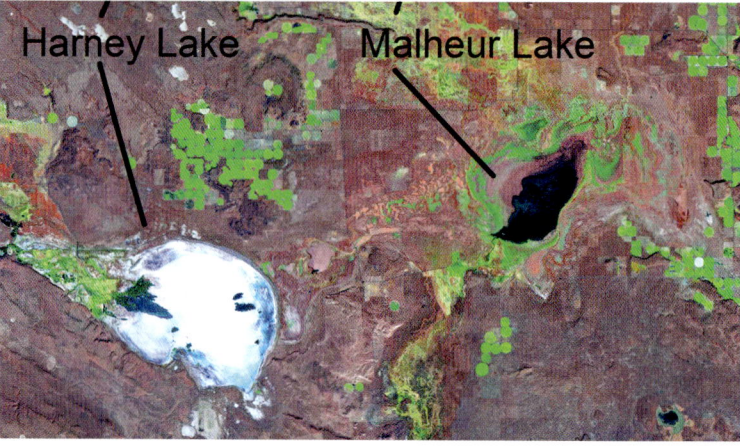

Center-pivot-irrigation system watering an alfalfa field in the Chewaucan Basin. Center pivots can irrigate a circle up to one half mile in diameter. The increased use of center pivots and increased amounts of alfalfa production in the Chewaucan Basin and elsewhere in the region are causing an unsustainable drawdown of the water table.

desiccation (Moore 2016; Larson et al. 2016). Researchers have determined that from 1926 through 1990, average inflow to the lake was 70 percent of the flow measured at Paisley, while from 2000 through 2014, only about 40 percent of the Chewaucan River flow measured at Paisley reached the lake (Keister 1992; Larson et al. 2016). This suggests that in recent years, water diversions between Paisley and the lake might have increased, resulting in much less water flowing into Abert.

Agriculture's increased use of water in the Chewaucan Basin has been key. Ranchers may be using inefficient flood irrigation to compensate for their fields' low soil moisture, which is itself caused by increasingly dry conditions. In addition, some farmers are turning to center-pivot sprinkler systems to irrigate alfalfa fields with groundwater. Although center-pivot irrigation systems are more efficient than flood irrigation, their water-saving benefits are likely negated by irrigating over a longer growing season, and the fact that alfalfa requires more water than grass hay does. And we know that farmers are irrigating more of the land: satellite images show crop circles in areas of the Chewaucan Basin that were not previously irrigated, evidence that agriculture is further depleting the already overstressed water supplies.

Use of groundwater to irrigate crops can lower the water table if the rate of consumption is greater than the rate at which water replenishes the soil. This is particularly problematic in Oregon east of the Cascades, where precipitation is low and variable and water use for crops is high. Data from the use of a well on the ZX Ranch

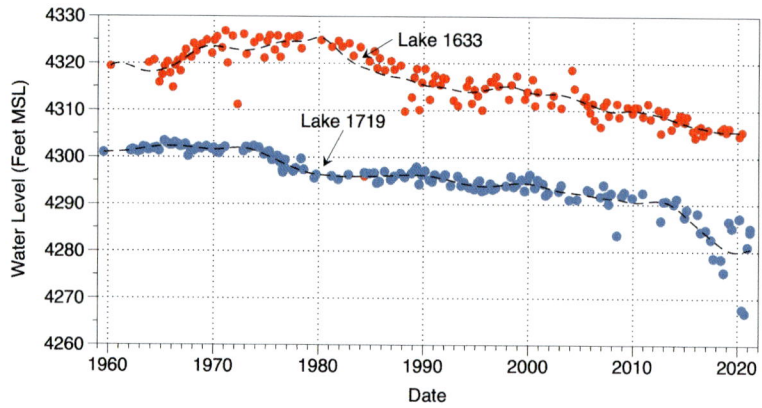

Changes in groundwater levels from 1960 through 2021 for OWRD wells #1633 (shown in red) and 1719 (shown in blue), which are located between Paisley and Valley Falls. Data courtesy of the Oregon Water Resources Department.

in the Chewaucan Basin clearly demonstrates this problem. For the first forty years of the well's operation, the groundwater's elevation slowly but steadily declined about five feet. Then in about 2000, the rate of decline increased, and more recently the drop has been even more rapid, especially in 2020, when the elevation plummeted twenty feet. The ranch's managers are clearly using more groundwater than is being replenished, and they won't be able to continue pumping the water for irrigation if this downward trend continues. Even more concerning is the fact that the substantially lowered water table could impact nearby wetlands, or even the Chewaucan River and Lake Abert, but no one is monitoring the adverse effects on these aquatic systems.

Additional new water uses in the Chewaucan Basin in recent decades include the construction of the River's End Reservoir in the 1990s. Developers built the reservoir with federal funding, which was justified as providing a beneficial resource to the public. In reality, the reservoir is little more than a private lake. Furthermore, during this same period, other projects were implemented to increase waterfowl habitat upstream of the lake. While these projects aimed to restore valuable wetland lost due to irrigation and drainage, they may have harmed the lake ecosystem by reducing inflows. While I generally support habitat restoration, I believe that our efforts to restore our lands and waters must avoid or mitigate adverse ecosystem effects. However, in all these cases, no one seems to have considered possible harmful consequences. Even more recently, PacifiCorp proposed building two pumped-storage hydropower facilities that would use water from the river to fill and maintain four reservoirs. If the government gives the go ahead, these facilities will exacerbate the current problems that the lake faces from too little inflow.

The history of the Chewaucan Basin is one of constant change, but most of the prehistoric transformations occurred slowly over centuries, or even over millennia, which gave species and ecosystems time to adjust. However, during the past 150 years, man's rapid pace of development has proved to be highly destructive to these aquatic environments.

CHAPTER 4

Life in Extreme Water

No fish can live in the water, nor any living thing except little brine shrimp. Chewaucan River, its principal feeder, is filled with fish. At the mouth of this stream there is a fall where fish that have ventured or fallen over these falls are there in evidence to show that nothing can live in Abert Lake. The shores of the lake at this point are composed of dead fish and fish bones. Tons of bones could be gathered up, and at certain seasons of the year the shores are lined with fish in all stages of decomposition. When the fish first strikes the water of the lake it makes for the shore and tries to flounder out, and if it fails, hugs the shore as closely as possible, with its head out of the water until it dies. The geese and ducks and other water fowls that abound in this section do not even light upon the lake, except at the mouths of freshwater streams.

—F. A. SHAVER et al., 1905

Historian F.A. Shaver's imaginary account of masses of dead fish spread along Lake Abert's shore, and of birds refusing to alight on the water, misses the true nature of what a vibrant and productive ecosystem the lake actually is much of the time. It is certainly not as hospitable as the typical freshwater lake, which offers a home to a wider variety of vertebrates, including many kinds of fish and amphibians, and to many invertebrate insects, mollusks, and worms. Salt lakes are typically harsh environments, and Lake Abert is no exception. In the distant past, the lake was deeper and less saline, with fresh enough water to support the organisms typical of freshwater ecosystems, but now it is home to a few specialized species that can tolerate its high and variable salinity, elevated alkalinity, and periodic drying.

Microorganisms dominate the lake, including single-celled archaea, bacteria, unicellular cyanobacteria (which form colonies), and such algae as the diatom (*Nitzschia frustulum*) and the filamentous, green *Ctenocladus circinatus*. The lake also hosts an immense number of brine shrimp (*Artemia franciscana*) and the alkali fly (*Ephydra hians*). Harsh environments like Lake Abert's generally have low biodiversity that consists of only a few species, ones that

Varied organisms that currently live in Lake Abert or in associated springs, as well as fossilized snails that once existed in the lake: *from top to bottom, left to right—Bembidion* sp. beetles; copepod and diatoms; *Artemia nauplius*; alkali-fly pupal cases; female *Artemia*; fossilized *Helisoma newberryi* snail shells; adult alkali flies; cyanobacteria filaments; *Dugesia* sp. flatworm; *Navicula* sp. diatom; and *Hyalella* sp. amphipod.

Looking across the south end of Lake Abert in July 2014, when the water had turned red from the presence of halophilic microorganisms. White alkali salts bordered the lake, covered the lake bottom, and also formed strange circular structures in the water.

have evolved the necessary behavioral, morphological, and physiological mechanisms to cope with the extreme conditions. However, some adaptive species not only tolerate these conditions: they actually require them.

Life at the Extremes

I will always remember the day in July 2014 when I arrived at Lake Abert and was shocked at what I saw. A broad border of brilliant, white alkali encircled the lake, and what remained of its water was bright red, almost as if it were filled with blood. I had never seen the lake in this condition, nor had I expected to, because no one had ever reported seeing it like this. As I later found out, this worldwide phenomenon is actually common for extremely salty lakes, although apparently this was the first time it had happened in a Pacific Northwest lake. Observers have seen red water in the Great Salt Lake in Utah and at Mono, Owens, and Searles Lakes and the Salton Sea in California, as well as in salt ponds around San Francisco Bay. Anyone curious about this phenomenon—who cares to see other-worldly scenes of red water contrasting with glaringly bright, salt-encrusted shorelines—can easily find "red salt lake" images online.

This red water is not due to the toxic, red-tide-causing organisms sometimes found in the ocean. The organisms that occasionally turn saline lakes red are salt-loving, or *halophilic,* ones that thrive in extreme environments, otherwise known as *extremophiles.* They are the only species that can tolerate, and which actually require, hypersaline conditions of nearly 25 percent for their survival. Unicellular, they include bacteria and the very primitive, bacteria-like archaea, as well as some cyanobacteria, fungi, algae, and a few protozoans (DasSarma and DasSarma 2012).

Scientists believe that the red color of Lake Abert in 2014 was due to the presence of salt-loving archaea called *Haloarchaea* (Larson et al. 2016). Archaea are among the most salt-tolerant of all known organisms and are often responsible for the red color of hypersaline environments. More complex inhabitants of Lake Abert, like brine shrimp and alkali flies, prefer salinities that range from 5–15 percent. In stark contrast, freshwater fish, mollusk, worms, and aquatic insects only thrive at salinities of less than 1 percent.

Haloarchaea produce red pigments under high-light intensities and low concentrations of dissolved oxygen, two conditions typical of most salt lakes. One of these pigments, called *bacteriorhodopsin,* gets energy from sunlight, as chlorophyll does in green plants. Some scientists believe that bacteriorhodopsin was the first photosynthetic pigment to evolve, about 3.5 billion years ago in cyanobacteria. Haloarchaea also produce red pigments called *carotenes,* which protect them against ultraviolet damage from sunlight (DasSarma 2007; DasSarma and DasSarma 2012).

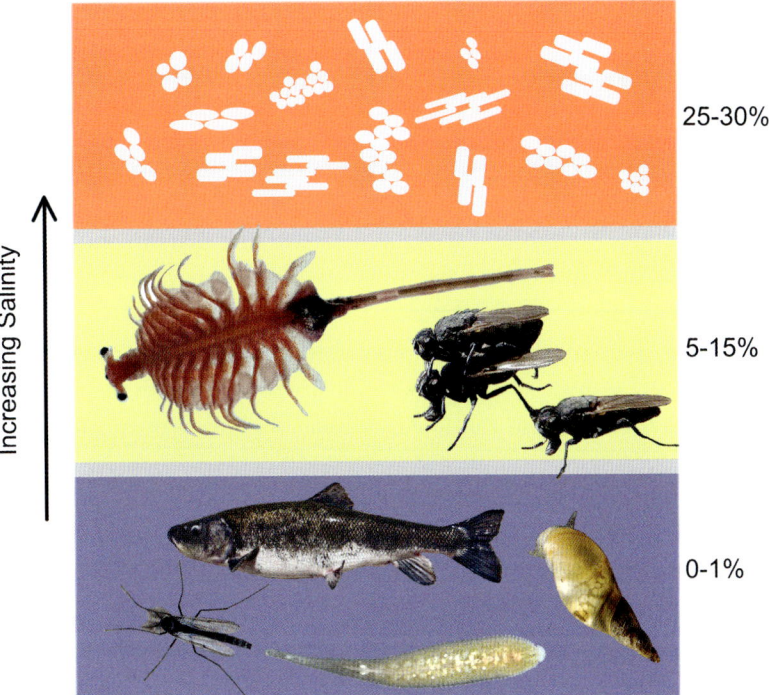

25-30%

5-15%

0-1%

Increasing Salinity

Effects of increasing salinity on aquatic organisms: the blue box on the bottom represents fresh water with a salinity of less than 1 percent that is inhabited by fish, mollusks, worms, and many aquatic insects; the yellow box in the middle represents water with an intermediate salinity of 5 percent to 15 percent that is tolerated by just a few macrobiota, including *Artemia* and alkali flies; and the orange box on the top represents hypersaline water with a salinity greater than 25 percent that is tolerated only by such single-celled extremophiles as archaea and bacteria.

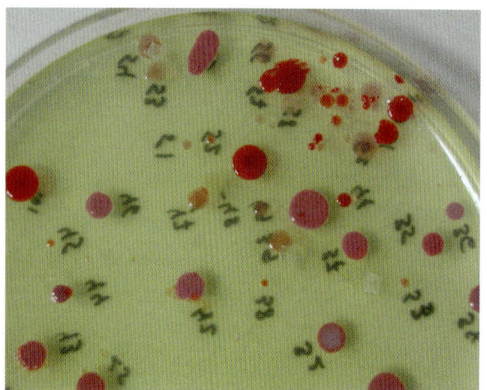
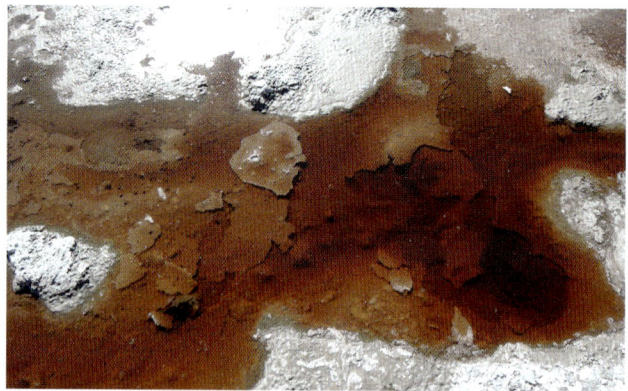

Red-pigmented halophilic microorganisms: *left*—red and pink spots, haloarchaea colonies, were cultured on an agar plate in a lab (photo courtesy of Bras del Port S. A., Spain); *right*—red-pigmented halophilic microorganisms living in a pool of brine were surrounded by salt that the lake deposited as it receded in 2014.

Two microbiologists from universities in Baltimore, Wolf Pecher and Shiladitya DasSarma, have conducted recent studies of water and sediment samples from Lake Abert, using DNA sequencing to focus on their microbiota. The researchers' findings show that the microbial community in the lake is extremely complex and diverse and that many different microorganisms find Abert to be a suitable habitat as salinities change (Larson et al. 2016). Their ability to quickly reproduce as conditions become more suitable enables them to benefit from the rapid environmental changes typical of the lake.

Extreme halophiles may thrive in salinities of up to 30 percent, the saturation concentration at which halite (table salt) forms and the highest level at which any species remains viable (DasSarma and DasSarma 2012). They can survive even when trapped inside salt crystals. Most organisms would die under such harsh conditions, which would quickly suck the water out of them, but extreme halophiles have specialized chemicals called *osmoprotectants* that allow them to maintain their internal balance of salt and water.

Most scientists studying halophilic microorganisms have taken samples directly from saline ecosystems, including the ocean and salt lakes. However, in the mid-nineteenth century, some researchers found *Halobacterium,* a rod-shaped species of archaea, growing on salted cod and salt-cured meat (DasSarma et al. 2010). This discovery has helped today's researchers and teachers: they routinely use one *Halobacterium* strain, called NRC-1, because they can culture it easily, it serves as a model organism for experimental studies, and it is suitable for use in classrooms because it is harmless, readily available, and would not be a threat to the environment if released (DasSarma et al. 2016).

Haloarchaea are primitive and able to survive extreme environmental conditions, including intense radiation and desiccation, and scientists have even isolated them from ten-million-year-old salt deposits. Some researchers suggest that these extremophiles are so unusual that they may serve as biological models for life from other planets (DasSarma 2007).

Microbial Mats—An Ecosystem in Miniature

Lake Abert's complex environment also includes *microbial mats,* communities of microorganisms organized into layers only a few millimeters thick and prone to develop under less-extreme salinity conditions, like those occurring near springs along the lake's eastern shore. Microbial mats look like thin, brownish, yellowish, or purplish blankets that cover the mud. The colors and textures of the mats vary a great deal, depending on the organisms, whose own variability hinges on their energy source, the amount of available oxygen, salinity, and other conditions. They can include: (1) an upper layer of photosynthetic, greenish cyanobacteria and golden-brown diatoms; (2) a middle layer of purple, photosynthetic

Microbial mats at Lake Abert: *top—brackish-water seep* with a golden-colored microbial mat, which was located along the northeastern shore of Lake Abert near a patch of three-square sedges; *bottom*—close-up of a microbial mat that probably contained gold-colored diatoms, green algae, and green cyanobacteria.

bacteria, or bacteria that can thrive in oxygen-rich environments during the day and oxygen-poor conditions at night; and (3) a bottom layer of anaerobic bacteria and archaea that use sulfur as an energy source (Caumette 1993; Caumette and Lucas 1994; Risatti et al. 1994; DasSarma and DasSarma 2012). Some of these organisms even migrate vertically as oxygen conditions change.

The microbial mats at Lake Abert mostly comprise masses of green, threadlike cyanobacteria and golden-brown diatoms; ciliate protozoans; and small, threadlike nematode worms. If you look at a sample of a mat under a microscope, you will see the diatoms glide, the cyanobacteria slide, the ciliates cruise, and the nematodes wiggle, and you will quickly come to realize that the microscopic world of these mats is alive with many unseen critters. This list does not even contain the more numerous, but mostly invisible, bacteria, many of which are also mobile.

Scientists believe that microbial mats are among the oldest ecosystems, consisting of primary producers that use sunlight to produce energy; consumers that eat the producers; and decomposers. Their fossils, known as *stromatolites,* date back to the Precambrian era, around 3.5 billion years ago (Bauld 1981; Dupraz et al. 2008). Stromatolites are the remains of some of the first ecosystems to release oxygen into an atmosphere that was otherwise essentially oxygen free (Hoehler et al. 2001; Abele 2002). The ancient microbial mats, therefore, created atmospheric oxygen that fueled vastly greater rates of biological productivity than was possible under anaerobic, or oxygen-free, conditions, which itself led to the evolution of complex, multicellular organisms that need oxygen.

Algae—The Base of the Food Web

One species of plant, *Ctenocladus circinatus*—a branched, filamentous, green alga that lives in saline habitats—is responsible for most of Lake Abert's primary production, or the increase in biomass

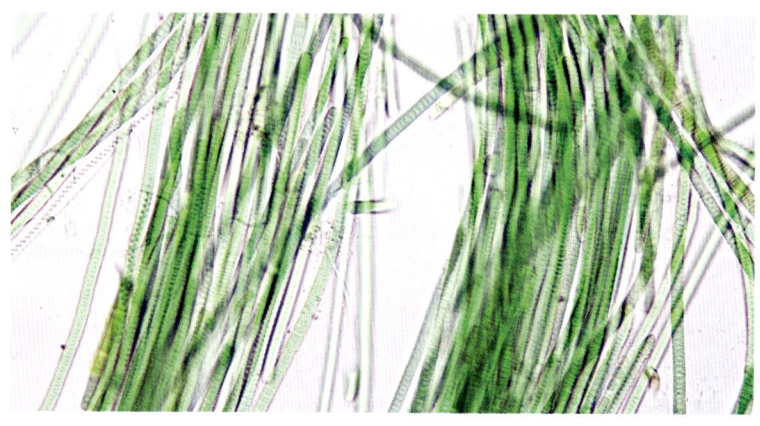

Cyanobacteria (highly magnified) from a microbial mat at Lake Abert. Most numerous are the green filaments of the cyanobacterium *Oscillatoria,* but a few *Spirulina* filaments, with spiral organelles inside them, are visible on the left side of the image. Both are approximately 10–20 μm in diameter (a μm is one millionth of a meter).

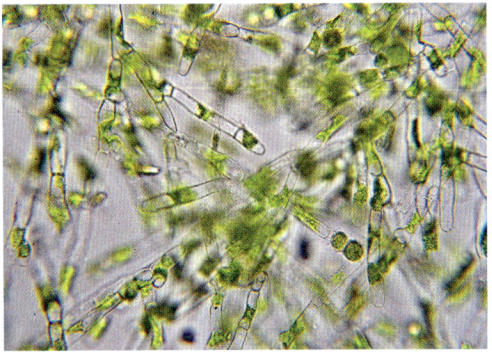

Ctenocladus circinatus, a type of filamentous green algae: *left*—floating *Ctenocladus* balls, approximately 0.5 inches in diameter, and scattered, dark-colored, alkali-fly pupae; *right*—transparent *Ctenocladus* filaments (highly magnified), which show individual cells and the small, green chloroplasts where photosynthesis occurs.

created by photosynthetic organisms. *Ctenocladus* attaches itself to rocks, lies on the lake bottom, or floats in the water. In late summer, waves roll the floating algae around, creating marble-sized balls that form a layer on the lake bottom or at the water's surface, where the wind blows them into concentrated piles near the shore. *Ctenocladus* is most prolific when salinities range between 2.5 percent and 7.5 percent. Its production continues all the way up to a salinity of about 10 percent, but at a reduced rate (Herbst 1986, 1994; Herbst and Castenholz 1994; Herbst and Bradley 2004). Researchers in the lab find that the alga stops producing when salinity goes above 10 percent. You may observe *Ctenocladus* in the lake at salinities as high as 15 percent, but likely only as part of a dying colony.

A variety of microscopic algae (including the planktonic diatom) and such cyanobacteria as *Oscillatoria, Anabaena,* and *Spirulina*

Microalgae from Lake Abert's brackish-water seeps: *top*—*Surirella striatula,* a diatom; *bottom*—*Closterium* sp., an alkali-tolerant desmid (a microscopic green alga). Both types of microalgae are approximately 10–15 μm in length.

are associated with the lake's sediment and rocks (Herbst 1994). Cyanobacteria likely play a role in the way tufa is deposited on rocks. Other organisms in brackish-water seeps and springs along the edge of the lake live in a kind of transition zone, where the salinity and pH are moderate. They include *Closterium,* a type of green algae, and the diatom *Surirella striatula,* which thrives worldwide in saline and marine habitats and is sometimes linked to microbial mats in salt lakes (Caumette and Lucas 1994).

Aquatic Invertebrates—Animals without Backbones

Few multicellular organisms can survive in Lake Abert's caustic, saline waters, although a variety of invertebrates serve as notable exceptions. These hardy creatures can even be abundant, provided that both salinity and food availability are optimal. On the other hand, they nearly vanish when conditions are not just right. Researchers have observed both extremes in the past three decades.

One such invertebrate, the alkali fly, nearly disappeared from 1983 to 1986, when Abert's salinity fell to about 2 percent, due to high water levels. Some other invertebrates prevailed during this same period, especially aquatic insects like the water boatman (*Corixidae*) and the water flea (*Moina*), which are adapted to living in fresher water. Drought years then followed, which greatly reduced lake volume and increased salinities, and by 1992, salinities reached as high as 18 percent. By 1994, ecologist Dave Herbst reported that the lake's aquatic community was "all but eliminated," except for small numbers of alkali fly larvae. Within a year, however, wetter conditions returned, lowering salinities to levels optimal for both alkali flies and brine shrimp, both of which proliferated. Other invertebrate species also reappeared in fair numbers (Herbst 1994). Research during the 1980s and 1990s, under varying salinity conditions, showed just how dependent aquatic fauna are on the lake's salt concentrations. And the startlingly high salinity of 18 percent that proved so problematic has since been overshadowed, first by

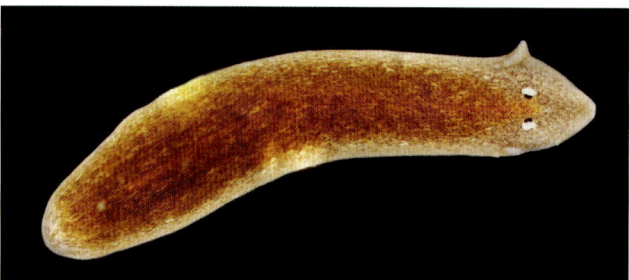

Two invertebrates collected from brackish-water seeps along the eastern shore of Lake Abert: *left—Dugesia,* a flatworm that has a pair of eyespots on its pointed head; *right—Hyalella,* a shrimplike amphipod. Both invertebrates are approximately 0.3–0.4 inches long.

even higher salinities observed from 2013 to 2016 and then again in 2021 and 2022, when the level reached 25 percent or possibly higher.

A few invertebrates also make their homes in the brackish springs and seeps near the lake, creatures including: *Dugesia* sp., a flatworm; *Rhyacodrilus* sp., a small aquatic worm related to the earthworm; and *Hyalella* sp., an amphipod crustacean. Large numbers of the latter burrow in the sand in the upper parts of the spring runs, swim over the substrate, or hide under rocks. Researchers who conducted experiments on *Hyalella* sp. at Pyramid Lake in Nevada found that the crustacean tolerates salinities as high as 10 percent for up to a year (Galat et al. 1988), an indication that it handles salty water relatively well.

During the low water levels that occurred between 2005 and 2016, marshes spread outward toward the lake and created freshwater pools that provided a habitat for aquatic insects like the western pondhawk (*Erythemis collocata*), a large, powder-blue dragonfly.

Between 2005 and 2016, low water levels in Lake Abert allowed marshes to spread, which provided a habitat for some aquatic insects that thrive in wetlands: *left*—the western pondhawk (*Erythemis collocata*); and *right*—a damselfly (*Enallagma* sp.).

Brine Shrimp

> There are no fish in Mono Lake—no frogs, no snakes, no polliwogs—nothing, in fact, that goes to make life desirable. Millions of wild ducks and sea-gulls swim about the surface, but no living thing exists under the surface, except a white feathery sort of worm, one half an inch long, which looks like a bit of white thread frayed out at the sides. If you dip up a gallon of water, you will get about fifteen thousand of these. They give to the water a sort of grayish-white appearance.
>
> —MARK TWAIN, 1872

Of all the interesting critters that live in Lake Abert, perhaps none is as amazing as the brine shrimp, the "feathery sort of worm" that humorist Mark Twain described after he visited California's Mono Lake. I first saw Abert's brine shrimp in June 1965, when my high school biology class took a bus tour around South Central Oregon that included a stop at the lake. We were there just long enough for me to get a water sample containing several ¼-inch-long, pale-pink crustaceans swimming continuously on their backs. I had grown up along the Oregon Coast, so I was familiar with a variety of crustaceans, including crabs, shrimp, and barnacles, but I had never seen anything like these little swimmers, which I later learned were brine shrimp. The *Artemia* struck me as strange looking, and their ability to thrive in a salt lake in Oregon's high desert amazed me even more. At that time, I had no idea that I would return to the lake half a century later to study these tiny creatures and so many other species that live in the lake.

Artemia are not true shrimp but instead are related to so-called fairy shrimp and other primitive crustaceans belonging to the class Branchipoda, which first appeared in the Triassic era, approximately one hundred million years ago. Today, ten species of *Artemia* swim in salt lakes and ponds worldwide (Asem et al. 2010). However, they normally do not live in the ocean or in other habitats where fish and predatory aquatic insects would eat them. In fact, we actually use brine shrimp to feed marine aquarium animals, ranging from coral to clown fish, that require live prey and are very happy to eat them. When I was doing marine biology graduate work in Puerto Rico, I offered newly hatched *Artemia* as food to many kinds of predatory invertebrates, including sea wasp jellyfish, that typically feed on plankton.

Brine shrimp normally make their home in highly saline water, which protects them from fish and aquatic insect predators that can't survive in such an environment. Under laboratory conditions, these little organisms can survive salt concentrations above 30 percent, but their optimal salinity is near 10 percent (Castro-Mejia et al. 2011). *Artemia* feed by filtering microalgae and bacteria from the water as they swim, which is one reason why they move continuously. Swimming also aids their respiration by circulating water around their flattened, feather-like appendages, called *pleopods*,

Pair of adult *Artemia* mating and swimming. The smaller male (on the bottom) holds onto the female. Adult brine shrimp normally swim on their backs. The approximate length of each *Artemia* is 0.4 inches.

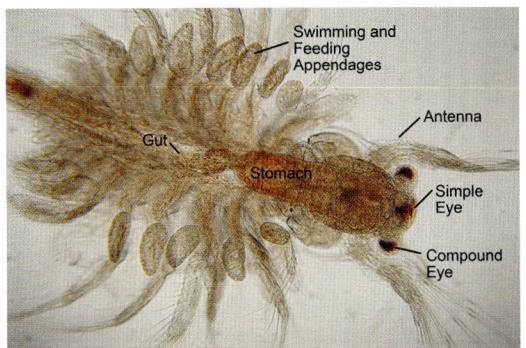

Adult *Artemia* morphology: *left*—the head and thorax of a female are pictured, but not the long tail (*Artemia* have nineteen body segments, eleven of which have flattened swimming/feeding appendages called *pleopods,* and their heads have two compound eyes, a simple eye, and two pairs of antennae); *right*—in males, the second pair of antennae are modified into large claspers for grasping the female during mating.

which serve as gills and for which the Branchipoda—or "gill foot"—class is named. Brine shrimp swim by moving adjacent pairs of pleopods sequentially, from front to rear, creating a wavelike motion called a *metachronal wave,* which propels the animal forward.

Scientists don't know why *Artemia* swim upside down, but they have found that the organisms reverse their orientation when a light source is placed under them. Their light-oriented movement, or *phototaxis,* causes them at times to swim toward the light—positive phototaxis—and at other times to swim away from the light—negative phototaxis. This activity varies according to environmental conditions, including the quality of the light and how hungry the brine shrimp are. Phototactic behavior helps explain why the brine shrimp are sometimes more numerous at the lake's surface at night than they are during the day (K. Kreuz, pers. com. 2014). When *Artemia* swim to a light in the lab, they swarm near

Several *Artemia* swarms—pictured as red patches—in Lake Abert. The swarms are 1–1.5 feet in diameter and inhabit water that is about 1–2 feet deep.

the light. Such swarms also occur in the lake during the day, but they don't appear to be phototactic. These natural swarms are one to two feet in diameter and consist of thousands of shrimp swimming together in a circle and forming a ball that I have called a "shrimp boil." We haven't yet figured out why or how these swarms form. However, I don't believe the behavior has anything to do with reproduction, because most of the *Artemia* in the swarms I have seen are immature.

Part of *Artemia*'s survival strategy in such stressful habitats as Lake Abert is to reproduce in several ways, depending on environmental conditions. Some populations of brine shrimp consist only of females that can reproduce without males. In populations that reproduce sexually, such as the ones at Lake Abert, the male has a pair of enlarged, modified antennae for clasping the female, so that during mating, he remains attached to the female. Under good conditions at the lake, such as when salinity is moderate and both food and dissolved oxygen are adequate, female *Artemia* produce many fertilized eggs, and the eggs hatch soon after being released into the water, thus increasing the population. However, when conditions are stressful, due to high salinity, low dissolved oxygen levels, or a lack of food, females produce "cysts." Although we often refer to these cysts as eggs, they actually are partially developed embryos, protected by waterproof shells, that are in a very early stage of development. The shells protect the embryos from desiccation and ultraviolet damage.

Cysts are metabolically dormant, meaning that they remain inactive—much like a plant seed—while suspended in a process called *cryptobiosis,* or "hidden life." Cryptobiosis allows the cysts to withstand extreme environmental conditions, such as heat of 200°F for several hours, severe cold of −300°F, high amounts of ultraviolet radiation, and prolonged desiccation, and the process can delay hatching for years, or sometimes for decades. Biologists have taken advantage of *Artemia*'s tolerance for harsh conditions, sending them to the moon in 1972, aboard Apollo 16, and into space on several Russian satellites, as part of experiments to study the effects of space travel on organisms.

"Sea monkeys," brine shrimp's other common name, are popular aquatic pets because they are easy to raise. They are also often used for school science projects because of their ability to survive harsh conditions. One of the projects, called "Abuse-a-cyst," directs students to test the survival of cysts under various stressors (http://teach.genetics.utah.edu/content/gsl/).

Artemia's cryptobiotic cysts not only allow brine shrimp to survive in unstable environments for years, but they also enable the little creatures to disperse. Currents and the wind carry and blow the cysts a short distance onto shore at Lake Abert. Scientists have also found evidence that birds disperse the cysts over longer distances when they migrate. Researchers have, for example, linked Lake Abert's *Artemia franciscana* to sites throughout the Americas, where they have located the same species. Additionally, they have discovered viable *Artemia* cysts attached to shorebirds and in the guts of waterfowl (Munoz et al. 2013).

The annual life cycle of *Artemia* in Lake Abert starts in early spring, and very occasionally in winter, when inflows raise water levels and dilute salinities, allowing the cysts that floated to shore during the previous summer and autumn to hatch. The first stage in their lifecycle is the *nauplius,* which is very small and has just a few appendages and a primitive eye spot. As the nauplius grows, it acquires additional segments and legs, and then it becomes a *metanauplius.* Eventually as the shrimp grows, it develops a pair of compound eyes and a full set of swimming legs. When it reaches a ¼-inch to a ½-inch in length, it is ready to reproduce. Reproduction often occurs in May, but in some years, it is delayed until August.

At first, any eggs that the shrimp produces will quickly hatch, increasing the population. By midsummer or late summer, however, stressful conditions often develop in the lake, due to high salinity and related low dissolved-oxygen concentrations, and also due to reduced food levels, and these harsh conditions trigger cyst production. During this period, you can observe the brownish cysts floating on the lake and washing ashore. As water temperatures cool in the fall, many *Artemia* die, but some do survive into winter, when water temperatures are suitable.

Cryptobiosis and the ease at which *Artemia* cysts can hatch under the right saltwater conditions make them highly valuable as food for tropical marine fish and other marine animals that need small, live organisms to eat. Commercial harvests of these cysts

Artemia nauplius (*left*) and metanauplius (*right*): both have a length of about 0.1–0.2 inches, or 3 mm.

High concentration of brown *Artemia* cysts along Lake Abert's shore.

have become big business at the Great Salt Lake and in salt ponds around San Francisco Bay. In the Great Salt Lake, planes are used to spot patches or "streaks" of floating cysts, and then boats speed out to these sites to set out booms that surround the cysts, similar to those used to confine oil spills. Other boats, called harvesters, cruise out to the booms and harvest the cysts using pumps; screens then separate the cysts from the water. Once onshore, workers clean, dry, and put the cysts into cold storage before shipping them out in vacuum-sealed containers. Annual permits to harvest cysts from the Great Salt Lake can cost fifteen thousand dollars, but these harvests have yielded revenues totaling thirty million dollars in some years.

One small business, a real mom and pop operation, has worked to harvest adult *Artemia* at Lake Abert, but the operation has been extremely low tech and labor intensive. Keith and Lynn Kreuz operated the Oregon Desert Brine Shrimp Company, based out of their cabin at nearby Valley Falls, for thirty-five years. In 2014, however, their longtime operation came to a standstill: they could no longer find any brine shrimp to harvest. In a February 2022 letter to *The Oregonian* newspaper, Lynn expressed her concern for the lake, after environmental reporter Rob Davis published an article titled "Oregon's Lake Abert is in Deep Trouble."

Lynn wrote, "As owners of the Oregon Desert Brine Shrimp Company, we operated a low-impact, sustainable commercial brine shrimp fishery on Lake Abert from 1979 to 2014…[when] the lake went dry and we lost our livelihood. Every summer…we harvested brine shrimp and as a result developed a special relationship with this remote, beautiful lake, appreciating the wonder of its uniqueness and the countless birds that depended on it. From

Left: Keith and Lynn Kreuz at Lake Abert. *Right*: Keith preparing to harvest *Artemia* from the lake. The two nets he used for harvesting the brine shrimp were located on either side of his tiny raft. A small outboard motor (not shown) powered the boat. Keith wore rubber clothing to protect himself from the caustic alkali.

the beginning, we were ambassadors for the lake bringing our concerns to state agencies over numerous projects.... In almost every case we would get only lip service. Now, finally, the plight of Lake Abert has been exposed."

Keith, a tall, youthful-looking man in his late 60s, harvested shrimp from a tiny, homemade raft powered by a three-horsepower outboard motor and pulling two fine-meshed nets. He did most of his harvesting at night, when the *Artemia* came near the surface. Keith has lots of stories about being out on the lake at night, including tales of having to fend off bats and phalaropes that were attracted to his headlight. It was very hard work and could also be dangerous. In fact, one night Keith was forced to swim to shore after his motor fell off. Lynn, a petite woman with a gentle smile, helped Keith process the shrimp each morning before putting it into cold storage units beside their cabin. Keith and Lynn sold their frozen shrimp as aquarium fish food or for the aquiculture of fish and shrimp. Keith's brine shrimp harvesting was featured on a video segment shown in February 1993 on Oregon Public Broadcasting's *Oregon Field Guide* TV program.

Lake Abert's *Artemia* have died off in some years as early as July, due to stressful water conditions, like those in 2010, 2013, 2015, and 2016. The cause of these mortality events appeared to be high salinities and associated low levels of dissolved oxygen. Under low dissolved-oxygen levels, *Artemia* produce hemoglobin to increase the transport of oxygen in their bodies, and the hemoglobin turns them red. However, some of this coloration, especially in females and in nauplii, is due to carotenoid pigments obtained from algae (Gilchrist 1954). *Artemia* turn bright red, usually, when they have

Brine shrimp: *left*—red water caused by hemoglobin-rich *Artemia* dying and being blown inshore at Lake Abert in August 2010; *right*—hemoglobin-rich *Artemia* swimming in an aquarium, mostly females, as indicated by the swollen egg pouches at the base of their tails.

inadequate dissolved-oxygen concentrations. In July 2010, when the *Artemia* were bright red, the lake's salinity was above 15 percent. Also at that time, hot weather and calm winds likely reduced oxygen levels in the water, which led to the stressful conditions.

Under good conditions, *Artemia* can be extremely numerous in the lake. In the late 1980s, researchers sampled Abert and estimated that it contained three billion individuals and a biomass of fifteen million pounds (Conte and Conte 1988). That assessment was just a snapshot of conditions present then, and the abundance certainly varies considerably from one year to the next; however, these data indicate that under optimal conditions, the lake can be home to vast numbers of *Artemia*. Keith Kreuz found considerable variability in his catches, from year to year, and research has backed up his personal experience: interestingly, he generally caught the most brine shrimp when the lake was at relatively low elevations, those between 4,248 and 4,250 feet, and he caught the fewest shrimp at a higher lake level of 4,260 feet (Senner et al. 2018). As lake elevations get too low and as salinities increase, conditions are more likely to be lethal, as happened in 2010. Catches may sometimes be higher when the lake is shallow, however, possibly because the shallower water prevents *Artemia* from swimming down deeply enough to avoid capture by nets.

Other Branchiopods—Water Fleas and Fairy and Tadpole Shrimp

Branchiopods, like the brine shrimp, are primitive crustaceans that occur in a wide variety of aquatic habitats, from oceans to small ponds. Besides *Artemia,* three kinds of branchiopods occur in and around Lake Abert: water fleas, fairy shrimp, and tadpole shrimp.

In the summer of 2017, the low, 3–4 percent salinity of the water allowed the water flea (*Moina hutchinsoni*) to invade the lake from the river, and it was abundant by midsummer. Researchers have seen it at the lake in other years too (Herbst 1980). Like *Artemia,* water fleas feed on microalgae and bacteria.

Unlike the water fleas, fairy and tadpole shrimp are two kinds of branchiopods directly related to *Artemia.* They usually live in vernal pools or playa lakes throughout South Central Oregon and adjacent areas of the Great Basin in years when these lakes and pools contain water. However, we know little about these organisms because most of these pools lie in remote areas that are difficult to reach when they do contain water, and most of the time, the pools are dry. Nevertheless, one type of ¼-inch-long fairy shrimp (*Branchinecta* sp.) has made its way to both Lake Abert and Summer Lake, when the salinity levels have been low enough (Keith Kreuz, pers. com.). You can also find this particular species in a small pond east of Highway 395 near River's End Reservoir, and it lives in most of the small lakes, including Skookum Lake, at the north end of Abert Rim. It likely makes its home in other similar playa lakes throughout the region.

Another fairy shrimp, the two-inch-long *Branchinecta gigas,* lives together with an unidentified *Branchinecta* sp. in a small, muddy, vernal pool—the color of coffee and cream—near The Narrows, west of Lake Abert; at one of the lakes on Abert Rim; and in a playa lake adjacent to Highway 395 in the Alkali Lake Basin north of Lake Abert. The small, seasonal pool near The Narrows is also home to a population of ½-inch-long tadpole shrimp (*Lepidurus lemmoni*),

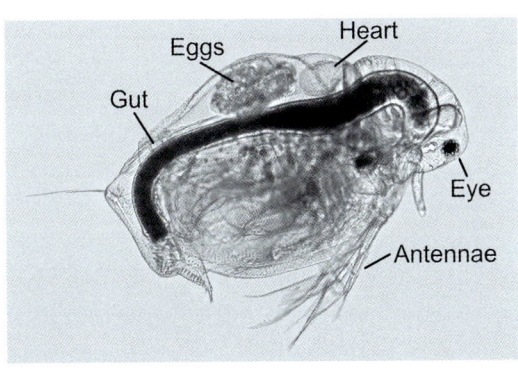

Moina hutchinsoni, the water flea (highly magnified), collected in Lake Abert during July 2017. The total length of the flea was approximately 0.05 inches, or 1 mm.

Fairy shrimp and a tadpole shrimp, *clockwise from top left*: a male *Branchinecta* sp., or fairy shrimp, which has large claspers on its head to hold onto the female during mating; *Lepidurus lemmoni,* the tadpole shrimp, which is about three quarters of an inch long, has a pair of eyes located at the front of its shield-like carapace, and has a pair of long tails; and the female fairy shrimp, which has a conspicuous egg sack (both male and female fairy shrimp are about one quarter inch long).

which look like trilobites, with broad, shield-like carapaces covering numerous legs, and which date back to the Cambrian period 500 million years ago. A tadpole shrimp has two bean-shaped eyes, a row of spines on its posterior, and a spiny, jointed abdomen terminating in two long, whiplike tails. Fairy shrimp swim upside down like *Artemia,* but tadpole shrimp lope over the muddy bottom or swim right side up, using their numerous feet.

Alkali Flies

> Then there is a fly, which looks something like our house fly.
> These settle on the beach to eat the worms that wash ashore—
> and any time, you can see there a belt of flies an inch deep and
> six feet wide, and this belt extends clear around the lake—a belt
> of flies one hundred miles long.
>
> —MARK TWAIN, 1872

The *Artemia* may be the dominant macroinvertebrate in Lake Abert when salinities are above 5 percent, but alkali flies—also called brine flies or *ephydrids*—are a close second and are the major benthic, or bottom-living, macroinvertebrate. In favorable years, dense masses of the adult flies swarm along the shore and out on the lake's surface. Their countless larvae crawl on the sediment and their pupae are attached to rocks. Like *Artemia,* alkali flies are highly adapted to life in lakes that are both salty and alkaline and thus show an amazing tolerance for both relatively high saline and pH

Adult alkali flies: *left*—a side view; and *right*—a top (dorsal) view. Both flies are about 0.3 inches long.

levels, but ephydrids have a much different life cycle than that of brine shrimp. *Artemia*'s dormant cyst stage allows them to survive the drying of their lake habitats, but alkali flies lack such an adaptation and thus must find suitable habitats under drought or high salinity conditions.

Twain was correct: alkali flies do look like other, more familiar, flies. They are only found, however, in salty and alkaline lakes in the United States, Canada, and Mexico, and they are different in other ways. Adult ephydrids are only present on the mud, around the lakeshore, and on the surface of a lake, where they feed on algae, bacteria, and organic detritus.

Alkali flies can be so numerous along the shore that several thousand can congregate in one square foot of mud. Seen from a distance, a mass of these flies turns the shore black, making it look as though it is covered by tar. When you draw closer to inspect the shifting blackness, you can see their movement, hear their buzzing, and then recognize what they are. These dense concentrations of flies, comprising as many as 80,000 per square meter, have been called *fly mats* for obvious reasons (Swarth 1983). On calm mornings, you can hear the buzzing sounds of millions of the adults some distance from the shore. They are present at the lake when water levels are between 4,248 and 4,258 feet, and they were especially numerous in 2002 and 2011, when the lake was at 4,254 and 4,251 feet and the salinity was at approximately 5 percent and 8 percent, respectively (Keith Kreuz, pers. com.).

Alkali fly larvae hatch from eggs and live on the mud, both under and above water, where they feed on algae, bacteria, and other organic matter. Each wormlike larva has a small head, three thoracic segments, and eight abdominal segments, and it actively crawls, using short legs equipped with hooklike appendages that can firmly grasp onto rocks and algae.

Although the larvae have a pair of posterior "snorkels," they mostly breathe through their skin, so they don't need to reach the water's surface to get oxygen. The larvae tolerate relatively high

Alkali flies: *left*—high concentrations of adult alkali flies swarming along the eastern shore and spreading out onto the surface of Lake Abert in September 2011, forming fly mats; *right*—cluster of adult flies feeding, using their proboscises to touch the mud.

salinities, due to their lime glands, which remove carbonate and bicarbonate ions from their blood as limestone. The resulting limestone pellets may act as ballast that prevents the larvae, which are less dense than the surrounding water, from floating to the surface.

After several weeks to months, an alkali fly larva develops into a pupa, encased in a protective, brown-colored, leathery sack that is attached to rocks or other substrates below the water. Pupae are also affixed to *Ctenocladus* algae and wash ashore and can float to the surface after being dislodged by the action of the waves. Keith Kreuz has reported that he was sometimes unable to harvest *Artemia* because alkali fly pupae were so abundant at the lake's surface. When the pupae are so numerous, waterbirds will feed on them. Anytime large numbers of gulls scatter over the lake and peck at the surface, they are likely feeding on the floating pupae. At Mono Lake in California, Northern Paiute Native Americans once collected the pupae for food, because the organisms are rich in fats.

Adult flies emerge from the submerged pupae within a few weeks during the summer and float to the surface inside a silvery envelope of air. The pupal cases that the adults leave behind wash ashore and are sometimes so numerous that they form brownish mats several inches thick and five feet wide, extending along the shore

Underside of an alkali-fly larva. The head is on the left, and the two appendages at the rear are the "snorkels." The larva has gas bubbles near its middle and eight pairs of tiny, black hooks in rows on its feet. It is approximately 0.3 inches long.

Alkali fly stages: *left*—masses of empty pupal cases that washed onshore at Lake Abert; *right*—alkali fly larva and a fresh pupal case, both about 0.3 inches long.

over long distances. Observers once saw an estimated seven billion pupal cases lying along six miles of the Great Salt Lake shoreline (https://learn.genetics.utah.edu/content/gsl/foodweb/brine flies).

After feeding and mating, the female fly crawls back down into the water, drawing oxygen from a bubble of air surrounding her. Once she finds a suitable rocky site on the lake bottom or on algae, she lays up to two hundred eggs, and the larvae emerge one to two days later (Herbst 1986). All of this happens very quickly: adult alkali flies live for less than a week.

According to Herbst, who studied the ecology and physiology of the flies at Lake Abert, all of these life stages are present throughout the year, and the flies can produce three to six generations annually. However, the larval stage predominates during the winter months. During the winter, most adult flies die. However, in some years, such as in 2018, a few fly mats decorate the shore as late as mid-November.

Herbst's research at Lake Abert, Mono Lake, and in the laboratory provides some important insights into the ecology of the alkali fly. His focus on the impact of salinity and food availability on the fly's growth and survival has been particularly revealing. Herbst has found that under conditions of either food scarcity or high salinities, larvae develop slowly, wind up as smaller adults, and experience low survival rates. High salinity likely forces the flies to use too much energy as they try to regulate their internal salt concentrations. Herbst discovered that young larvae experience lower survival rates when salinity rises above 15 percent and that older larvae's maturation is inhibited at a salinity of 20 percent. He also found that small pupae produced under these stressful conditions have lower energy reserves, which increases the adults' risk of starvation. High salinity also negatively impacts the growth of algae, the flies' main food source, which further reduces their chances of survival. These effects likely explain why few adult flies are present at the lake when salinities are high.

At Lake Abert, and in similar habitats where they live, small numbers of adult flies frequently make short-distance flights to new feeding areas multiple times per day. Sometimes vast numbers of adult flies also make longer-distance flights. I got to witness this one afternoon in September 2011, when the flies flew north. As far out over the lake as I could see, unimaginably large numbers of these insects flew for over forty-five minutes. In July 2017, I saw another mass movement of flies heading south along the shore and then southwest out across the lake. Keith Kreuz has also reported seeing huge swarms of the flies journeying north from the south end of the lake, which suggests that these mass movements are common. We don't know the purpose of these longer flights or how the flies coordinate their mass movements, but we can hypothesize that their goal is to find new feeding areas. Because the flies can be extremely abundant, they would likely deplete their food supplies if they remained at one location for too long.

Alkali flies, in all their life stages, serve as a critical food source for migratory waterbirds: the flies are abundant and nutritious and have about twice the caloric content of adult *Artemia* (Herbst 1988). Because the flies are packed with protein and fat, they are the perfect high-energy food for fowl preparing for long-distance migrations. All waterbirds that visit Lake Abert—including avocets, stilts, gulls, phalaropes, sandpipers, Willets, and blackbirds of the Brewers, Red-winged, and Yellow-headed varieties—probably eat fly larvae and pupae (Boula 1986; Jehl 1988, 1999). Avocets sweeping along the lakeshore are likely feeding on these tasty treats. Birds find the adult flies more difficult to catch, but they still make many an ill-fated attempt, sometimes with humorous results.

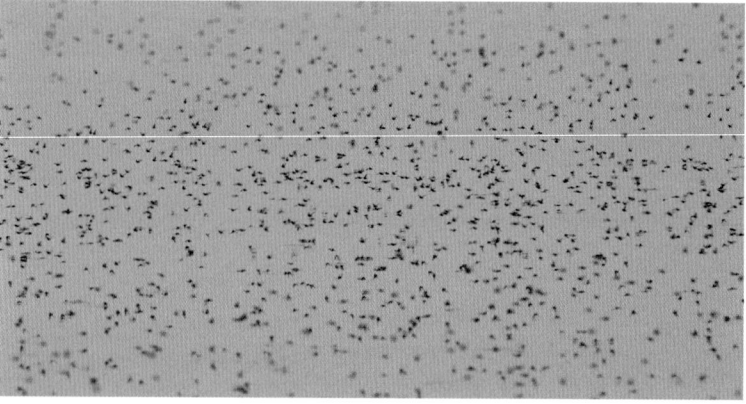

Migrating alkali flies flying over the lake in July 2017.

Black-necked Stilt
trying to catch
adult alkali flies at
Lake Abert.

Among the waterbirds that feed on the flies, Wilson's Phalaropes are the most skilled hunters and use considerable stealth when pursuing them. Like a cheetah stalking a gazelle, a phalarope crouches low and approaches a fly sitting on the water's surface by slowly moving toward it. The bird lowers its head and points its bill forward, almost touching the water. When it gets within striking range, the bird quickly lunges forward, using its long bill like a pair of forceps to snatch the fly. This technique works best on individual flies sitting on the water; however, when the flies gather in masses, they are more difficult to catch because they can better sense the presence of the birds and quickly fly away. Once alarmed, the flies closest to the bird fly up and away, and this causes more distant ones to flee, resulting in a cascade of flies moving outward.

Lake Abert's Other Aquatic Flies

Alkali flies have some cousins at Lake Abert, including *Mosillus bidentatus,* a small, black insect, about one-half the size of the alkali fly but similar to it. *Mosillus* flies also live at Mono Lake, the Great Salt Lake, and numerous other lakes in the West (Swarth 1983; Mathis et al. 1993). They feed on organic matter, particularly on algae that washes up along the lakeshore. *Mosillus* flies gather in small areas and in high concentrations, but they are never as plentiful as the alkali fly.

Another common aquatic fly at Abert is the long-legged fly, *Hydrophorus plumbeus,* which skates over the lake's surface like a water strider, supported by surface tension, and jumps from the water to avoid predators (Swarth 1983; Herbst 1988; Burrows 2013). Long-legged flies, like water striders (the common name for a family of long-legged, aquatic insects), feed on other bugs trapped in the surface tension. At Lake Abert, I have watched them eating

Lake Abert flies, *clockwise from top left*: long-legged fly (*Hydrophorus plumbeus*) skating on the water's surface (approximately 0.3 inches long); *Mosillus bidentatus* feeding on algae growing on a rock (approximately 0.2 inches in length); horsefly (*Tabanus* sp., whose huge eyes have numerous facets, giving it excellent vision); and side view of a horsefly, approximately 0.5 inches in length.

the cysts of brine shrimp. Long-legged flies also skate over Mono Lake's waters.

In 2017 and 2018, the deerfly (*Chrysops sp.*) and the horsefly (*Tabanus sp.*) became numerous at Abert, and the females inflicted a nasty bite on anyone walking near the lake. Large *Tabanus* flies, with big, colorful, iridescent eyes marked by horizontal lines, predominated. As I was watching birds at the edge of the lake in July 2017, I nervously kept an eye on several dozen of these flies as they continuously circled me, planning their painful attack. The aquatic larvae of the flies likely live in the marshes that fringe the lake. These flies are also common at Summer Lake and elsewhere in similar habitats.

Lakeshore Beetles

Some insects do not easily fit into either an aquatic or a terrestrial category. Lake Abert hosts two such beetles in its littoral zone, an area located between the lake and the uplands. Water covers this zone in some seasons and when there is a wind-blown surge. I discovered one of these beetles when I was at the north end of the lake, again watching birds. A south wind began blowing, and it forced water levels to rise along the shore. As the water advanced, I noticed small beetles scurrying over the mud ahead of the water, just as their burrows flooded.

Beetles from around the Lake Abert shore, *clockwise from top left*: rove beetles (*Bledius* sp.), dorsal and ventral views, approximately 0.3 inches long; adult rove beetles swarming at the south end of Lake Abert; Oregon tiger beetle *Cicindela oregona,* approximately 0.5 inches long; and carabid beetles (*Bembidion flohri*), approximately 0.3 inches long.

Afterward, I determined that these small insects were burrowing rove beetles (*Bledius* sp.), common on the shorelines of salt lakes around the world (Garcia and Niell 1991; Gerdes et al. 2008). Their burrows reach estimated densities of over one hundred per square meter in some areas around Lake Abert. On another occasion at the lake, this time in the evening, Steve Sheehy, a friend and fellow naturalist from Klamath Falls, and I were fortunate enough to witness the rarely seen aerial swarming of the adult *Bledius* at the south end of Abert. Innumerable flying beetles surrounded us, their wings glowing in the golden light of the setting sun. It reminded me of the nuptial flight of winged termites. Recent low water levels at the lake have increased the habitat for these beetles. This proliferation of *Bledius* beetles may also benefit the snowy plover, a shorebird known to feast on them (Purdue 1976).

The second insect that does not fit neatly into either terrestrial or aquatic categories is *Bembidion flohri,* a small, predatory ground beetle that lives at the lakeshore. I have seen these insects crawling on the damp mud in the same areas where the rove beetles burrow, and they are sometimes numerous under rocks at the edge of the lake. *Bembidion* beetles live in a variety of habitats, and some species dwell along saline shorelines, where they survive just fine, even when they are temporarily submerged in salt water (Desender 2005).

In August 2016, while watching birds at Juniper Point on the eastern shore of the lake, I noticed a strange-looking insect scurrying over the wet shoreline. Looking closer, I determined that I was actually seeing two beetles mating. They had large bulging eyes and were a dark brownish-purple, with green and red iridescence over their heads, thoraxes, and abdomens. The colorful and active beetles turned out to be predatory Oregon tiger beetles (*Cicindela oregona*), which make their homes in many parts of the Pacific Northwest on ocean and inland beaches. A related species, the hairy-necked tiger beetle (*Cicindela hirticollis siuslawensis*), has been recently proposed for protection under the Endangered Species Act because it has disappeared from many coastal sites (Federal Register 2021).

Mollusks

The various insects at Lake Abert certainly face adverse effects when the salinity get too high. Most mollusks, however—including clams, snails, and mussels—simply can't live in the lake's highly concentrated salt and can only survive in the environment of freshwater lakes and streams. Nevertheless, you can find evidence of a few kinds of mollusks that have survived in the lake at various times, during rare, low-salinity events. If you walk along the shore at the south end of the lake, you will find shells of a small clam, *Psidium* sp., lying scattered there. Lake sediments exposed by road cuts along Highway 395 reveal even more abundant fossilized mollusk shells. Additionally, archaeologists have unearthed shells of four snail and clam species, and of one freshwater mussel species, along the east side of the lake (Pettigrew 1985).

Researchers have determined that fossilized snail shells found in Paleolake Chewaucan range in age from 13,000 to 14,000 years old (Friedel 2001; Licciardi 2001). I did some further sampling at two road cuts along Highway 395, at elevations of approximately 4,295 feet at the north end and 4,280 feet at the south end of Abert, and I found some of the same fossils, as well as shells of the small springsnail, *Pyrgulopsis robusta,* which still makes its home in XL Spring at the north end of the lake.

The most numerous fossilized mollusks in this area, found in the sediments located fifteen to twenty feet above Highway 395 at both ends of the lake, are from the snails *Vorticifex effusa* and *Helisoma newberryi.* Researchers have found other Paleolake Chewaucan shell fossils southeast of Summer Lake near Paisley Cave, including from the snails *Gyraulus* sp., *Menetus* sp., and *Vorticifex effusa,* as

Fossilized snail shells once buried in lake sediments before being uncovered by a road cut at the north end of the Lake Abert basin: *left*—fossilized shells exposed in sandy sediment (the largest ones are Great Basin rams-horn snails, *Helisoma newberryi,* which are about 0.5 inches across, and the smaller ones are lined ramshorn, *Vorticifex effusa,* which are about 0.2 inches across, and the presence of so many shells in one place could be the result of a mass-mortality event); *right*—close-up of fossilized robust springsnails, *Pyrgulopsis robusta,* which are about 0.2 inches long.

well as from the small clam *Psidium* sp. (Negrini et al. 2001). *Vorticifex effusa* is the dominant mollusk in these deposits, just as it is in the sediments above the eastern shore of Lake Abert. The fossils found near Paisley remain undated, but scientists have estimated that a sediment layer above the fossils consists of fifty-thousand-year-old volcanic ash from Mount St. Helens in Washington, so they appear to be much older than the fossils mentioned above.

Several of the fossilized mollusks found in the sediment deposits near Lake Abert likely come from aquatic creatures that lived in the Chewaucan River. One of these, the robust springsnail, also left its remains at several sites in Harney and Malheur Counties; along the Columbia River; and in tributaries of the Snake River in Wyoming and Idaho (NatureServe 2011). These scattered fossils indicate that the snail was once widespread in the West and, almost certainly, much more abundant than it is today. Scientists now consider it to be critically imperiled, due to its rarity. The springsnail genus, *Pyrgulopsis,* is one of the most diverse and widespread aquatic invertebrates in the Great Basin and includes eighty species (Hershler and Sada 2002).

Another snail present as a fossil in Paleolake Chewaucan sediments is the Great Basin ramshorn, *Helisoma newberryi,* which was named for John Newberry, who served as a doctor and naturalist on the 1854–1855 Abbott and Williamson Pacific Railroad Survey. This relatively large snail made its home throughout much of the interior West near the end of the Pliocene epoch, approximately three million years ago. Proof of its prevalence in the Great Basin lies not only in ancient fossils but also in evidence of recently wiped-out populations in Wyoming, Utah, Nevada, California, and Oregon. In modern times, we still find this snail, but only in a handful of

Live Great Basin ramshorn snail, approximately one inch long, from Upper Klamath Lake. The snail lives buried in sediment and has hemoglobin in its blood to aid in respiration, which turns its body a reddish color.

sites in Shasta and Lassen Counties in California, as well as in Oregon's Upper Klamath Lake and in several of its tributaries (Frest and Johannes 1998; NatureServe 2020). Therefore, we consider the ramshorn to be rare and critically imperiled in Oregon.

Based on the diversity of the mollusks that lived in Paleolake Chewaucan during the Pleistocene, the lake was likely similar in some respects to present-day Upper Klamath Lake, which still supports many of these species or related ones (Frest and Johannes 1998).

Fish

Like mollusks, fish don't find Lake Abert's environment suitable, due to its high salinity. Once again, however, a little excavation unearths fish bones in sediments near the lake. Researchers have discovered fossils of tui chubs, dating from approximately one hundred thousand years ago, in Paleolake Chewaucan sediments near Summer Lake (Gobalet and Negrini 1992). The east side of Lake Abert includes other archaeological treasures, both fossilized bones of tui chubs and a single catfish bone (Pettigrew 1985). Such findings lead us to believe that fish swam in the lake earlier in our Holocene era, when Native Americans lived along the lakeshore.

Scientists weren't surprised to find tui chub bones in archaeological sites above the lake because these fish still live in the Chewaucan River. The catfish bone was much more unusual, however, because fossilized catfish are extremely rare in Oregon and are only found in much older deposits than those at Lake Abert. The bone does not allow us to identify the species, but it might be the only evidence that catfish were native to Oregon in the not-too-distant past. Brown bullheads (*Ameiurus nebulosus*), which are a type of catfish, are now widespread in Oregon streams east of the Cascades. People likely introduced them because they are a good food source (Markle 2016; pers. com.).

I found fossilized fish vertebrae in the same lake deposits that contained mollusk shells, those exposed by road cuts along Highway

Close-up of fossil tui chub vertebrae, 0.1–0.2 inches long, from Lake Abert shoreline deposits at a 4,280-foot elevation. Radiocarbon dates of fossil snails by Licciardi (2001) from this deposit indicate that snails are approximately 13,000–14,000 years old.

395 at the north and south ends of the lake. The tiny vertebrae, only about 0.2 inches long, are likely from tui chubs that lived in the lake until it became too saline.

The Chewaucan River upstream of Abert Lake contains native fish, including Great Basin redband trout (*Oncorhynchus mykiss newberryi*), tui chub, and speckled dace (*Rhinichthys osculus*), as well as some fish more recently introduced to the river, such as the largemouth bass (*Micropterus salmoides*) and the brown bullhead (Snyder 1908; Tinniswood 2007).

Redband Trout

When we think of trout, we usually envision them living in fast-flowing rivers and streams that surge down from the Cascade Mountains, but redband trout have adapted to life in high-desert streams that experience relatively broad temperature ranges and can exist in summer only as isolated pools fed by ground-water. Redbands are the Chewaucan River's largest native fish, are related to the coastal rainbow trout, and are endemic to Oregon east of the Cascades (USFWS 2000; Dambacher et al. 2001; Currens et al. 2009).

Living in their Great Basin habitats, redbands have evolved an amazing ability to survive in small brooks by finding shaded pools that are fed by cool groundwater during droughts. Also, redbands can mature to the small size of less than one foot, but if they have access to a river or lake where food is more plentiful, they can grow much larger. Upper Klamath Lake redbands, for example, commonly weigh more than ten pounds.

The genetics of redband trout are complex and show considerable diversity. Their DNA reveals evidence of ancient river connections, likely dating back to the Pliocene epoch, 2.6 million to 5 million years ago (Currens et al. 2009). Redbands from different

Six-inch-long Great Basin redband trout (*Oncorhynchus mykiss newberrii*), photographed in Rock Creek at Hart Mountain National Wildlife Refuge.

interior basins show nearly three times the genetic divergence of coastal rainbow trout (*Oncorhynchus mykiss*). We can conclude that the redband populations now present in South Central Oregon's isolated basins have been separated from other trout species for long enough to acquire significant genetic divergence. Researchers have found that redbands from South Central Oregon are genetically distinct from, but still related to, Sacramento Basin redband trout, *O. mykiss stonei* (Currens et al. 2009). Neither the close relationship nor the genetic divergence of these two trout populations is surprising: in the distant past, Goose Lake drained south into the Sacramento River via the Pit River. However, the Chewaucan Basin has been isolated from the Sacramento River for a very long time, perhaps for several million years, and so we should expect that Chewaucan Basin redbands would show great genetic divergence from the Sacramento River trout.

Redbands occupy only about one-third of their potential range in the Great Basin, and large, healthy populations live in only 10 percent of that range (Currens et al. 2009). The US Fish and Wildlife Service named the redband trout a "species of concern" in 2000 because of its limited range and small populations. This designation warns land managers that the species is vulnerable, but it does not provide any special protection.

The only other trout native to southeastern Oregon is the Lahontan cutthroat trout, *Oncorhynchus clarki henshawi*, which swims in several small streams in the far southeastern corner of the state. This species once lived in Paleolake Lahontan during the Pleistocene, and it successfully adapted to the increasing alkaline conditions that occurred at smaller remnant lakes, including Pyramid and Walker, that replaced the paleolake during the Holocene. More recently, the range of the Lahontan cutthroat trout has substantially contracted as livestock ranching and other development-related factors have damaged streams and their banks. The Lahontan has also faced negative impacts from interbreeding and competition with introduced trout. Currently, self-sustaining Lahontan cutthroat trout populations occupy less than 1 percent of their former habitat, and much of their current range is confined to areas where their populations are sustained by hatcheries. Consequently, the federal government now lists the native Lahontan as threatened.

Tui Chub—Desert Minnows

Tui chub minnows are one of just a few fish species still living in interior closed basins in Oregon and elsewhere in the Great Basin, and they are widespread in the West, ranging from the Columbia River Basin to the Mohave Desert in Southern California (Sigler and Sigler 2014). Their fossils in the Great Basin date back to the late Miocene or Pliocene eras, approximately five million to ten million years ago (Smith et al. 2002), so they have been present in the region for a very long time. Tui chubs live in the Chewaucan River and in several South Central and southeastern Oregon basins. The closest tui chub habitat to Lake Abert is XL Spring, north of the lake, and this home to the minnows was first described by J. O. Snyder, an ichthyologist and a professor at Stanford University.

Snyder characterized the tui chub as a new species when he found them in the XL Spring in the early twentieth century. In his 1908 report on the lake fishes of southeastern Oregon, he wrote:

XL Spring, as it appeared in April 2010. The spring is located near the XL Ranch house, north of Lake Abert, and is home to the tui chub and the robust springsnail.

At the north end of Abert Lake, on the XL Ranch, is a remarkable spring. Its water has a temperature of 61°F., is said to be constant in volume, clear, and fresh. It pours at once into a boggy pool 100 feet in diameter and about 17 feet deep, from whence an outflowing stream spreads over a marsh of tules and rushes. The water of the pool is clouded with algae and swarms with fishes (*Rutilus*). A handful of crumbs thrown out on the surface attracts great numbers, causing the water fairly to boil, the food disappearing almost instantly. One haul of a seine net enclosed hundreds of specimens measuring from a few inches to nearly a foot in length. Their stomachs were stuffed with vegetation and numbers of a minute gastropod [likely the robust springsnail]. Whether the fishes derive their entire support from the pool was not learned. A few were seen in the stream leading from the pool, and it is reported that during wet weather great numbers pass out to the marsh, where they are left to die as the dry season approaches. The fishes are no doubt natives of the spring; the species having been left by the retreat of the desiccating lake.

Snyder's belief that tui chubs are native to the spring seems plausible, especially because the robust springsnail—a key food source—makes its home there. However, some researchers have argued that tui chubs may have been introduced into XL Spring from elsewhere. The fish do share genetic similarity with tui chubs from Hutton Spring and with Lahontan Basin minnows that swim in waters one hundred miles distant (Remple 2013). Others have argued for and against the possibility that Native Americans introduced them into the spring (Sada and Vinyard 2002).

I believe that it is highly unlikely that Native Americans would have transported live fish so far, although they certainly included tui chubs in their diet (Aikens et al. 2011). Furthermore, because the robust springsnail is very likely native to XL Spring, and has lived there since the time of Paleolake Chewaucan, it seems probable that the tui chub is also native to the spring.

The last time I saw XL Spring was in May 2020, and it was reduced in size to a small pool only about thirty by ten feet. That

Drawing of an Oregon lake tui chub (*Siphateles bicolor oregonensis*) from J. O. Snyder's 1908 article titled "Relationships of the Fish Fauna of the Lakes of Southeastern Oregon," published in the *Bulletin of the Bureau of Fisheries*.

was prior to the drought conditions of 2021 and 2022, so I am concerned about the future of both the tui chub and the robust springsnail, given that even drier conditions are likely to occur in the future. It seems almost unbelievable that species that have survived in this spring since the Pleistocene might soon be gone.

Wetland and Shore Plants

Flowering plants normally don't grow along Lake Abert because the salinity is too high for them to survive. However, some hardy, flowering wetland plants live in its shoreline marshes, where there are fresh or brackish springs. These marshes began expanding out toward the lake in 2010, as the lake's elevation started decreasing, because the marsh plants were no longer flooded and killed by salt water.

The most widespread plant in the marshes is a type of bulrush or sedge called the common three-square (*Schoenoplectus pungens*), which forms green islands and verdant borders along parts of the lake's east side, especially at the two-mile-long Mile-Post 74 Springs complex at the northeast end of the lake. Several other native, wetland plants coexist with the common three-square, including: cattails (*Typha latifolia*), swordleaf rush (*Juncus ensifolius*), seaside arrow grass (*Triglochin maritima*), seacoast bulrush (*Bulboschoenus maritimus*), hardstem bulrush (*Schoenoplectus acutus*), and spike rush (*Eleocharis* sp.). In addition, the non-native teasel (*Dipsacus fullonum*) has invaded the upper parts of these marshes and the lower reach of Poison Creek below the highway. A small, prostrate, grasslike plant called horned pondweed (*Zannichellia palustris*) grows at the outer edge of the marsh in muddy areas where water seeps in from the springs.

The higher springs and seeps around the lake provide a habitat for a variety of other flowering wetland plants, including: the

A verdant, brackish-water marsh—comprised mostly of the common three-square (*Schoenoplectus pungens*) sedge—surrounding one of the seeps at the Mile Post 74 Springs along the northeast shore of Lake Abert.

Flowering plants from fresh and brackish-water wetlands adjacent to Lake Abert, *clockwise from top left*: cutleaf water parsnip (*Berula erecta*); foxtail barley (*Hordeum jubatum*); yellow monkeyflower with red-dotted lip (*Mimulus guttatus*); horned pondweed (*Zannichellia palustris*); and basin goldenrod with fly (*Solidago spectabilis*).

attractive, yellow-bloomed monkeyflower (*Mimulus guttatus*), the white-flowered water parsnip (*Berula erecta*), the willow dock (*Rumex salicifolius*), the basin goldenrod (*Solidago spectabilis*), and the brightly colored scarlet paintbrush (*Castilleja miniata*). Higher up on the shore, the silvery seed heads of foxtail barley (*Hordeum jubatum*) wave in the wind and are quite striking. Gray rabbitbrush (*Ericameria nauseosa*) grows above the marshes in gravel that marks the former beach lines and also farther upslope on Abert Rim. Its bright-yellow flowers attract a wide variety of nectar-seeking insects, including bumblebees and butterflies like the Great Basin wood nymph (*Cercyonis sthenele*; see chapter 5).

These hardy and beautiful plants are part of a rich ecosystem at Lake Abert, although certainly one that lacks the biological diversity present at some other lakes in the region. The species that do live in and around Abert have, nonetheless, evolved amazing adaptations that enable them to thrive under extreme conditions and even to become abundant when conditions are good. Invertebrates in this delicately balanced community, such as brine shrimp and alkali flies, attract thousands of birds to the lake each year. The brilliant marsh flowers provide sustenance to pollinating insects.

We now realize that the lake, in its less saline days, was an even richer ecosystem than it is today. The fossil remains of mollusks and fish found around the Chewaucan Basin indicate that diverse, freshwater fauna were once profuse and that they had an affinity with fauna as far away as those in Wyoming. In addition, evidence proves that Lake Abert's former biological diversity once attracted settlers, who, for a time, lived and thrived there.

Today, our observations of these normally robust animals and plants, now living under the stresses caused by drought, too-high salinity, and water diversions, serve as a wake-up call to us all.

CHAPTER 5

Life in an Arid Landscape

The northern corner of the Great Basin, southeastern Oregon and northern Nevada, is a drift of sagebrush country the size of France. The raw landforms incessantly confront us with both geologic time and our own fragility. The rims were built over eons; we can see the layers, lava-flow on lava-flow. Shadows of clouds travel like phantoms across the white playas of the alkaline wet-weather lakes. But the endlessness of desert is not so intimidating if you focus on the beauties at your feet, red and green lichens on the volcanic rocks, tiny flowers.

—WILLIAM KITTREDGE, 1993, *Hole in the Sky: A Memoir*

Living in the Cascades' rain shadow—the region of reduced rainfall sheltered by the mountains—makes life difficult for plants and animals around the shores of Lake Abert. This area, on average, gets less than one foot of precipitation per year, and half of that in some years, so any species living there must be highly adapted to drought, a requirement getting only tougher as climate change makes the area even drier. Adaptations, therefore, are key to surviving in this landscape. Perhaps no organism has evolved the capacity to withstand the rigors of the Great Basin environment so well, and in such diversity, as the intriguing lichen.

Lichens—Slow Life on Hot Rocks

Fifteen million years ago, after the flood basalts covering much of South Central Oregon had cooled, the wind carried spores and fragments of crustose lichens, dropping them on the rocks' surfaces, where they gradually gave life to what had been a barren landscape. Lichens are true pioneers, among the first visible life forms to settle on bare rock, and they can live on little more than sunlight and the water provided by dew and occasional rainstorms. They consist of an alga or cyanobacterium in a symbiotic relationship with a fungus, and they grow very slowly, often on the side of a rock. They most likely die only when their rock substrate succumbs to the effects of weathering.

Lichens occur in two basic forms. *Crustose lichens,* which consist of barely more than thin, hard coatings that cling tightly to their

Sunburst and other colorful lichens and cushion mosses richly displayed on Steens Basalt boulders on the lower slope of Abert Rim.

rocks, have adapted themselves perfectly to high-desert environments that are hot and dry in summer and bitterly cold in winter, and their viability makes them both diverse and abundant. *Foliose lichens,* which look more leaflike, adhere less completely to the substrates and are less numerous in deserts, but they are abundant in the forests west of the Cascades.

Nineteenth-century geologist Israel Russell described the cobblestone lichen (*Pleopsidium flavum*)—a yellow-tinted crustose that paints the near-vertical fault scarp of the uppermost ramparts of Abert Rim—in his writing: "The palisade rising abruptly from the eastern shore of Abert Lake has a deep brownish tone, relieved by a growth of bright yellow lichens, and is grand in its barren ruggedness, especially towards evening, when its inherent richness is heightened with sunset tints."

Cobblestone lichens grow throughout the world, even in Antarctica. In the northern Great Basin, they mostly adhere to the faces of vertical basalts, and they are especially abundant and conspicuous on the five-hundred-foot-high rock wall of Abert Rim near the

Colorful and diverse lichens from Abert Rim, *from top to bottom, left to right*: rock tripe (*Umbilicaria hyperborea*); tile lichen (*Lecidea* sp.); sunburst lichen (*Xanthoria elegans*) and other lichens; rockbright lichen (*Rhizoplaca melanophthalma*), *Lecidea fuscoatra,* and other lichens; moonglow lichen (*Dimelaena oreina*); cobblestone lichen (*Pleopsidium flavum*); sunburst lichen (*Xanthoria elegans*) and a woodrat marker (white deposit, five-cm- or two-inch-long scale); and wolf lichen (*Letharia vulpina*), which primarily grows on trees.

southeastern end of the lake. There, the lichens bake in the full after-noon summer sun, enduring extremely hot and dry, seasonal conditions. Even in spring, the steep surfaces on which they live are rarely dampened by rain or snow.

These cobblestone lichens have adapted to life in unforgiving environments and under challenging conditions of widely varying temperatures, high light, intense UV radiation, and very little moisture. Consequently, one of them—*Pleopsidium chlorophanum*—has attracted the interest of astrobiologists, scientists who study life on and outside of Earth and who want to understand what life forms may have existed in the distant past on Mars or elsewhere in our solar system. In fact, they have used *P. chlorophanum* in an experiment to determine whether lichens could adapt to Mars-like conditions (de Vera et al. 2014). In a lab simulation that lasted a month, the cobblestone lichen not only survived, but it also actually adapted to extreme conditions that mimicked those on Mars. The experiment's results suggest that life forms similar to lichens could have existed on the planet when liquid water flowed on its surface.

The imposing, basaltic palisade of Abert Rim and the countless boulders below it provide homes to multitudes of other crustose lichens. Displaying varied hues of red, orange, brown, gray, and black, the lichens create a multi-tinted palette on rocks and boulders located just above the lakeshore and strewn across the landscape all the way up to the top of the rim. Scientists have not totaled the exact number of lichen species that live around the lake, but my friend Steve Sheehy, who is a lichenologist, has identified more than seventy, including the striking, reddish-orange sunburst lichen (*Xanthoria elegans*), a type of crustose lichen that adorns the north sides of waterworn basaltic rocks on former beaches and also thrives on upslope boulders. Sunburst lichens, another crustose species, subsist on nitrogen and other nutrients transported as aerosols from the lake. They also grow on woodrats' "signposts"—my name for the limy deposits excreted by these soft-furred rodents when they urinate—because the deposits are rich in nitrogen. A similar, orange-colored lichen, the desert firedot (*Xanthomendoza trachyphylla*), lives predominately on calcareous tufa.

Mosses and Ferns

Unlike the hardy lichens, very few mosses and ferns can survive in the dry uplands and in the salty and alkaline soils around the lake. If you are a very good detective, and very lucky, you may find cushion mosses, forming dark-green mounds that range in size from the diameter of a dime to several inches across. Cushion

Mosses and ferns from around Lake Abert, *clockwise from top left*: cushion moss (*Coscinodon calyptratus*), with white hairs and pointed sporangia, about one inch across; star moss (*Syntrichia* sp.), also with white hairs and about one inch across; and fragile fern (*Cystopteris fragilis*), with fronds approximately six inches long.

mosses belong to the family Grimmiaceae, and they grow on the north and east sides of boulders above the lake and throughout the Great Basin. When these plants are dry, they are nearly black and less visible than when they are wet and green. Like the lichens that cling to rocks, cushion mosses grow very slowly under dry conditions, and it is not unusual to see lichens growing over them, because of just how gradual the mosses' development is. Even bryologists, scientists who study mosses, find it nearly impossible to identify these cushions in the field. One variety however, *Coscinodon calyptratus,* is distinctive because it has silver threads running through its pointed, bulblike sporangium, the structure in which its spores are produced. Other mosses common around the lake, such as the star moss (*Syntrichia* sp.), grow near the ground in shady sites, such as below boulders located on north-facing hillsides.

The only fern that I have found near Lake Abert is the fragile fern, *Cystopteris fragilis,* which is usually less than eight inches long. You can see it most often during the spring, when its delicate, light-green fronds reach out into the sunlight from under boulders or near streams; by summer the fronds are shriveled, brown, and inconspicuous.

Upland Plants—Living with Water Scarcity

> Sterility, on the contrary, is the absolute characteristic of the val-
> leys between the mountains—no wood, no water, no grass, the
> gloomy artemisia the prevailing shrub.
>
> —CAPTAIN JOHN FRÉMONT, 1845

Although the area around Lake Abert is indeed desertlike—due to limited rainfall, hot and dry summers, and soils that are both salty and alkaline—the uplands have many different habitats that support a variety of interesting and colorful flowering plants, despite Captain Frémont's gloomy assessment.

Above the lake and its associated wetlands, the soils are dry and alkaline, and plants growing in these soils must have adapted to these tough conditions, especially at the north end of the lake, which gets average precipitation of eight inches or less annually. Shrubs located north of Poison Creek are highly drought toler-ant and have therefore spread throughout much of the drier Great Basin. These hardy shrubs include: big sagebrush (*Artemisia tri-dentata*); greasewood (*Sarcobatus vermiculatus*); spiny hopsage (*Grayia spinosa*); spineless horsebrush (*Tetradymia canescens*); and shadscale (*Atriplex confertifolia*).

In July and August, when much of the vegetation is brown, temperatures exceed ninety degrees, and dust devils whirl across the snow-white playa, this area might seem like a desert, espe-cially if you are used to living west of the Cascades, where the dominant color of plant life is green. People often refer to this upland region as Oregon's high desert, but technically, it is semi-arid, not a true desert like the Mojave or the Sonoran in the

Close-up of a sagebrush leaf densely covered by a feltlike mat of tiny white hairs, which gives it a woolly appearance. The hairs are thought to reduce heat loading and to mini-mize water loss. Scale: 1 mm = 1/25 inch.

Southwest. Nevertheless, plants growing in this part of the Great Basin have adaptations similar to those of true desert plants: small leaves covered by silvery hairs; leaves with waxy coatings, which minimize water loss; leaves that are deciduous (fall off) when water deficient; long roots to reach for water deep underground; tolerance for salty soils; spines to reduce herbivory (animals feeding on the plants); and other modifications.

Grasses

Besides shrubs, the only other plants that grow plentifully above the shore are grasses, mostly bunchgrasses that grow in clumps. These plants are different from the sod-forming grasses that build our lawns: they have numerous, tightly packed, growing points instead of blades that develop from a network of roots. These high-desert grasses die back each winter and then form new leaves in the spring.

One native, drought-tolerant bunchgrass growing on the lower slope of Abert Rim is bluebunch wheatgrass (*Pseudoroegneria spicata*), the state grass of Washington, which flourishes on slopes near the south end of the lake but which is scarce or nonexistent farther north along the lake. Other common bunchgrasses include squirreltail (*Elymus elymoides*) and Sandberg's bluegrass (*Poa secunda*),

Grasses growing along the lower slope of Abert Rim, *clockwise from top left*: squirreltail (*Elymus elymoides*), which has seed heads that break off and roll over the ground, as the wind disperses them; close-up of Indian rice grass (*Achnatherum hymenoides*) and its open seed capsules, which resemble tiny bird beaks; basin wildrye (*Leymus cinereus*), a large and conspicuous bunchgrass; and crested wheatgrass (*Agropyron cristatum*) seed heads.

which is probably the most common variety in the area, but because it is small, it is easily overlooked. Basin wildrye (*Leymus cinereus*) is the largest of these bunchgrasses and can grow to be waist high. Less common are beardless wildrye (*Leymus triticoides*), western needlegrass (*Achnatherum occidentale*), and Indian rice grass (*Achnatherum hymenoides*). The latter of these has open seed capsules resembling tiny bird beaks and once served as an important seasonal food for Native Americans.

The aptly named needle-and-thread grass (*Hesperostipa comata*), one of the more unusual plants growing on Abert Rim's lower slopes, has seeds with attached appendages called *awns* that are over seven inches long (the awns of most grasses are less than one inch in length). Salt grass (*Distichlis spicata*), as its name suggests, is highly tolerant of salt and often grows close to the lake. Land managers are responsible for one additional common but non native plant, the crested wheatgrass (*Agropyron cristatum*), which they have planted widely around the Great Basin to improve forage for livestock.

Sagebrush—The Bush That Makes a Sea

> These bushes shade the ground and hold the snow, build up humus, bind the soil, conceal the sage grouse and young antelope, and provide choice fuel for the campfire. Their pleasant odor is one of the charms of the desert, and the smell of a dried spray brings back the memory of broad valleys and clean wholesome air.
>
> —VERNON BAILEY, 1936

Big sagebrush is the most abundant shrub near the south end of the lake, as well as on the west side of the lake in areas that have not recently burned and above the highway and higher up on the rim to the north. It is also the most numerous and widely spread sagebrush species, growing in all of the western US states, and, in fact, it is one of the most abundant shrubs in the world (Schultz 2006). Big sagebrush comprises four subspecies, each of which occupies a specific habitat. The one growing at low elevations around Lake Abert—the Wyoming big sagebrush (*A. tridentata wyomingensis*)—is the most drought-tolerant subspecies. Big sagebrush benefits other nearby plants by using its roots to pump water into the upper soil.

Sagebrush shrubs are members of the Asteraceae, or aster, family, and therefore are not related to the true sage genus, *Salvia,* which is in the Lamiaceae, or mint, family. Sagebrush is probably the most widespread and ecologically important dryland shrub in the intermountain West, growing in a wide range of habitats

Sagebrush species from around Lake Abert, *clockwise from top left*: a five-foot-high, Wyoming big sage-brush (*Artemisia tridentata wyomingensis*) in bloom; close-up of big sagebrush leaves covered by many small, reflective hairs that reduce solar heat loading and conserve water; close-up of big sagebrush seeds, about 1/16 inches, or 1.75 mm, in length; bud sage (*A. spinescens*), a species tolerant of very dry sites; flower of a low sagebrush (*A. arbuscula*); and low sagebrush plant.

from basin bottoms to above tree lines and across nearly 160 million acres. Because of its prevalence, this prolific plant has given its name to a vast landscape nicknamed "the sagebrush ocean," the title of a book by writer and naturalist Stephen Trimble. Although still relatively widespread, sagebrush habitat has decreased by half in the past 50 years.

A second type of shrub, the low sagebrush (*Artemisia arbuscula*), more commonly grows on top of Abert Rim, where the soils are thin, rocky, and have a high clay content. Yet another species, the gray sagebrush (*A. ludoviciana candicans*), grows on lower slopes of the rim in seasonally moist, riparian meadows and has one- to two-inch-long leaves divided into lobes with two–three points. A fourth variety, the bud sage (*A. spinescens*), roots itself in rocky outcrops and on sand dunes near the XL Ranch, north of the lake. It is a very short, spiny, deciduous, and extremely drought-tolerant shrub, and it even grows where the only soil lies in cracks between the rocks. The presence of each of these sagebrush types often indicates the condition of its soil.

These habitats are sometimes called "sagebrush steppes." *Steppes,* however, more commonly refer to Eurasian grasslands in cold, semiarid climates (Bone et al. 2015), which differ markedly from the botanically rich, intermountain West of the sagebrush (Cronquist et al. 1972). Therefore, the term *steppe* seems more appropriate for the shortgrass prairies of North America, which are dominated by grasses and are seasonally cold and dry, much like the steppes of Eurasia.

Sagebrush flowers are so inconspicuous that many people probably never notice them. The shrubs, lacking showy flowers to attract pollinators, instead produce numerous, tiny, wind-pollinated flowers in the late summer. Each blossom produces a minute seed that is only about one millimeter long, but each plant can produce many thousands of these seeds. If you have ever been in sagebrush country, you know that the evergreen shrubs' gray-green leaves are intensely aromatic, especially in early spring. Their fragrance comes from volatile oils that have a complex chemistry and which probably evolved to dissuade ruminant animals, like deer, from feeding on the shrubs. Some of the aromatic compounds produced by the sagebrush, such as terpenoids, are bitter and discourage insect grazing; at high concentrations, they can even kill the beneficial bacteria that live in the rumen (or first stomach compartment) of cattle and deer, thus making it difficult for these animals to eat sagebrush. As a result, few large animals feed extensively on the plant, except for pronghorn. One smaller creature also safely gets its nourishment from the shrub: the Greater Sage-Grouse (*Centrocercus urophasianus*), which depends almost entirely on sagebrush leaves for food and cannot exist without them (Rosentreter 2005).

Fascinatingly, sagebrush shrubs' fragrant oils actually help the plants communicate with each other. Scientists have shown that as animals graze on sagebrush plants, the leaves release aromatic compounds that serve as cues for nearby, related plants to increase

Greater Sage Grouse hen, displaying its characteristic earth tones of cream, gray, and brown and watching out for predators from a boulder.

their chemical defenses (Shiojiri et al. 2015; Karban et al. 2016). We don't yet know which specific volatile cues are involved in this communication or how plants detect them, but we do know that as many as fifty species of plants worldwide appear to use similar chemical messaging (Karban et al. 2016). In addition, because of its volatile oils (or oils that vaporize easily), sagebrush has religious or spiritual significance for many Native Americans, who burn small bundles of the leaves, sometimes called smudge sticks, during special ceremonies.

Besides providing food for select animals, sagebrush also offers cover for birds, smaller mammals, and reptiles. In fact, at least eight vertebrates depend on sagebrush, including grouse, pronghorn (Paige and Ritter 1999), Sage Thrasher birds, and sagebrush lizards. Unfortunately, the widespread destruction of this key shrub, as part of rangeland "improvement," has imperiled some sagebrush-dependent species, including the grouse and the pygmy rabbit (*Brachylagus idahoensis*). It is surely not an overstatement to say that the sagebrush once was ecologically the most important plant in the Great Basin. Regrettably, however, the non-native cheatgrass may now have that distinction.

Three Yellow-Flowered Great Basin Shrubs

Rules for Life in the Oregon Desert

1. The color must be predominantly gray. It can be gray-green, or gray-blue. Sagebrush is mostly dusty gray-green, junipers are gray-blue. One kind of rabbit brush is bright gray-green, the other kind has so little green that it is called gray rabbit brush.

—E. R. JACKMAN, 1964

Three shrubs from the aster family, annually resplendent in bright, yellow flowers, grow on the lower slope of Abert Rim. The first of these to bloom—in May and June—is the spineless horsebrush, which has white, hairy stems and produces clusters of egg-yolk-yellow flowers. Because it blooms so early, the horsebrush likely avoids competition with other plants for such pollinators as bees, beetles, flies, and moths.

Gray and green rabbitbrush (*Ericameria nauseosa and Chrysothamnus viscidiflora*) are the two other aster family plants that grow around the lake. Gray rabbitbrush (also called rubber rabbitbrush) is a four-foot-tall, gray-green shrub with bright-yellow flowers, which begin to open by late July and continue to bloom into the autumn. This species is conspicuous near the lake and in areas throughout the region that have been altered by development and

Close-up of the flowers of the spineless horsebrush (*Tetradymia canescens*).

Rabbitbush flowers, pollinators, and seed predators, *clockwise from top left*: gray rabbitbrush (*Ericameria nauseosa*), blooming; Hunt's bumblebee (*Bombus huntii*); Great Basin wood-nymph butterfly (*Cercyonis sthenele*); yellow-pine chipmunk (*Tamias amoenus*), eating seeds of a gray rabbitbrush; and fritillary butterfly (*Speyeria* sp.).

overgrazing. Its twigs and leaves sport a dense, feltlike mat of tiny hairs, which give the plant a grayish-green cast and its name.

Green rabbitbrush, as its name implies, displays a brighter green color and has glandular hairs that make it sticky. The narrow, linear leaves of this species are noticeably twisted. Green rabbitbrush grows abundantly on the lower to middle slopes of the rim.

Jackrabbits, mule deer, and pronghorn feast on rabbitbrush, and the shrub also provides cover and food for smaller mammals and birds. The blossoms of all rabbitbrush species readily attract nectar-feeding insects, including bees, wasps, flies, beetles, and butterflies, because they flower later in the season after most other flowers have finished blooming.

Both the gray rabbitbrush and the green rabbitbrush grow extensively throughout the Great Basin and in much of the West, from British Columbia all the way to Mexico. Gray rabbitbrush tends to become weedy, due to fine, hair-tufted seeds that disperse readily in the wind. Its presence in large stands often indicates that cattle have overgrazed the land or that humans have disturbed other native plants enough to make room for such a fast-spreading, weedy plant, and this is why you can often find numerous gray rabbitbrush shrubs growing on highway shoulders. Desert landscapers also sometimes intentionally plant the shrubs because they require very little water and bloom attractively.

The sap of the gray rabbitbrush plant contains natural rubber in small concentrations, perhaps an evolutionary feature that reduces herbivory. The University of Nevada's Agricultural Station has studied the shrub to assess its usefulness as a source of natural rubber (USDA 2006).

Other Shrubs in an Arid Land

> There was very little grass [along the Lake Abert shore] for the animals, the shore being lined with a luxuriant growth of chenopodiaceous shrubs [greasewood], which burned with a quick flame and made our firewood.
>
> —JOHN FRÉMONT, 1845

Abert's nearby landscape serves as home to three additional, relatively common shrubs, especially at the lake's north end. Two of these—spiny hopsage and shadscale—belong to the goosefoot subfamily, Chenopodioideae. Frémont touted the third of these plants, greasewood (*Sarcobatus vermiculatus*), for its usefulness in making a campfire. Until recently, botanists had included greasewood in the goosefoot subfamily, but they now categorize it as belonging to

Arid-land shrubs from around Lake Abert, *clockwise from top left*: Spiny hopsage (*Grayia spinosa*) flowers and foliage; flowers of shadscale (*Atriplex confertifolia*); close-up of scalelike shadscale leaves; male flowers of the greasewood (*Sarcobatus vermiculatus*); and greasewood female flowers.

its own family, Sarcobataceae. More broadly, both goosefoot and greasewood are small shrubs that grow in dry, salty, and alkaline soils, and thus they are common in the Great Basin and in deserts around the world.

One of the lake area's most common goosefoot plants, the spiny hopsage is a dryland shrub that also grows throughout southeastern Oregon, as far north and west as Bend, and which is scattered across the West. It grows to about four feet in height and has gray-green, elliptical leaves. Its wind-pollinated flowers are small, but the spiny hopsage shows off in other ways, with colorful, winglike leaves—purplish-pink, reddish-yellow, or white—that resemble hops. By midsummer, the leaves turn reddish-brown, and in winter, they fall off. Deer, pronghorn, bighorn sheep, rabbits, and some birds feast on the leaves, as well as on the spiny hopsage's fruit.

The Great Basin's shadscale is also widespread and drought tolerant, reaches a height of two feet, and has a rounded crown. You can recognize it easily for its spiny twigs and its dense cluster of silvery-green leaves that resemble fish scales and which are oriented in various directions that tightly cover the branch tips. At the lake, shadscale grows primarily near the north end on southwest-facing slopes that are hot and dry in summer. Hugh Mozingo, author of *Shrubs of the Great Basin,* describes it as a true desert plant that thrives in the driest locations, surviving on only four to eight inches of annual precipitation and tolerating salty soils. Shadscale's flowers are small and wind pollinated. Wildlife not averse to a salty flavor nibble on its fruit and leaves. In Oregon, shadscale grows as far west as Lakeview and as far north as Fossil, and it is widespread across the West.

The third of these prolific, woody plants is greasewood, sometimes referred to as black greasewood, and it is the most common shrub between Lake Abert and the highway, as well as a short part of the way upslope. Its distribution around the lake might mirror the area once covered by Paleolake Chewaucan, thriving there now, perhaps, because of the alkaline soils the ancient lake left behind. In Oregon, greasewood grows east of the Cascades, north to La Grande, and west to Klamath Falls. Throughout the Great Basin, greasewood is one of the most plentiful shrubs along the basin bottoms that the paleolakes once covered, where the soil is alkaline, rich in clay, and marked by a deep water table that the plant tolerates well.

Greasewood grows to about six feet in height and has small, gray-green, fleshy, fingerlike leaves. The tiny male flowers are set in

A greasewood plant's extensive roots, exposed by wind erosion at Alkali Lake dunes.

conelike structures near the end of its branches, and the female flowers are dish shaped and occur at the junction of the leaves and stem. The small fruits are winglike.

According to Mozingo, greasewood has roots that can extend for more than fifty feet, which enables the plant to grow where the water table is deep. In the Alkali Lake Basin north of Lake Abert, greasewood successfully thrives in sand dunes, despite the fact that its deep and extensive root system is sometimes exposed by the blowing sand. The shrubs also tolerate flooding, so they can grow in intermittently wet playa habitats, where they capture windblown soil and build mounds that allow them to keep growing upward. Greasewood leaves contain oxalates, acidic chemicals produced by plants that can be toxic to animals if the leaves are their only food source. Some wildlife can eat the leaves as part of their diet, however, and varied animals seek shelter in the shrubs in winter (Mozingo 1987).

Streamside Plants

Riparian (bank-dwelling) plant communities thrive along the three stream channels that flow from Abert Rim. They comprise mostly tall shrubs, including: willows (*Salix* spp.); interior rose plants (*Rosa woodsii*); red-osier dogwoods (*Cornus sericea*); service-berries (*Amelanchier alnifolia*); golden currants (*Ribes aureum*); and chokecherries (*Prunus virginiana*). All of them, except for the willow, produce edible fruits.

Nearby habitats support the white-flowered Klamath plum (*Prunus subcordata*), which attracts bees in spring and has edible, Ping-Pong-ball-sized, red fruit in late summer (see chapter 7). The sweet-tasting fruit produced by these shrubs in late summer was likely welcome food for the people who lived along Abert's lakeshore in the distant past, and, in fact, several former Native American village sites lie in areas where the Klamath plum still grows. Many species of wildlife continue to eat this fruit, including Western Scrub-Jays (*Aphelocoma californica*) and coyotes.

Colorful, pink-flowered, sticky geraniums (*Geranium viscosissimum*) grow in the seasonally wet meadows located near creeks and springs. Purple-flowered horsemint (*Agastache urticifolia*) blooms in May and June and attracts many types of bees and other insect pollinators. Nearby aspen groves flourishing along Poison Creek are home to the attractive, red-and-yellow-flowered western columbine (*Aquilegia formosa*) and to the white-flowered starry Solomon's seal (*Smilacina stellata*). Sometimes a great horned owl (*Bubo virginianus*) or a mule deer (*Odocoileus hemionus*) seeks cover in these shady groves.

Flowers from around Lake Abert and Abert Rim, *from top to bottom, left to right*: fragrant evening primrose (*Oenothera caespitosa*); lava aster (*Ionactis alpina*); sticky geranium (*Geranium viscosissimum*); Applegate's paintbrush (*Castilleja applegatei*); Wyoming paintbrush (*Castilleja linariifolia*); dwarf onion (*Allium parvum*); sagebrush mariposa lily (*Calochortus macrocarpus*); wooly-pod milkvetch (*Astragalus purshii*); longleaf phlox (*Phlox longifolia*); Klamath plum (*Prunus subcordata*); and interior rose (*Rosa woodsii*).

Abert Rim and Its Flora—The High Life

The imposing Abert Rim stands like a giant wall protecting Lake Abert. Any account of the natural history of this area would be incomplete without mentioning this impressive, looming part of the landscape.

Abert Rim is a block of the earth's crust, a horst created by numerous basaltic lava flows that originated near Steens Mountain. The rim resembles a table top that tilts up to the west and slopes down to the east and the north. At its highest point, above the south end of Crooked Creek Basin, the rim's elevation is 7,500 feet, and at its lowest point, east of the north end of the lake, its elevation is only 5,000 feet. Thus, in about 25 miles, the elevation drops about 2,500 feet, a large enough drop to have a pronounced effect on the plants growing there.

An array of colorful, flowering plants do thrive high on the slopes of the rim, especially above the south end of the lake. They include: arrowleaf balsam root (*Balsamorhiza sagittata*); western puccoon (*Lithospermum ruderale*); hoary aster (*Machaeranthera canescens*); spiny skeleton weed (*Stephanomeria spinosa*); slimpod milkvetch (*Astragalus filipes*); silvery lupine (*Lupinus argenteus*); yarrow (*Achillea millefolium*); Applegate's (*Castilleja applegatei*), desert (*Castilleja angustifolia*), and Wyoming paintbrushes (*Castilleja linariifolia*); and the pale-blue or purple sagebrush lily (*Calochortus macrocarpus*).

Just below the top of the rim, the vegetation is dense and nearly impenetrable because of the lush growth of shrubs, including mountain snowberry (*Symphoricarpos rotundifolius*), mountain

Abert Rim near the south end of Abert Lake after a light snowfall.

spray (*Holodiscus dumosus*), and wax currant (*Ribes cereum*), the latter of which has slightly sweet, reddish fruit in midsummer. A variety of wildflowers grow there too, poking up among the boulders, including sagebrush bluebells (*Mertensia oblongifolia*), pale-pink-flowered western boneset (*Ageratina occidentalis*), and the white-flowered, round-leaf alumroot (*Heuchera cylindrica*).

The top of Abert Rim, like its lower slope, is a land of contrasts. Dense forests of white fir, ponderosa, and lodgepole pine trees grow at an elevation of six thousand to seven thousand feet on the rim's higher and better-watered south end, situated in the Fremont National Forest. Stands of aspens that turn golden in autumn are rooted near springs, where there is more moisture. In May, masses of bright-yellow-flowered heartleaf arnicas (*Arnica cordifolia*) grow in leaf mold under the trembling aspens and are surrounded by the verdant foliage of white-flowered, starry Solomon's seal plants.

Farther north, the rim is treeless, and rock is often the dominant substrate. Plants growing there are mostly relegated to pockets of soil tucked between the rocks. From a distance, the top of the rim looks flat and easy to traverse. In reality, the abundance of small boulders scattered over the ground makes hiking difficult and laborious for people, although pronghorn gallop along effortlessly and at amazing speeds. Frost wedging—when the pressure of freezing water causes boulders to crack or break down—and weathering of the underlying basaltic bedrock have created this jumbled and rocky landscape.

The wax currant (*Ribes cereum*) has pale-pink or purplish flowers in spring and slightly sweet, red, edible fruit in midsummer.

Pocket groves of aspens and stands of white firs and ponderosa pines also thrive in a few north-facing habitats below the rim, where snow accumulates and creates a moist microclimate. Clusters of firs and pines stand out most visibly above the south fork of Poison Creek, nestled just below the rim on a northwest-facing slope. These conifer stands and aspen groves might be relicts from the Pleistocene, when the climate was cooler and the trees likely

Views along Abert Rim: *top*—looking southwest to a grove of white firs and ponderosa pines, with some aspens and willows mixed in, growing on a talus slope near the rim above the south fork of Poison Creek; *bottom*—looking north along the crest of Abert Rim south of Poison Creek (numerous low sagebrush shrubs and a few scattered western junipers grow along the rim, yellow cobblestone and other lichens spread across stable parts of the cliff face, and below the cliff, stands of white fir, aspen, and willow trees, as well as chokecherries and other shrubs, dot the landscape).

grew much farther downslope, perhaps even as far down as the lakeshore. However, as the Holocene progressed and the climate became warmer and drier, the trees could only survive in moist microhabitats along the creeks and at some north-facing, higher-elevation sites.

The face of Abert Rim remains a geologically active place, where large boulders are on the verge of caving off and tumbling down. In some areas, such as east of Valley Falls, the scarp is nearly vertical for about five hundred feet. At Poison Creek, the upper scarp ranges in height from a few vertical feet to more than fifty feet. In places along the edge of the rim, the rocks look like they are being stretched by gravity, and they form a downward-sloping edge that is highly fractured, unstable, and likely to give way during the next earthquake. Just south of where Poison Creek cascades over the rim, arresting petroglyphs decorate large boulders hanging onto the cliff-edge (see chapter 7). Here, too, fearless woodrats leave their signposts on the sharp edges of the boulders.

Poison Creek marks the approximate boundary between the higher-elevation areas of the rim to the south, with their stands of firs, pines, and junipers, and the lower areas to the north that are mostly treeless. If you head south from the creek and move up in elevation, you see the number of western junipers steadily increasing, until you arrive at a point about five miles southeast of Valley Falls, where the junipers join the near-continuous stands of ponderosa pines and white firs. If you head north from Poison Creek, you don't see another substantial cluster of junipers until you travel twenty miles to Juniper Mountain, east of Alkali Lake.

Colvin Timbers, an isolated, one-square-mile cluster of old-growth ponderosa pines, white firs, and western junipers, lies about one mile southeast of where Poison Creek cascades over the rim. This stand of old-growth trees is part of the twenty-three-thousand-acre Abert Rim Wilderness Study Area. If you travel southeast of Colvin Timbers, you encounter the lovely, forty-acre Colvin Lake, one of the few permanent water bodies on the rim. Hikers may be tempted to swim and cool off in this enticing spot, but leeches lurk beneath the waters. The lake was apparently an important camping area for Native Americans, who carved numerous petroglyphs on the nearby rocks.

Colvin Timbers is also notable because it is the northernmost array of ponderosa pines on the rim and, at this latitude, for some distance east. No pines grow on Poker Jim Ridge or even on the much-higher Steens Mountain (Mansfield 1999). A similarly isolated stand of old-growth ponderosa pines lies at the southeastern end of Hart Mountain in an area called Blue Sky Camp.

Colvin Lake and Colvin Timbers, *clockwise from top left*: Google Earth satellite image, showing the south end of Lake Abert; remarkable stand of old-growth ponderosa pine trees at Colvin Timbers; an aerial view of Colvin Lake, looking south to the North Warner Mountains, taken in May 2018 when water levels were high.

East of the lake, the landscape on top of the rim differs markedly from anything below. You find very few trees north of Poison Creek, and nearly all the plants that you see are low growing. Even many of the western junipers look more like multi-trunked shrubs with rounded crowns than like actual trees. The soil is very thin and dries out quickly, and the wind near the edge of the rim—blowing from the west and the southwest—can be quite strong and likely damages tall-growing plants with stiff limbs, including typical junipers. A few scattered, curl-leaf mountain mahoganies (*Cerocarpus ledifolius*)—small, evergreen trees with leaves that curl when dry—thrive along the edge of the rim, preferring the rocky habitat, where fires rarely burn. These trees likely earned their name from their distinctive, very hard, brown wood, which resembles tropical mahogany but is unrelated. Native Americans once used the wood to create digging tools.

When the seeds of the mountain mahogany trees mature in July, they make the trees look frost covered. The seeds have a long, curled, fuzzy "tail," a distinctive characteristic responsible for the tree's Greek genus name, *Cerocarpus,* which means "tailed fruit." The tail likely aids in wind dispersal and is actually part of the flower, which stays attached to the seed. Mountain mahogany trees grow in rimrock habitats throughout the Great Basin and as far

Curl-leaf mountain mahogany: *left*—ten-foot-tall tree growing on the rocky crest of Abert Rim; *right*—mountain mahogany seeds with fuzzy, curled, taillike appendages that likely aid in dispersal.

Hood's phlox (*Phlox hoodii*): *left*—carpet of phlox growing on a rock pavement on top of Abert Rim; *right*—close-up of blossoms and tiny leaves.

south as Mexico's interior. You can often see them in lines or as clusters of low trees on rocky ridges. They grow slowly because of low soil fertility and inadequate moisture, but they can live for centuries, possibly for one thousand years or more (Schultz et al. 1990).

On top of the rim, the ground is especially rocky, forming a kind of pavement. In its cracks, the clay-rich soils are thin and dry by midsummer, making it difficult for plants to get established and to grow. However, this area has diverse spring flora that have adapted to these demanding conditions, and these plants put on a show in May and June that is worth a climb to the rim. Common flowering plants there are: low sage; Pursh's milkvetch (*Astragalus purshii*); crag aster (*Ionactis alpina*); dwarf onion (*Allium parvum*); Hooker's balsamroot (*Balsamorhiza hookeri*); tapertip hawksbeard (*Crepis acuminata*); Bloomer's daisy (*Erigeron bloomeri*); false agoseris

(*Nothocalais troximoides*); mat buckwheat (*Eriogonum caespito-sum*); desert paintbrush; tansy-leaved evening primrose (*Camisso-nia tanacetifolia*); long-leaved phlox (*Phlox longifolia*); and Hood's phlox (*Phlox hoodii*). In many sites among the rocks, the ground looks snow covered in May because of the masses of low-growing, white-flowered Hood's phlox.

Taking a Hike

Hiking to the rim on a sunny May or June morning is delightful. Countless bees and flies buzz around, pollinating the profuse carpet of wildflowers, many of which are in full bloom. The smell of sage permeates the air, and songbirds sing their spring tunes from the cover of the riparian vegetation. From the rim, clear skies offer an impressive, long-range view of distant snow fields on the North Warner Mountains to the south, on the Hart and Steens Mountains to the east, and on the Winter Ridge to the west.

Far below, the lake spreads out like a blue mirror nestled between Abert Rim and Coglan Buttes. Later in the afternoon, looking to the west over Coglan Buttes, you can see dust devils as they form on the Summer Lake playa and carry a white plume of alkali high into the sky. Even if you hike to the rim in September or October, you won't be disappointed, because the south fork of Poison Creek should still be flowing, so you will hear its tumbling water, one of the best sounds in the desert. (Poison Creek may have gotten its name from the milky color of its water, which contains suspended clay particles, but the water is not toxic.)

The view from Abert Rim in May 2010, looking north from a point near Poison Creek: the rim's top is flat and nearly treeless; Highway 395 is just visible as a thin ribbon near the water's edge; and the lake is relatively high.

Travel along Abert Rim looks easy from a distance, because the landscape appears to be flat. Hiking on foot, however, is slow because of the abundance of exposed rocks, so you need to take the tricky terrain into consideration, especially when you plan longer hikes. One way to access the rim is from Highway 395 at Poison Creek, via a historical shepherds' trail that Native Americans likely used even earlier. This trail originates east of the highway, about one quarter mile south of where Poison Creek crosses under the highway. It remains clear and usable in some places, and, in fact, you will be astonished when you see the hulking size of the boulders that people once moved off the path. However, over time, rocks have fallen onto parts of the trail, and vegetation has overgrown and obscured other parts.

The trail heads northeast along the slope. About halfway up to the rim, the path finally disappears completely, and from there you have to find your own way up, which is not that difficult. Regardless of which final route you take to the rim, however, you should avoid hiking over the large talus slopes, consisting of boulders that are dangerous to cross. From a distance, the boulders look small, but some are actually eight to ten feet across, and there are deep holes between them, so they are far too risky to traverse.

The Hiking Project has an online map showing a route to Colvin Timbers from Highway 395 and from the Fremont National Forest (https://www.hikingproject.com/trail/7016576/oregon-desert-trail -seg-8-abert-s-to-colvin-timber), and Andy Kerr's *Oregon Desert Guide* describes the southern and Poison Creek routes. The southern route begins at the end of Fremont NF Road #3615, which you access from Highway 140, east of Lakeview. From there, a jeep trail leads to Colvin Timbers and to nearby Colvin Lake. In May or June, this is an enjoyable trip because of the abundance of wildflowers, scenic views of the North Warner Mountains and their remaining high-elevation snow fields, and abundant pronghorn.

From the north you can access the rim by traveling on an improved gravel road that comes off Hogback Road (Lake County # 3–10), about 3.4 miles east of the highway, and which parallels the high-voltage powerline running north to south. That road joins the rim at a spot where the powerline crosses over the rim above the northeastern corner of the Lake Abert playa; however, for most of its length, the road is up to 4 miles away from the rim. About 9 miles after leaving Hogback Road, you can head out on foot west for about 1 mile, and you will then see the Twin Lakes and some amazing petroglyphs, as well as an extensive, hand-built stone corral located on the south side of North Twin Lake.

Going farther south, you can camp at Rabbit Creek meadow and, once refreshed, go on foot for several more miles, before heading west for an additional two miles to reach the rim and nearby Mule Lake. And don't forget, as Andy Kerr suggests, "to dangle your tootsies over the rim." While cooling your feet and admiring the spectacular view, you might see Violet-green Swallows (*Tachycineta thalassina*) and White-throated Swifts (*Aeronautes saxatalis*) gracefully gliding back and forth in the updraft along Abert Rim's face.

The Twenty-Five-Thousand-Acre Abert Rim Wilderness Study Area

In 1992, the BLM designated twenty-five thousand acres along Abert Rim, east of the lake, as a wilderness study area (WSA). The BLM awards such designations as a means of protecting the wilderness characteristics of a landscape until Congress decides whether or not to designate it formally as wilderness. To earn this temporary

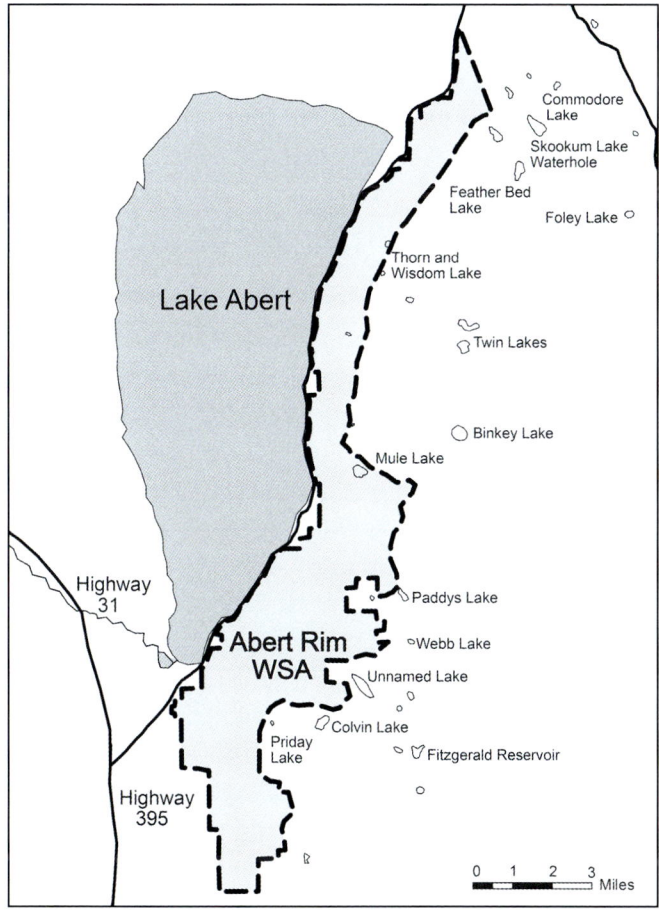

Abert Rim Wilderness Study Area.

protective label, a region must be (1) larger than five thousand acres and without developed roads; (2) natural; and (3) rich in outstanding, primitive recreational opportunities.

Currently there are over 500 WSAs in the United States, and Oregon has 89 such protected areas, totaling 2.7 million acres. Only Utah has more WSA acres. Although the Abert Rim WSA represents a sliver of land, which in some spots is just a mile wide, the designation does help protect the rim—one of the most scenic and remarkable areas in Oregon—from development. Unfortunately, the WSA only includes 2 of the 25 picturesque and biologically vital playa lakes on the rim. And some of the unprotected ones, such as Colvin Lake and the Twin Lakes, have outstanding petroglyphs, artifacts that should be preserved. Furthermore, North Twin Lake, which also deserves protection, has an incredible stone corral that is about 0.2 miles long, probably constructed by shepherds more than a century ago. Only Mule Lake (30 acres in area), 9-acre Thorne and Wisdom Lake, and several smaller lakes are within the WSA.

I am really puzzled about why BLM has excluded most playa lakes from the WSA. One possible explanation is that the lakes have lost some of their wilderness value because, ironically, the BLM has allowed too many cattle to graze and to denude areas around these lakes, clearing away vital, wild habitats and replacing them with invasive plants. To prevent more such degradation, and before Congress finally designates Abert Rim as a wilderness, BLM should propose protective designations for as many of the playa lakes as possible.

Flowering Plants Growing at the North End of the Lake

Yet another rich habitat lies north of the lake and west of the XL Ranch house, near Lake County Road #3–09. There, dunes made of sand and clay host several plants that don't grow anywhere else near the lake, including the yellow-flowered desert dandelion (*Malacothrix glabrata*) and two small, low-growing herbs—corrugated spring parsley (*Cymopterus corrugatus*) and an unusual relative of lettuce, carveseed (*Glyptopleura marginata*), which blooms with white flowers and once served as an early spring food for Paiutes in the region. Corrugated spring parsley and carveseed grow more abundantly in other areas of the Great Basin, and in Oregon, these plants are at the northern and western limits of their ranges.

The most conspicuous wildflower in the dunes west of the XL Ranch house is the attractive evening primrose (*Oenothera caespitosa*), which has surprisingly large and fragrant, four-petaled blossoms that are white with yellow-tinted centers. This primrose was first described in Western writing by botanist Frederick Pursh, who

Flowering plants growing on the dunes at the northwest end of Lake Abert near the XL Ranch house: *left*—carveseed (*Glyptopleura marginata*); *right*—corrugated spring parsley (*Cymopterus corrugatus*).

examined plants that were collected in 1806 by Meriwether Lewis along the Missouri River in Montana.

Invasive Plants—Mediterranean Sage, Cheatgrass, and Crested Wheatgrass

Unfortunately, a variety of invasive weeds have also taken hold in the area and likely have upset the ecology of the land around Lake Abert. Three of these non-native plants deserve special attention because of their abundance. The first of these, Mediterranean sage (*Salvia aethiopis*), is a two-foot-high, herbaceous plant with white flowers that form a candelabra shape and with a basal rosette of silvery leaves covered with wooly hairs. By late summer, the sage's brittle stem breaks off, and the upper part of the plant blows away like a tumbleweed, dispersing its seeds, which can number up to one hundred thousand from a single plant. The Mediterranean sage is native to areas around the Mediterranean and in north Africa.

This non-native plant initially invades sites altered by development or overgrazing, but once it takes root in such an area, it then quickly spreads into other habitats. In Oregon, botanists first noticed the plant in the 1920s, and they now classify it as a "B-rated weed of economic importance," meaning that its control could be costly. It has spread east of the Cascades and is most abundant in Klamath and Lake Counties. Mediterranean sage first likely got a foothold around Lake Abert along the highway or in nearby rangelands, and from there it was dispersed by the wind. Unfortunately, the plant has spread abundantly throughout the area, including on top of the rim.

A biological control agent—a species of root weevil—is available and should be considered as an option to limit its spread. Because

Mediterranean sage is distasteful and foul smelling, it is not useful for most wildlife, although bees do get nectar from its flowers. The strong odor of this plant is due to an aromatic chemical called aethiopinone that has antimicrobial properties (Hernández-Pérez et al. 2008). Surprisingly, Mediterranean sage has taken on one positive ecological role in Abert Lake. Because the flower stems of the plant are dispersed by the wind, tens of thousands of them end up in the lake every year, where they provide substrates on which alkali fly pupae develop.

The second invasive plant's importance is summed up aptly in the title of a BLM publication, *Cheatgrass: The Invader That Won the West* (Pellant 1996). No other invasive plant in the West is so widespread or so damaging. Accidentally introduced to the West in about 1890, it covered 160,000 square miles by 1930, and now there is hardly a square foot of rangeland habitat that does not contain this highly invasive grass (Grayson 1993; Young and Clements 2009). In 1994, cheatgrass dominated over 3 million acres of public land and infested or threatened to invade another 76 million acres (Pellant 1996). Cheatgrass apparently got its name because it cheated wheat farmers of their crop in parts of Oregon and Washington. Anyone who has hiked in cheatgrass-dominated land in the summer or fall has undoubtedly experienced the pain imparted by its sharp-pointed seeds, fitted with thousands of tiny, rearward pointing barbs, clinging to their socks, shoe laces, and pants. With every movement, the barbs work the seeds inward, which makes their removal all the more difficult. Even worse, the seeds can harm animals by piercing their feet, noses, and eyes.

Cheatgrass is a winter annual, meaning it germinates in the fall

Mediterranean sage (*Salvia aethiopis*): *left*—blooming plant; *right*—close-up of flowers.

or early winter after the first rain, grows during the winter and spring, then sets its seeds and dies by summertime. Its spikelets, which contain its seeds, turn red when they mature in June and can color entire hillsides if the grass is abundant. The plant's seeds can lie dormant in the soil for up to five years, and its ability to produce seeds under nearly all moisture conditions gives it an advantage over native grasses. Consequently, observers have reported seeing cheatgrass densities of over ten thousand plants per square meter.

Besides having a competitive advantage over native grasses, cheatgrass can alter rangeland ecosystems: after it dies and dries up, early in the summer, it can readily catch fire and burn, spreading flames across the landscape. Consequently, fires can return to cheatgrass-dominated habitats every five years, while they tend to burn native sagebrush habitats only a few times per century. If a habitat has evolved while experiencing very few fires, and then starts burning much more often, its plant cover and diversity will substantially decrease, which, in turn, will provide room for cheatgrass and other non-native plants to dominate. The resulting cheatgrass-wildfire cycle relies on the accumulation of fine fuels from the growth of the grass and other plants during wet years, which allows wildfires to burn in dry years (Pillod et al. 2017).

Unfortunately, the cattle now grazing on approximately 270 million acres of public land exacerbate this problem by reducing the competitive cover provided by native bunchgrasses (Reisner et al. 2013). Scientists stress that we have few good options for controlling cheatgrass, so we must manage our rangelands properly, particularly the way that we graze them, before cheatgrass completely takes over (Pellant 196; Reisner et al. 2013). Three recommended practices may minimize damage to rangelands and restore habitats impacted by cheatgrass: (1) reducing grazing pressure; (2) increasing native bunchgrass cover and diversity; and (3) using biological

Seeds of cheatgrass (*Bromus tectorum*): *left*—close-up of four spikelets, each about one inch long, that contain the developing seeds; *right*—highly magnified image of the awns' numerous, rearward-pointing barbs, which cause the seeds to penetrate and cling tightly to clothing or to animal fur (scale, 1 mm = 1/25 inch).

soil crusts (communities of beneficial organisms) to slow cheatgrass establishment (Reisner et al. 2013). Sadly, we may already be too late; we have woefully overgrazed public rangelands in the Great Basin, and we also must contend with climate change, which is further drying the landscape and causing more frequent fires. Urgent restoration actions may be the only solution.

We humans likely introduced both cheatgrass and Mediterranean sage to our landscapes by accident, but we intentionally planted the third common, invasive plant: crested wheatgrass. Land managers sowed its seeds throughout much of the Great Plains and the intermountain West to improve livestock forage and to reduce soil erosion (Rogler and Lorenz 1983; Fansler and Mangold 2010; Grant-Hoffman et al. 2012). Native to Asia, crested wheatgrass adapts extremely well to cold and semiarid climates and now grows widely across the West and in Canada. It does provide good forage for cattle in some areas that have been damaged by overgrazing, and some state agencies still advocate for its use in some situations.

However, crested wheatgrasss spreads too rapidly, forming monocultures that crowd out native grasses and forbs (herbs other than grass), increase bare ground, and reduce soil nutrients and organic matter, all of which reduce biodiversity and adversely impact ecological integrity. Consequently, federal land managers who need to meet multiple-use objectives have taken measures to control it (Fansler and Mangold 2010; Grant-Hoffman et al. 2012). Around Lake Abert, crested wheatgrass grows in scattered clumps. Fortunately, I have not seen much evidence of harmful monocultures forming, except in small areas damaged by overgrazing.

Terrestrial Invertebrates—Small and Often Overlooked Critters

The area around Lake Abert not only features a wide array of plants: it also supports a thriving, diverse community of insects, spiders, and other land-dwelling invertebrates. One of the more unusual of these is the northern scorpion (*Paruroctonus boreus*), which is native to the western United States and parts of nearby Canada and is the most widespread scorpion in the United States (Tourtlotte 1974). Northern scorpions are dark brown and can be up to two inches long, although they are usually smaller.

Interestingly, scorpions give birth to live babies, which ride on their mothers' backs. Northern scorpions are most active in the summer, just after twilight, when they forage. They are ambush predators: they hide, waiting for insects and other small invertebrates to approach, and then they seize their unwitting prey.

Although it has adapted to a variety of habitats, the northern

Northern scorpion (*Paruroctonus boreus*): *left*—photographed at night, using an electronic flash; *right*—photographed using a black light. Both scorpions are about two inches long.

scorpion is most abundant on dry hillsides with rocks that provide daytime cover (Tourtlotte 1974), which is exactly what the lower slopes of Abert Rim offer. One night in early August, I searched for scorpions along the lower slope of Abert Rim, using a black light. I knew that I would have trouble finding them with a standard flashlight at night, and I also knew that a blacklight casts a blue glow on them in the dark and makes them highly visible. As I hunted, I kept in mind that northern scorpions are venomous and should never be handled.

Frogs and Toads—Desert Amphibians

No amphibians can survive in Lake Abert because of its high salinity. However, the marshes along the eastern shore of the lake provide a welcoming habitat for the western chorus frog (*Pseudacris regilla*), also commonly called the western or Pacific tree frog.

These hardy amphibians are the most common frogs throughout the region, and they seem to thrive in almost every habitat. I have even seen them in small spring-fed ponds on the Lake Abert playa. How they can get onto the playa without having the alkali

A western chorus frog (*Pseudacris regilla*) adopting a cryptic color pattern to match a lichen-covered rock.

Great Basin spadefoot toad (*Spea intermontana*): *left*—large larvae, about two inches long; *right*—a young, quarter-sized toadlet.

burn their sensitive skin puzzles me. I am also amazed every time I see these frogs living miles away from any possible breeding site. Their ability to cover so much ground and to find their way back to a pond to breed is marvelous. Although they are small, when they make chirping, croaking sounds together at their breeding sites, the sound is nearly deafening.

I thought that I would find Great Basin spadefoot toads (*Spea intermontana*), which are much larger than chorus frogs, close to the lake, but for a long time, I could not find a trace of them. I knew that areas with sandy soil, suitable for burrowing, are likely the preferred habitat for adult spadefoot toads (Buseck et al. 2005). One such area consists of sandy dunes near the XL Ranch cabin at the north end of the lake, so I did some investigating there. Finally, in May 2019, I found hundreds of *Spea* larvae concentrated in a shallow, drying, roadside ditch near the cabin.

My very first encounter with *Spea* happened when I was collecting branchiopod shrimp in the Warner Basin. Actually, the big tadpoles found me by swimming against my bare feet. These tadpoles are unmistakable, resembling iridescent-green, tiny-eyed Ping-Pong balls with tails. Later, I found *Spea* toadlets in Guano Basin's wetlands, fifty miles southeast of Lake Abert. The miniature toads had recently metamorphosed from larvae and hid in mud cracks left by the receding water.

Scientists don't believe that the Great Basin spadefoot toad is threatened in Oregon, but I can find no current information on their status. A review done in Wyoming several years ago suggested that there had been very few sightings of the toad across its range there and that, therefore, its true conservation standing was unknown (Buseck et al. 2005). We cannot easily verify their status here in Oregon because the adults are nocturnal and much of their habitat is remote.

Lizards

Dry shrublands encircling Lake Abert include an abundance of boulders, the ideal habitat for lizards. The most common of these reptiles is the western fence lizard (*Sceloporus occidentalis*), a widespread species throughout the West. At Lake Abert in May and June, the lizards frequently sun and boldly display themselves on the rocks. Perhaps not surprisingly, numerous, old lizard petroglyphs, based on their real-life counterparts, decorate boulders around the lake.

Western fence lizards are about eight inches long and have spiny scales in colors ranging from gray or light-brown to nearly black, and their bodies appear to be especially dark early in the day, when they are first warming themselves in the sun. Their scales sport dark chevron-, wave-, or blotch-shaped designs, and the critters have light-colored undersides, usually white or yellow. In addition, males have two bright-blue patches along their bellies and one on their throats, making them distinctive; they also have yellow patches on the back of their thighs, but these are more difficult to see.

These reptiles primarily eat arthropods, including insects and spiders, and they may also consume scorpions, but they mostly nibble on beetles in southeastern Oregon (Whitaker and Maser 1981). Females lay eggs in the spring, which hatch two months later.

The fence lizards have close relatives in the area, sagebrush lizards (*Sceloporus graciosus*), which are smaller, just a few inches long, and are less conspicuous, seemingly shy about being out in the open. However, they seem to be more widespread, and I have seen small numbers of them in many locations, even on Abert Rim. They share similar colors to those of western fence lizards, but they are usually not so dark. Furthermore, male sagebrush lizards often have blue flecks on their backs and orange markings on their sides, something not seen on western fence lizards. Scientists believe that

Great Basin lizards (*Sceloporus* spp.), *left to right*: male western fence lizard (*Sceloporus occidentalis*), showing off the blue patches on his throat and belly; another western fence lizard, sunning itself on a boulder; and a male sagebrush lizard (*Sceloporus graciosus*), resting on a lichen-covered boulder.

Horned lizards from around Lake Abert, *clockwise from top left*: desert horned lizard (*Phrynosoma platyrhinos*) with cryptic coloration and a rough skin that mimics rocky desert soils; juvenile pygmy short-horned lizard (*Phrynosoma douglasii*), which is the size of a quarter; and an adult pygmy short-horned lizard, which is 1.5 inches long.

most fence and sagebrush lizard populations are secure, due to how broadly dispersed they are throughout the West.

An unusual and less-common lizard from around the lake—the desert horned lizard (*Phrynosoma platyrhinos*)—is a distinctive species that is characterized by its broad, flattened belly, a row of fringed scales along its sides, numerous pointed scales on its back and tail, and six larger, hornlike scales at the rear of its head. The abundant armor and flattened body make members of the *Phrynosoma* genus unmistakable, and their distinctive appearance has led casual observers to mistakenly call these reptiles "horned toads." Zoologists Eric Pianka and William Parker, writing about these reptiles in 1975, said: "*Phrynosoma* are characterized by a relatively bizarre constellation of morphological, behavioral and ecological adaptions that tend to set members of this genus apart from most other lizards." The genus includes thirteen species in North America.

Desert horned lizards reach a length of about five inches. Their coloration includes a mixture of orange, brown, and gray, and it is similar for both sexes. They live in South Central and eastern Oregon, and their habitats extend east to Utah and all the way south to northern Mexico. Although desert horned lizards eat a variety of insects and spiders, they most enjoy dining on ants (Pianka and Parker 1975; Whitaker and Maser 1981). You can often find them near harvester anthills, where the lizards sit without moving and wait for their prey. Because harvester ants (*Pogonomyrmex* spp.) are small and mostly consist of indigestible chitin, their predators must eat lots of them. This requirement might account for the desert horned lizard's large belly and its sticky tongue. The lizards' flat-tened bodies have large surface areas that might help them more effectively absorb warmth from the sun on cold mornings.

Unlike the quick western fence lizard, which can sun itself in the open and then run away quickly when it is threatened, the slower desert horned lizard relies on camouflage and armor as its primary defenses, but it can also inflate itself with air, becoming larger and more difficult to swallow. However, if these defensives fail, this spe-cies, and some of its relatives, can squirt blood from a pouch under their eyes. Interestingly, they only seem to use this effective defense when predators like coyotes and bobcats—which apparently don't like the taste of blood—attack them.

Another horned lizard that makes its home around the lake, and in other parts of Oregon east of the Cascades, is the appropriately named pygmy short-horned lizard (*Phrynosoma douglasii*). You can easily recognize it for its small size, snub nose, short legs, and short tail. This lizard reaches a maximum length of about three inches, but it is usually much smaller, and its young are only about the size of a quarter. This species produces live young, which is unusual for lizards. Unlike the widespread desert horned lizard, this small rep-tile has a more limited range, most of which is in Oregon east of the Cascades but which also includes eastern Washington, south-ern Idaho, northern Nevada, and a small part of northeastern Cali-fornia. Its conservation status is likely secure in most of these areas, but it is listed as vulnerable in Washington State and is thought to have vanished from British Columbia, where the last reliable sight-ing was in the 1950s.

Two additional reptiles that live around the lake are the leop-ard lizard (*Gambelia wislizenii*) and the side-blotched lizard (*Uta stansburiana*). Both these reptiles are widespread throughout the western United States. The leopard lizard grows to a length of about five inches and sports dark spots on lighter-colored skin, and it lives in open, sandy sagebrush and greasewood habitats, such as those north of the lake. Side-blotched lizard adults are about six inches long, and, as their name suggests, they display a dark blue or black spot on each side, behind their front legs. These latter lizards are notable because of their unusual mating strategies and associated color patterns. Zoologists believe that these special characteristics

Adult leopard lizard (*Gambelia wislizenii*) with conspicuous dark spots.

could lead to the formation of a new, distinct species (Corl et al. 2010). Side-blotched lizards live near the tufa mounds north of the lake.

Many lizard species throughout Oregon do, unfortunately, face threats from the conversion of sagebrush habitats into grasslands and from the loss of vegetative cover, due to overgrazing by cattle (Werschkul 1982; Newbold and MacMahon 2008).

Snakes

At least two species of snakes crawl around Lake Abert, the Pacific gopher snake (*Pituophis catenifer*), which is relatively common, and the Great Basin rattlesnake (*Crotalus oreganus lutosus*). The gopher snake grows to about six feet in length, and its back is yellowish to light brown; its sides have blotches ranging from dark brown to black, but its coloration varies a great deal. Its range extends from southwestern British Columbia to northern New Mexico. You won't find a more plentiful snake in the upland environments east of the Cascades in Oregon, and it thrives in nearly all habitats. I often see Pacific gopher snakes in residential areas around my home in Klamath Falls. In fact, one year in June, a young gopher snake crawled into our house, much to the displeasure of my wife. Happily, I caught the young snake and released it into our backyard, unharmed. At Lake Abert, the Pacific gopher snake mostly inhabits the upland edges of the marshes.

Gopher snakes are solitary homebodies, mostly staying within a small range. They are constrictors that feed primarily on small mammals—hence their name—but they also eat insects, lizards, small snakes, and small birds. When alarmed, these snakes coil themselves into a striking posture similar to that of a rattlesnake, and they will even wiggle their tails to appear more threatening.

Unfortunately, this defensive behavior causes them to be mistaken for rattlesnakes and makes them targets for frightened people, who kill them. Coyotes are probably their primary natural predator; however, many of the snakes are killed by vehicles on highways because they like to warm themselves on the pavement. Gopher snakes reproduce by laying eggs. Because of their wide distribution and abundance, most Pacific gopher snake populations are likely secure.

Rattlesnakes are far less common throughout the West, and in fact, I rarely see them at Lake Abert, even though suitable rocky habitat abounds. Despite its smaller numbers, the western rattlesnake inhabits a vast region that extends from British Columbia all the way south to northwestern Mexico and east to Kansas. Only two of its subspecies live in Oregon: the northern Pacific rattlesnake (*Crotalus oreganus oreganus*), which dwells in the southwestern part of the state, as well as in the southern Willamette Valley and the Columbia Plateau; and the Great Basin rattlesnake, which makes its home in South Central and southeastern Oregon and suns itself on the rocks around Lake Abert.

This rattlesnake has a broad, triangular head, is relatively heavy-bodied, and can mature to a length of four feet, although it averages three feet. Its coloration varies a great deal and changes over the snake's life. Typically, it has a lighter background color on its back, topped by a series of darker saddles that resemble diamonds, and it has smaller blotches along its sides. There is also a dark-brown blotch on its head. The last two rings at the base of its tail are dark.

The western rattlesnake is Oregon's only venomous snake that poses a threat to humans, but it rarely strikes, unless it is provoked. In fact, most of the western rattlesnakes I have seen lie coiled on rocks or on the road, trying to warm themselves and intending no harm. However, one day, while I was staying with Keith and Lynn Kreuz at their home in Valley Falls, a Great Basin rattlesnake

Snakes that live around Lake Abert: *left*—Pacific gopher snake (*Pituophis catenifer*); *right*—Great Basin rattlesnake (*Crotalus oreganus lutosus,* photographed at Valley Falls with a small, reddish tick attached to the back of its head).

ventured to within a few feet from us, slithering alongside their cabin. Even when I followed it in an attempt to take some photos, it did not coil up threateningly or try to strike; it instead only wanted to escape. The snake was likely attracted to the Belding's ground squirrels (*Urocitellus beldingi*) that were plentiful around the house.

As pit vipers, western rattlesnakes use heat-sensing organs on the sides of their heads to locate small mammals. However, younger rattlers mainly eat insects and lizards. Mature females bear live young. Their habitats are dry, feature only sparse vegetation, and include rock piles suitable for hibernation. The snakes engage in activity mostly during the warmer months; in the winter, they hibernate. Humans have wiped out western rattlesnakes from some parts of their Oregon range, such as in the northern half of the Willamette Valley, and consequently, ODFW classifies them there as a sensitive species (https://oregonconservationstrategy.com/strategy-species /western-rattlesnake/). In my personal experience, I have rarely seen these snakes in much of South Central Oregon.

Mammals

According to B. J. Verts and Leslie N. Carraway, authors of the 1998 book *Land Mammals of Oregon,* nearly ninety mammal species inhabit the Basin and Range province where Lake Abert lies. Most of these mammals are rodents, including mice, gophers, kangaroo rats, woodrats, voles, chipmunks, squirrels, beavers, and porcupines, and they constitute the second most diverse group of mammals for any region in the state.

The area around the lake supports a wide variety of warm-blooded animals, including the mountain cottontail (*Sylvilagus nuttallii*), the black-tailed jackrabbit (*Lepus californicus*), the white-tailed antelope squirrel (*Ammospermophilus leucurus*), the yellow-bellied marmot (*Marmota flaviventris*), and the bushy-tailed woodrat (*Neotoma cinerea*), in addition to many smaller and less-conspicuous ones, such as shrews, voles, jumping mice, and bats. Yellow-bellied marmots frequently sit on boulders not far from the highway. The appealing, white-tailed antelope squirrel, one of the small rodents, lives in sagebrush habitats on the west side of the lake, is relatively tame, and often doesn't shy away from human observers.

The larger mammals around the lake include the pronghorn antelope (*Antilocapra americana*), mule deer, bighorn sheep (*Ovis canadensis*), coyote (*Canis latrans*), cougar (*Puma concolor*), bobcat (*Lynx rufus*), and badger (*Taxidea taxus*). In my explorations, I have most frequently seen mule deer, especially does, because they often feed in the marshes along the eastern side of the lake. I have more rarely seen bighorn sheep, but on several occasions, I have seen small herds high on the rim and even near the highway, where, unfortunately, they show little fear of humans. Pronghorn often graze in the alfalfa fields south of the lake and on top of the rim.

Rabbits and Hares

If you hike around the lake, you will most frequently see rabbits and hares, specifically mountain cottontails and black-tailed jackrabbits, among the small mammals living there. When I walk in the area, I often catch glimpses of these animals darting away, especially jackrabbits, which move quickly and live in nearly every vegetated habitat near the lake, except in the marshes.

Rabbits and hares from around Lake Abert: *top*—mountain cottontail (*Sylvilagus nuttallii*); and *bottom*—a black-tailed jackrabbit (*Lepus californicus*), which is a hare.

Mountain cottontails are the smaller of the two, less than a foot long. Their fur has a mix of gray, black, and tawny colors, and their tails are noticeably white. Black-tailed jackrabbits are nearly twice as large and have long, sometimes black-tipped ears. Although commonly called rabbits, they are really hares, and so they do not burrow but instead depend a great deal on sagebrush and other shrubs for cover.

Because jackrabbits do not live in burrows, they make easy targets for human hunters. In earlier times, people hunted them by using long nets. In fact, archaeological evidence from a site called Buffalo Flat Bunny Pits, located in the Fort Rock area of Oregon, shows that Native Americans participated in "rabbit drives" about 11,500 years ago, community hunts followed by mass processing of the hares. Archaeologists have found thousands of jackrabbit bones in the remains of what apparently were ovens (Oetting 1994a).

Black-tailed jackrabbits undergo periodic population explosions, which can have a significant economic impact on the farmers whose crops they start eating. These burgeoning populations once forced county governments to issue bounties. For example, in 1915, Harney County paid five cents per jackrabbit scalp, and in that year, the bounty hunters provided the government with over one million scalps (Bailey 1936). Rabbit drives were another means of reducing their numbers.

One other hare lives around the lake, the white-tailed jackrabbit (*Lepus townsendii*), which is larger than its black-tailed cousin, has shorter ears, and sports lighter-colored fur, especially in winter, when it turns mostly white. It is very rarely seen.

White-Tailed Antelope Squirrel

Yet another white-tailed critter makes its home around the lake, the white-tailed antelope squirrel, although the lake represents nearly the northwestern limit of its range. These animals are the most widespread of all antelope squirrels, and their range extends from Oregon all the way east to Colorado and as far south as the interior of Baja California (Zeveloff and Collett 1988; Hartson 1999). In Oregon, you will only find them in the three South Central and southeastern counties—Lake, Harney, and Malheur (Verts and Carraway 1998).

This desert-adapted squirrel lives in habitats with sparse vegetation, and, in fact, its genus name, *Ammospermophilus,* means "lover of sand and seeds." White-tailed antelope squirrels dwell in colonies, but they have individual burrows. They are, at most, just eight inches long, about the size of a chipmunk, and they are the smallest of Oregon's ground squirrels. You can recognize this small

White-tailed antelope squirrels (*Ammospermophilus leucurus*) from the west side of Lake Abert, holding their tails against their backs, a characteristic behavior.

mammal from the light gray fur on its body, its tawny legs and forehead (the latter of which has a white stripe on each side), and the dark stripe along its otherwise white tail.

A distinctive behavior of this species is that it holds its tail low over its back when it is stationary, perhaps as a sunshade, but holds it vertically when it runs, displaying that distinctive, white underside. This squirrel dwells on the west side of the lake in the same dry, big-sagebrush-dominated habitats where the desert horned lizard lives.

Yellow-Bellied Marmot

The most visible medium-sized mammal that lives around the lake, weighing in at about ten pounds, is the yellow-bellied marmot, or rock chuck, a member of the woodchuck tribe of rodents. Actually, its belly color is more cinnamon than yellow but varies a great deal among different populations. Marmots' backs and faces are brown, and their fur features numerous, white guard hairs over much of their bodies, making them look grizzled or hoary. They live in burrows that they often dig between or under boulders, which protect them from the badgers or coyotes that attempt to haul them out.

Through the summer, marmots put on a lot of weight, mostly as fat. In winter, they use that fat for energy, losing up to 50 percent of their body weight during hibernation, which can last six months or longer. Mammologist Chris Maser says that marmots eat almost constantly when they are active and that by the time they hibernate, they are nearly round. Their food consists of low-growing vegetation, including grasses, flowers, and seeds (Maser 1998).

You can frequently see them sunning themselves, or sitting alertly and vertically, while on the lookout for such key predators as coyotes and Golden Eagles. Marmots chirp or whistle shrilly when they are alarmed. They can live singly or in small colonies,

Life of the yellow-bellied marmot (*Marmota flaviventris*): *left*—adult stands at attention; *right*—a marmot's scat, 1.2 inches or 3 cm long.

and they are widespread throughout the West and the Great Basin, dwelling in the mountains and basins. You will often see them on the slopes of Abert Rim, due to the abundance of boulders and talus there, which make perfect burrowing grounds.

In earlier eras, Native Americans likely relied on marmots as an important food source, particularly in late summer, when the animals' substantial fat deposits would have made them especially nourishing. In fact, the Northern Paiute people called the Kidütökadö, or the Gidu Ticutta, and who lived in the region between Goose Lake and the Warner Basin in South Central Oregon, as well as in the Surprise Valley area of northeastern California, earned the nickname "Yellow-bellied Marmot Eaters" because the mammals were such a common food for them.

Bushy-Tailed Woodrat

In Oregon their favorite haunts are the rimrock cliffs or broken lava beds and caves, where they revel in safe retreats and comfortable winter or cool summer quarters. They are noted builders and endeavor to fortify their rocky caverns by piling them full of rubbish, sticks, chips, stones, bones, thorny branches, dried manure, refuse food material, and anything they can find to carry and fill up the vacant spaces and hide or protect their nests and young.

—VERNON BAILEY, 1936

Although I have never seen one alive, I know that bushy-tailed woodrats are fairly common around Lake Abert. I also know that they dwell in especially large numbers higher on the rim, based on the abundance of their signposts among the boulders and rocky cliffs that they call home. These animals live in the Great Basin wherever there are numerous crevices and caves, as naturalist Vernon Bailey noted in the first monographic treatment of Oregon's mammals in 1936. Vernon and his wife, Florence Merriam Bailey, who was herself a noted nature writer and ornithologist, often traveled and published their research together. In 1902, she published a book on western birds titled *Handbook of Birds of the Western United States.* In 1992, the Oregon Geographic Names Board honored the couple's contributions to the natural history of the state by naming a mountain after them, Mount Bailey, in the southern Oregon Cascades.

Florence and Vernon Bailey around the time of their marriage in 1899. With permission of University of Wyoming, American Heritage Center.

In his monograph, Vernon Bailey described the limy deposits—or signposts—deposited by woodrats along rock edges. The two minerals, calcium oxalate and calcium carbonate, that form these deposits come from their urine and are derived from the plants that they eat (Bailey 1936; Emerson and Hoffman 1978; Finley 1992; Verts and Carraway 1998). Some plants synthesize and store oxalates that reduce herbivory because the chemicals can cause gastric hemorrhaging and even death in unadapted mammals (Miller et al. 2014). However, woodrats can eat oxalate-producing plants without being harmed because their stomachs contain symbiotic gut bacteria that detoxify the chemicals.

Woodrats' white, linear leavings are unmistakable, and that's why I call them signposts. They almost always comprise several distinct zones, including white calcareous deposits that the woodrats leave, which later become bordered by bright, reddish-orange sunburst lichens. The lichens apparently benefit from this association because they likely use the nitrogen in the urine.

I believe these deposits—which are visible even at night, when the highly territorial, male woodrats are most active—serve as signposts because the animals leave them behind to mark their territories. At some sites, I have found dozens of them left together on rocks, an indication that multiple generations of the same woodrat family may occupy these territories. For example, on Modoc Rim in Klamath County, I counted over fifty visible markers on one highly jointed, basaltic cliff about 50 feet high and 150 feet wide. These markers are conspicuous not only around Lake Abert and on Modoc Rim but also on the basaltic cliffs forming the south wall of Little Blitzen Gorge on Steens Mountain. In addition, they lie on rimrock throughout South Central and southeastern Oregon and in adjacent parts of California and Nevada. If these actually are

Two examples of bushy-tailed woodrat "signposts": *top*—two-foot-long deposit on the edge of a boulder; *bottom*—deposit on the edge of a basaltic rock. The ruler is five cm (two inches) long. Signposts are conspicuous because woodrats, with the help of a lichen, create several distinct zones on the edge of rocks: here, for example (particularly in the top photo), the woodrats have made a linear, white, calcareous deposit in the center area, and reddish-orange sunburst lichens have spread along each side of it.

signposts, they are an amazing adaptation, the use of a toxic waste product for a behavioral purpose. The deposits also represent an unusual form of symbiosis between a mammal and a lichen.

When I hike around the area, I always look for woodrat leavings on the sharp edges of cliff rocks, on the top of fault scarps that have lots of crevices, in talus, and even below boulders. They usually appear as a narrow band a few inches wide and oriented horizontally, but not always. You can differentiate woodrats' deposits from the more diffuse, uric-acid residues left by raptors, which you will most often see on prominent rock perches or on roosting or nesting ledges used by the large birds.

As their common name suggests, woodrats—also called pack-rats—can live in trees, and they collect almost anything that they can carry to build their nests up in branches or under boulders. I have even found one nest constructed in the end of a protective steel barrier along Highway 395, which skirts the lake. Woodrats build most of their nests using small sticks, and over time, they add to the nests, gradually making them quite large.

Signposts left by bushy-tailed woodrats (*Neotoma cinerea*): *left*—deposited on the south wall of Little Blitzen Gorge, Steens Mountain, displayed along with yellow cobblestone lichens; *right*—displayed on the rim in the Guano Basin, along with orange-colored sunburst lichens.

Bushy-tailed woodrat nests: *left*—abandoned nest, several feet across, containing sticks, rocks, and bones and built under a tufa mound, north of Lake Abert: *right*—large nest, about four feet across, built against a greasewood bush that mostly consists of sticks but which also contains dried cow patties.

Woodrats' behavior in collecting vegetation and storing it in dry places has led to the discovery that ancient woodrat middens, where the animals defecate and leave other refuse, hold important clues to how vegetation has changed over time in the Great Basin and in other arid areas (Betancourt et al. 1990; Grayson 1993). Some middens are so old—even older than fifty thousand years—that we can't accurately date them using radiocarbon. Scientists have studied the contents of more than one thousand of these middens, including some in southeastern Oregon, and their contents provide a unique window into the past environment, with some going back nearly one million years, a window that augments analyses of pollen taken from lake cores. Middens have also provided other information, including evidence of changing insect fauna, of retreating cliffs, and of changing climate (https://www.climate-policy-watcher.org/ecological-footprints/pack-rat-middens.html).

Pronghorn—Life in High Gear

We bay them down from the feeding-ground,
We fend them back from the pool,
And ever we raise the hunting howl
When the sun-warmed mesas cool.
And well they need both wind and speed
When the gray coyote pack,
By twos and threes from the hidden hills,
Breathes hot on the pronghorn's track.

—MARY AUSTIN, *The Rhyme of the Pronghorns*

Pronghorn, popularly called pronghorn antelopes, are the quintessential mammals of the wide-open spaces of the northern Great Basin's high-desert landscape. Among the swiftest of hoofed animals, they can quickly disappear into the distance, even over the rocky terrain so typical of the back side of Abert Rim. Moreover, they exhibit unusual behaviors, showing more curiosity about humans than other large herbivores do and acting in sometimes unpredictable ways.

I once saw a pronghorn herd, consisting of a buck and about two dozen does and younger animals, feeding several hundred feet away from me, when I was driving slowly at the north end of the rim along Hogback Road (Lake Co. Road # 3–10). I stopped to take pictures and to watch them, but I stayed in my truck so as not to alarm them. The buck seemed mostly uninterested in me and continued to feed, although he would occasionally lift his head to watch me. Some of the does and fawns, however, were obviously curious and

Pronghorn (*Antilocapra americana*), *clockwise from top left*: buck standing amid big sagebrush shrubs; doe and fawn keeping watch; part of a small herd on Abert Rim (the black face masks of several of the immature males are starting to develop, and their horns, which they shed annually, are short but will grow larger).

seemingly unsure about whether or not I was a threat. At times, they cautiously moved closer to me, but then they suddenly rushed back, only to move forward again, timidly. Finally, after fifteen minutes of this game, they all ran off, followed by the buck.

On another occasion, I had an opportunity to see pronghorn behavior of a very different kind. That particular day in May, I climbed to the top of the rim at Poison Creek and saw several Turkey Vultures (*Cathartes aura*) circling low, not far away, so I headed in their direction to see what was attracting them. As I approached more closely, a Golden Eagle (*Aquila chrysaetos*) and a Bald Eagle (*Haliaeetus leucocephalus*) flew out of the sagebrush. Seeing the eagles, I knew that they had likely been feeding on something, so I walked toward the spot where they had been. Then, I noticed a pronghorn doe about two hundred yards away, watching me (I can always recognize a doe because its horns lack prongs and are smaller than those of a buck, and because it doesn't have a dark patch extending from its nose to its horns). This doe was obviously alarmed because her white rump hairs flared dramatically, shining brightly in the morning sun.

Pronghorn up close: *left*—bereft doe stands watch on Abert Rim, just after her fawns have died (her flared rump patch signals her alarm); *right*—close-up of a buck, displaying pronged horns, bold facial markings, and a stiff-haired mane.

Once I got to where the eagles had been feeding, I immediately understood why the doe was distressed: she stood there, guarding a pair of dead fawns. I suspected that the fawns had died that morning, either during birth or shortly afterward, perhaps killed by one of the eagles. Golden Eagles are known to prey on fawns, although coyotes are considered to be their main predators (Byers 1997).

As I watched the doe, she approached within one hundred feet of me. Her rump hairs remained distinctly flared, indicating her fear. I understood that I had stumbled upon a common event, although one not often witnessed by people, due to its brevity and remote location. Based on what I saw, I felt conflicted, both sad for the doe's losses and yet also privileged to have witnessed something that few of us ever get to see, as disconnected from nature as we often are.

Research biologist Arthur Einarsen wrote eloquently about these beautiful animals in his book, *The Pronghorn Antelope and Its Management,* which he published in 1948 during his field work for the Oregon Cooperative Research Unit (located then at Oregon State College). He clearly enjoyed his research, even in winter when it was bitterly cold:

Arthur Einarsen (1897–1965), posing in 1949 at Oregon State College with his recently published book on pronghorns and their management. Courtesy of Harriet's Photographic Collection, Item HC1314, Oregon State University Special Collections and Archives Research Center, Corvallis.

> Few may feel that winter days of study on the pronghorn range could be pleasant, but I assure you it was a happy occupation. The smell of sage and the throaty, high-pitched *ka lank* of the raven lingers on. How dull will be the world when isolation is no more and we cannot watch these wildlings in their natural habitat. A lifetime cannot be better spent than to defer this day.

Einarsen's book was the first major publication devoted to the pronghorn and is still the most detailed account of the animal's biology in the Great Basin. He wrote it in the hopes of improving pronghorn management because information about best practices was scarce then. What I like about the book is his evocative writing style: he makes you feel as though you are sitting by a campfire with him, seeing him smile as he talks about his glorious days on Oregon's Hart Mountain, watching these marvelous animals:

> We arrive at Spanish Lake. Since the spring has been dry, the lake bed is already like a sunbaked race track and only a few green shoots emerge from the soil. The feeding pronghorns hurry over the rim to disappear in the distance and we settle ourselves in the blinds prepared long ago on the sloping hillside above the lake, where we can watch the animals on the plain and lake bed. The sun, which had given promise of warmth to the earth early in the day has disappeared behind scudding clouds with fingers of mist reaching down toward the prairie. A shrill cry pierces the air and near at hand a red-tailed hawk flies by with a ground squirrel firmly gripped in his talons, heading for a favorite escarpment a short distance away. Near its base is a dead yellow pine where he alights, a favorite roosting place of the hawks, and where many an unwary ground squirrel is eaten.

Einarsen lived from 1897 to 1965. His headstone, fittingly, has a pronghorn engraved on it, as well as an apt quote from nineteenth-century poet Lord Byron: "I love not man the less, but nature more."

Pronghorn are extreme-endurance athletes, the swiftest four-footed animals with hooves, and they can sustain running speeds of more than forty miles per hour (Heinrich 2001). Their need for speed clearly drove their evolution; they have a variety of interrelated adaptations, including a large windpipe and lungs, an oversized heart, large leg muscles, and a small stomach. Plus, their blood has a higher capacity to absorb oxygen than the blood of most animals. Modern pronghorn may seem to be overly built for speed, but their ancestors evolved in this way because of a dangerous suite of powerful Pleistocene predators, possibly including a wolf-sized dog (*Borophagus*), two species of cheetahs (*Acinonyx*), a lion (*Panthera leo atrox*), three species of sabertoothed cats (*Megantereon, Smilodon,* and *Homotherium*), and a hyena (*Chasmaporthetes ossifragus*), all of which were extinct by about ten thousand years ago (Byers 1997; Grayson 2016). All species, including humans, faced selective evolutionary pressures in the distant past, but those pressures—and the resulting adaptations—have been pronounced in the pronghorn.

Oregon's subspecies, *oregana,* is one of four that constitute the entire *Antilocapra americana* species. The family to which pronghorn belong, called the Antilocapridae, evolved in North America and comprised a successful and diverse group of ruminant mammals in the Miocene and the Pliocene (23 million–2.6 million years ago). During the late Pleistocene, three other antilocaprid genera coexisted with the pronghorn, but they are extinct. These pronghorn ancestors would seem bizarre to us because they had four horns instead of two. However, what is perhaps more surprising is that today's pronghorn are more closely related to giraffes and to African okapis than they are to antelope, making its popular name a real misnomer.

Pronghorn herds throughout the Great Basin, unfortunately, have much smaller populations than they did at the time of the Euro-American settlement of the West, when they numbered in the tens of millions (Zeveloff and Collett 1988). In Oregon, pronghorn populations have varied, from fewer than 2,000 in 1915 to nearly 20,000 in 1939, but by 1945, the population had declined again to fewer than 10,000 (Einarsen 1948). ODFW states that the current population size is about 25,000 (https://myodfw.com/big-game-hunting/species/pronghorn-antelope). Currently, the total US pronghorn population is estimated at somewhere between 500,000 and 1 million, with the largest number living in Wyoming (https://animaldiversity.org/accounts/Antilocapra_americana/). The other three pronghorn subspecies, Sonoran pronghorn (*A. a. sonoriensis*), Peninsula pronghorn (*A. a. peninsularis*), and Mexican pronghorn (*A. a. mexicana*), are considered imperiled in all or part of their range because of their small population sizes (https://www.iucnredlist.org/species/1677/115056938).

Bighorn Sheep—High Climbers

Another iconic animal of the West—the bighorn sheep—makes its home on the rimrock and comfortably dwells along the steep cliffs that form the ramparts of Abert Rim. One day, I was busily looking at native plants at the lake, while my knowledgeable naturalist friend, Steve Sheehy, was on the lookout for sheep on the upper slope of the rim. Within a few minutes, Steve called out to me, saying that he saw two bighorn sheep moving on the steep slope, over a quarter mile away in the distance. I thought he was kidding and looked in that direction, but I saw nothing moving. Then he described exactly where he noticed the sheep, so I looked again, using my binoculars, and I finally saw two light-brown specks slowly walking over a steep, grassy slope. I was amazed that Steve could see them unaided, because they were so far away. This experience

Bighorn sheep, *clockwise from top left*: ram, identified by the large size of its horns; ram skull found along the lower slope of Abert Rim; ewe and kid; and ewe #22, fitted with a radio collar (she was part of an ODFW transplant effort to increase the genetic diversity of the Abert Rim herd).

made me realize that these animals are likely more numerous than I had previously thought, based on the few that I had seen near the highway.

Bighorn sheep are named for the large, curled horns of the mature ram. These mammals fall into an uncertain number of subspecies, two of which occur in Oregon. The Rocky Mountain subspecies, *Ovis canadensis canadensis,* is confined in Oregon to the Wallowa Mountains and Snake River Canyon, in the northeastern part of the state. The Sierra Nevada subspecies, *Ovis canadensis sierrae,* occurs in rimrock habitats of South Central and eastern Oregon, and was previously called the California bighorn sheep.

Unlike mule deer and pronghorn, which successfully use human-altered habitats like alfalfa fields, bighorn sheep live almost exclusively in natural habitats (Buechner 1960), which typically consist of steep, broken cliffs with traversable terraces (or loose talus) for escape, and with bluffs for bedding (Van Dyke et al. 1983). Much of Abert Rim provides just such an environment.

Bighorn sheep historically dwelled throughout much of the West and were numerous. By 1900, however, they were wiped out from much of their range, which was the result of a whole host of factors: excessive hunting; a disease spread by domestic sheep called "scabies"; competition from domestic sheep; and loss of winter range (Bailey 1936; Buechner 1960). Helmut Buechner, a zoologist and ecologist tracking these herds in the mid-twentieth century, reviewed the historical records of bighorns in Oregon and in three other states where they had been wiped out, and he warned, "That they could disappear completely from the wildest country in Oregon should be an object lesson in conservation."

Vernon Bailey, chief field naturalist for the US Bureau of Biological Survey in the early twentieth century, advocated for the reintroduction of bighorn sheep in Oregon, stating, "Who would not enjoy living for part of each year where a magnificent old bighorn could be seen on a cliff above a band of ewes and young following a heavy-horned leader up a terraced wall, bounding upward from ledge to ledge to look back from the skyline above?"

According to ODFW surveys conducted between 2010 and 2014, the Abert Rim's bighorn population varied from 135 to 150 individuals and was similar in size to the Hart Mountain herd. The total size of the Oregon population of the California subspecies is believed to be 3,700 animals found in 32 herds located in central and southeastern Oregon; this subspecies is also the rarest game that Oregon hunters are allowed to pursue. Only about 100 hunting tags are available annually and are legal for use just once in a lifetime (https://myodfw.com/big-game-hunting/species/bighorn-sheep).

These small numbers did motivate repopulation efforts, as well as a concerted effort to track the animals. During visits to Lake Abert in 2013 and 2018, I photographed the same ewe, a female fitted with a radio collar and designated as #22. Jonathan Muir, ODFW wildlife biologist in Lakeview, reported that she was one of nine ewes transplanted from the Deschutes River Canyon in 2012 to increase the genetic diversity of the Abert Rim's bighorns. In November 2018, Steve Sheehy and I saw ewe #22's herd. They were feeding in the marsh near the lake and later crossed the highway and headed upslope. The group included eight ewes, three kids, and two large rams. They all looked robust and healthy and had dark-brown, winter coats.

In addition to reintroducing bighorn sheep to some areas, scientists have been working to improve their habitat, in part by adding man-made water-supply sources called *guzzlers,* which collect snow and rain. The Oregon Federation of North American Wild

Sheep has built at least three guzzlers on the lower slopes of Abert Rim so far.

All Oregon bighorn sheep populations now include animals introduced from other areas, mostly from British Columbia. Fortunately, the species now seems secure in Oregon, and the Abert Rim's population seems to be healthy. I believe Vernon Bailey would appreciate the fact that his dream of seeing bighorns return to their rimrock habitat in southeastern Oregon has been achieved.

Coyote—The Desert's Dog

In spite of this steady, persistent effort of thousands to banish him from the prairie, if you sit outside on a quiet evening at the edge of any of North America's deserts, you will soon hear, from the rimrock, or distant hill, the eerie, long talk of the coyote. He must be talking. It isn't howling or screaming; it isn't a mating call, for it goes on all year. I know of no other sound so descriptive of the desert. It is wild, remote, something elemental from the beginning of time, out of harmony with civilization.

—E. R. JACKMAN, 1964

No discussion of mammals from the Lake Abert ecosystem would be complete without mentioning the coyote, our most numerous, adaptable, and misunderstood carnivore.

There can be no doubt that the coyote is the top predator around the lake, due to its incredible adaptability. You will readily find coyote tracks when you walk along the lakeshore, a prime spot for these predators to search for birds and their eggs. One day, as Steve Van Denburgh and I were looking down to the lake from the highway at the Mile Post 74 Spring complex, we suddenly saw

Coyotes (*Canis latrans*), in the Great Basin: *left*—relaxed coyote casing the area, displaying a thick winter coat (the pelage consists of white and brown fur, as well as longer, black guard hairs); *right*—coyote standing on a boulder on the lower slope of Abert Rim, watching me.

a coyote walk out of the marsh below us and then start trotting west toward the far shore, located five miles across shallow water and sandy mud. As we watched, we thought it would soon stop and turn around because of poor footing, but instead, the coyote easily continued crossing the playa. After about fifteen minutes, we could barely see it in the distance. This confirmed for me just how at home coyotes are at the lake and made me realize that they must be viewed as an important part of the lake ecosystem, along with all the other diverse mammals that continue to thrive in this environment.

Thinking Like a Mountain—Predator Management in the Age of Climate Change

Aldo Leopold, in *A Sand Country Almanac,* coined the phrase "thinking like a mountain" in an essay to help readers understand that everything in nature is interconnected and that predator control, therefore, seldom has the intended results. This idea relates directly to the management of coyotes, as well as to that of such predators as wolves and cougars.

In 1996, the US Fish and Wildlife Service proposed killing coyotes at Hart Mountain National Wildlife Refuge in Oregon, in an effort to increase the survival of pronghorn fawns. Some studies had shown that coyotes had killed a high proportion of these fawns, and so the agency concluded that the predators were responsible for declining pronghorn populations. The agency's subsequent proposal drew considerable criticism, not only from the public but also from professional wildlife biologists, and so it was withdrawn.

If Leopold had been asked about the plan, he undoubtedly would have told the agency to "think like a mountain." He clearly understood that ecosystems are complex and that we usually don't understand the specific causes of wildlife population changes. This is especially true for the management of ecosystems now that climate change is rapidly occurring, especially because the resource agencies tasked with monitoring and managing public lands have insufficient budgets and have unrealistic pressures placed on them to fix perceived problems immediately and with limited data.

In the case of Hart Mountain, many experts insisted, rightly, that the exact causes of the pronghorn declines were not known. Scientists also emphasized that choosing to eliminate or reduce the coyote population might have unintended consequences, especially because the data were so inconclusive. Arthur Einarsen, who perhaps studied pronghorn antelope more extensively than anyone, wrote these prescient words back in 1948: "Where remnant herds exist and an increase in numbers is desired, the control of the coyote is obviously necessary. But under normal conditions predation perhaps is rarely a factor in determining survival."

Perhaps we will never know the root cause of the Hart Mountain pronghorn decline. We can, however, predict that climate change will have a ripple effect throughout all such ecosystems. This ripple effect is known as a *trophic cascade*: a small change in the plant community, for example, can reverberate all the way to the top of the food web, affecting even predators at the very top. Fortunately, the process is reversible, as we witnessed at Yellowstone National Park, when wolves were reintroduced and the aspens then began regenerating.

If there is a lesson here, it might be that although climate change will deeply impact our future, we can mitigate its effects by making our ecosystems as resilient as possible. In 1991, Hart Mountain Refuge's wildlife managers took their own controversial step to improve the local habitat: they removed cattle that had grazed there for a century. It turns out that this far-reaching plan was exactly the kind of management needed to reduce pressures on the ecosystem coming from climate change. A 2015 report documented the multiple benefits that resulted from this policy decision, including increases in streamside vegetation and improved watershed function (Batchelor et al. 2015) and provided evidence that the cattle's removal did, in fact, improve the environment for all the interconnected species living there. The kinds of innovative policies that wildlife managers and scientists have put in place at Hart Mountain and at Yellowstone are providing a roadmap for us all.

CHAPTER 6

Birds

Feathered Abundance in
a Harsh Landscape

The Great Basin is no avian Garden of Eden. Living conditions
are often stressful. Yet a fascinating array of birds inhabits the
lands and waters of the basin. Some birds meet the fiercest
challenges of climate and land head on—and feed and court
and even nest or overwinter in the harshest of environments. To
know these birds, we must first meet the land and feel its muscles and become aware of the stresses which buffet the birds
from all sides. Only then can we appreciate how very well birds
have made this their land.

—FRED A. RYSER JR., 1985, *Birds of the Great Basin*

Birds drew me to start visiting Lake Abert, and I have continued
to be fascinated by them ever since, particularly by the flocks of
small *peeps,* sandpipers rapidly twisting and turning in flight with
astounding synchrony—their white bellies and darker top feathers
causing them to appear alternately light and dark—and flying with
incredible speed for their small size. I have had many rewarding
days sitting at the edge of the lake, just watching them, and I am
always thrilled when they fly so close that I can hear the air pass-
ing over their small wings. Often, they land nearby and begin foraging with such focus that they don't seem to notice me, or they
rest nearby with their heads tucked under a wing, bird napping. If
they never returned to the lake, the vista would still be beautiful,
but the pulse of the place would vanish, becoming dead for me.

The Great Basin can be a tough place to live, as Ryser eloquently
pointed out in his 1985 book about the region's avian abundance.
Lake Abert's birds, which are the most conspicuous animals on
the lake and in the surrounding shrublands, grasslands, and rimrock habitats, nonetheless manage to thrive there. When you visit,
you will see all sizes, ranging from the tiny Calliope Hummingbird
(*Selasphorus calliope*), the smallest bird in the United States, to the
majestic Golden Eagle, which has a wingspan of over six feet. Birds

also occupy a broad range of habitats; for example, Rock Wrens (*Salpinctes obsoletus*) and Canyon Wrens (*Catherpes mexicanus*) search for insects and spiders hidden under the boulders that have tumbled down from the rim. American Avocets, which use their gracefully curved bills to forage for aquatic invertebrates on the muddy shores of the lake, even nest on the mud.

It might come as a surprise that in such a low-precipitation area, some of the most diverse fowl are waterbirds that spend all or most of their time feeding and/or breeding near water. These waterbirds include avocets, ibis, plovers, sandpipers, phalaropes, gulls, waterfowl, and many others. Most of them migrate to the lake to feed on the seasonally high abundance of aquatic invertebrates. No other place in the Great Basin, except for the Great Salt Lake and the Lahontan Valley wetlands, has attracted such large concentrations of shorebirds, and in fact, it is not unusual to see over one hundred thousand birds at the lake in midsummer. Sadly, however, climate change has started to impact the lake and the birds that use it.

Habitats around the lake also attract migratory land birds, such as the Sage Thrasher (*Oreoscoptes montanus*) and the Loggerhead Shrike (*Lanius ludovicianus*). These birds nest in the arid greasewood- and sagebrush-dominated uplands around the base of Abert Rim. You can also occasionally see the Sagebrush Sparrow (*Artemisiospiza nevadensis*), named for its primary habitat, although it also spends time in the saltbush of the central Great Basin. Another sparrow that you might sometimes see flying or hear calling from the lower slope of Abert Rim is the Black-throated Sparrow (*Amphispiza bilineata*), but it rarely visits the northern Great Basin, spending the bulk of its time in the desert Southwest.

Other birds, like the colorful Yellow Warbler (*Setophaga petechia*) and Wilson's Warbler (*Cardellina pusilla*), glean insects from the thick growth of willows and dogwoods or pluck insects from the air, in hawklike fashion, along Juniper Creek as it tumbles down from Abert Rim. Golden Eagles, Prairie Falcons (*Falco mexicanus*), and Common Ravens (*Corvus corax*) nest on rocky ledges high on the rim and range over vast areas, searching for food to feed their hungry young. Ravens are the most common large upland birds at the lake (I almost always see at least one pair circling or flying along the edge of the rim, when I visit). The National Audubon Society identified Lake Abert as an Important Bird Area because of the abundance and diversity of its birds, and the lake is one of the key stops on the Basin and Range Birding Trail.

When I consider the large number and diversity of birds using Lake Abert seasonally, I find it surprising that researchers did not publish a report on the lake's regular avian visitors until the 1980s.

Adult Wilson's Warbler, resplendent in its distinctive black cap, stopping along Juniper Creek during spring migration.

Adult Black-throated Sparrow, unmistakable with its bold, white stripes and black throat. Observers have reported seeing a few of these sparrows from May through August on the lower slopes of Abert Rim. This adult was photographed in southern Arizona during the winter, when it is a common visitor there.

This seems particularly anomalous because naturalists tend to notice birds first and did, in fact, research birds in other parts of Oregon much earlier. For example, in the early twentieth century, zoologists working for the US Bureau of Biological Survey collected data throughout the state, which eventually led to the 1940 publication of *Birds of Oregon* by Ira Gabrielson and Stanley Jewett. The apparent lack of interest in Abert's birds was in marked contrast to enthusiasm for fowl living just west of the lake, in the Klamath Basin, as well as in the Malheur Basin to the northeast and in the Warner Basin to the east. Multiple researchers wrote about birds in these locations throughout the late nineteenth and twentieth centuries.

William Finley, a resident of Portland, finally brought broader attention to Oregon waterbirds and their habitats in the early 1900s, through his popular documentaries and in articles and photographs published in books and in magazines, including *National Geographic* (Mathewson 1986). Finley ardently defended birds and their ecosystems and fought against their decimation by market hunters and by those harvesting egret plumes, and he decried the loss of wetland birds' habitat resulting from water diversions. His tireless efforts persuaded President Theodore Roosevelt to designate Lower Klamath Lake in California and Malheur Lake in Oregon as national wildlife refuges in 1908. Finley also deserves credit as an early leader in environmental activism in the western United States. We owe a debt of gratitude to him and to his assistant, Herman Bohlman, for their dedication and perseverance on behalf of Oregon's birds.

Researchers including Kathryn Boula, Frank Conte, and David Herbst went on to provide valuable insights into the links between aquatic invertebrates and waterbirds at Lake Abert, helping to show the productivity of its environment. Ornithologist Joseph Jehl, who was especially interested in Eared Grebes (*Podiceps nigricollis*) and Wilson's Phalaropes (*Phalaropus tricolor*), published several scientific papers about waterbirds based on his own observations at the lake in the 1980s and 1990s. Scientists including Kurt Kristensen started focusing on the number of breeding birds at the lake, a data-gathering effort that required considerable time and perseverance. BLM wildlife biologist Walter Devaurs also made an important contribution to our knowledge of Abert's birds through an extensive counting effort he made in the 1990s (Bureau of Land Management 1995).

More recently, USGS staff at Oregon State University's Forest and Range Ecosystem Science Center investigated how waterbirds use multiple Great Basin wetlands during breeding and post-breeding periods, depending on water levels and other environmental conditions. Based on all these studies, we now know with certainty that during the spring and summer nesting and post-nesting seasons, waterbirds do, in fact, use multiple Great Basin wetlands, including those at Abert and at other salt lakes. We also know that we need to take a regional approach to managing these birds and their environment, due to both annual and seasonal variability of the water and food conditions, which affect waterbird movements. This became especially evident during recent desiccations of the northwestern Great Basin lakes, which forced many of these birds to fly off in search of suitable habitats elsewhere (Larson et al. 2016; Moore 2016).

Now that we understand that waterbirds use multiple wetlands regionally, we have moved on to conduct broadscale shorebird surveys in the western Great Basin, including at Lake Abert. Researchers have concluded, crucially, that the lakes support large enough populations to make them important not just regionally, but also hemispherically.

Starting near the end of the Pleistocene Epoch, about fifteen thousand years ago, a drier and warmer climate caused immense changes in the amount and distribution of water in the Great Basin.

An aggregation of American Avocets, Black-necked Stilts, and Wilson's Phalaropes taking advantage of Lake Abert's summertime bounty; large numbers of such shorebirds migrate to the lake annually to feed on the abundant brine shrimp and alkali flies there.

Such changes must have affected the waterbirds' habitats substantially, but not always in negative ways. As the deep, cold, and unproductive lakes receded and the climate warmed, more marshes and shallow water habitats developed in some areas, and these were increasingly productive, providing more food and offering diverse nesting habitats for waterbirds. In the past ten thousand years, as lake levels have continued to decline, salinity has increased, and

Snowy Plovers (*Charadrius nivosus*) in breeding plumage, standing on alkali-covered sediment.

Upland birds from around Lake Abert, *top to bottom and left to right*: Red-winged Blackbird (*Agelaius phoeniceus*); Song Sparrow (*Melospiza melodia*); Western Meadowlark (*Sturnella neglecta*); Black-billed Magpie (*Pica hudsonia*); Yellow-rumped Warbler (*Setophaga coronata*); Townsend's Warbler (*Setophaga townsendi*); Chukar (*Alectoris chukar*); Ruby-crowned Kinglet (*Regulus calendula*); Western Kingbird (*Tyrannus verticalis*); California Quail (*Callipepla californica*); White-crowned Sparrow (*Zonotrichia leucophrys*); and California Scrub-Jay (*Aphelocoma californica*).

highly productive, fish-free salt lakes, such as Lake Abert, have become good environments for brine shrimp and alkali flies and have provided much-needed food for migrating shorebirds, gulls, Eared Grebes, and waterfowl. Such increased productivity also attracted humans for a time: thousands of years ago, Native Americans settled near these water bodies to avail themselves of the plentiful wildlife resources.

For simplicity's sake, we can broadly divide Lake Abert's feathered community between upland birds and waterbirds. Of course, this is an artificial distinction because many of them—including the upland Red-winged and Brewer's Blackbirds (*Agelaius phoeniceus* and *Euphagus cyanocephalus*)—use both habitats, and some shorebirds, including the Killdeer (*Charadrius vociferus*) and the Western Willet (*Tringa semipalmata inornate*), nest or feed in upland areas. The most adaptable of all waterbirds—and perhaps of all Lake Abert birds—are the gulls, who find food in parking lots, landfills, and agricultural fields, but who return to the interior wetlands and lakes to nest.

Upland Birds

In early spring, the lower slopes of Abert Rim fill with the sweet songs of Western Meadowlarks (*Sturnella neglecta*). Somewhat later in the season, a variety of songbird migrants, including the distinctive Yellow-rumped Warbler (*Setophaga coronata*), forage for insects in the willows and dogwoods along Poison Creek. Song Sparrows (*Melospiza melodia*) and Red-winged Blackbirds take up territories in the marshes along the eastern shore. Pairs of Common Ravens dive and spin in midair, high above and near the crest of the rim, and a Golden Eagle soars past, searching for jackrabbits or fat yellow-bellied marmots. Although not as numerous as waterbirds, these upland birds are richly diverse, with nearly one hundred species dwelling around the lake. The diversity is especially high in the spring and fall when migrants pass through.

Golden Eagle

The Great Basin—with its richness of eagles in unobstructed skies—is an unexcelled stage upon which to view eagle courtship. The best viewing occurs in late winter and early spring. Much circular soaring occurs over the nest site. While soaring, one bird or the other may engage in a series of undulations— diving downward and swooping upward. Sometimes when the male dives toward the female, she will roll over on her back at his approach.

—FRED A. RYSER JR., 1985

It was probably not chance that led Fred Ryser to choose a photo of a Golden Eagle for the cover of his 1985 book, *Birds of the Great Basin: A Natural History*: the regal bird is utterly symbolic of the wide-open spaces of this vast region. Seeing one of these birds around Abert Rim is a matter of luck, however, because while they are not uncommon, they do forage over a large area and don't stay in any spot for very long, except near their nesting sites.

Golden Eagles have a prolonged nesting season that starts in February and can extend to early August. For some years, a pair nested on a cliff at the north end of Abert Rim. The site was exposed to southwest winds that buffeted the precariously placed nest, which sat high on a ledge above the highway. Early one spring, I checked the nest and was fortunate to see a pair of young eaglets, still covered with snowy-white down and barely able to stand. When I revisited the site in June, I saw that they had grown larger and that they were surrounded by many discarded jackrabbit legs left from previous meals.

The Golden Eagle, an important apex predator in the Great Basin. No other bird is more emblematic of that vast, arid landscape.

I had just one other close encounter with one of these magnificent predators, in May 2010, when I was hiking on Abert Rim. This was when I saw both a Golden Eagle and a Bald Eagle at the spot where a pair of pronghorn fawns lay, no longer breathing, with their distraught mother standing by.

Oregon's eagles and their conservation have been monitored by Frank Isaacs, cofounder of the Oregon Eagle Foundation and a researcher at Oregon State University for several decades. The foundation's most recent report documented over one thousand Golden Eagle nest sites in the state and estimated a minimum population of about six hundred nesting pairs (Isaacs 2018). These birds have produced nearly five hundred eaglets each year. They nest almost exclusively in areas east of the Cascade Mountains (Isaacs 2018), and they

Golden Eagle young: *left*—pair of downy eaglets just a few weeks old and just starting to move around in their nest in April; *right*—same pair of eaglets in June, about eight to ten weeks old, crouching in a nest and surrounded by the remains of previous meals (likely from black-tailed jackrabbits, which are well-documented prey of Golden Eagles). These birds probably fledged a month or so later.

Juvenile Golden Eagle showing off its spectacular adaptation, the ability to soar at high altitudes. The young eagle has white wing patches and a white tail band, which will disappear when it matures.

require the following ecological conditions: cliffs with isolated ledges for nesting; vast areas of open habitat for foraging; and medium-sized mammals, especially jackrabbits, to eat. Researchers, based on sightings of banded Golden Eagles in Oregon, know that the birds disperse widely as juveniles, even flying as far away as Saskatchewan, Canada, and Baja California, Mexico, each nearly one thousand miles away. However, once they reach breeding age, they tend to return to the places of their birth, where they remain.

Scientists believe that western Golden Eagles experienced a long-term decline, starting with the wave of Euro-American settlement in the mid-nineteenth century, because of habitat loss and humans' misguided efforts to control the birds as predators. The Oregon Eagle Foundation and its volunteers have been monitoring the status of these birds in the state, and their data shows that Golden Eagle populations in Oregon are currently stable, or are possibly even slowly increasing, which is encouraging. More-concerning statistics from the Breeding Bird Survey (BBS, https://www.pwrc.usgs.gov/BBS/PublicDataInterface/index.cfm), however, show a slight downward trend, and other data sources from elsewhere in the country suggest an ongoing, gradual population decline (Millsap et al. 2013; US Fish and Wildlife Service 2016). (The BBS started tracking North American bird populations in 1966; each spring, its volunteers count birds along selected routes and report sightings to the Patuxent Wildlife Research Center operated by the USGS.)

Humans cause over 50 percent of the birds' deaths. While Golden Eagles suffered less harm from DDT use in the mid-twentieth century than some raptors did—due to not feeding on

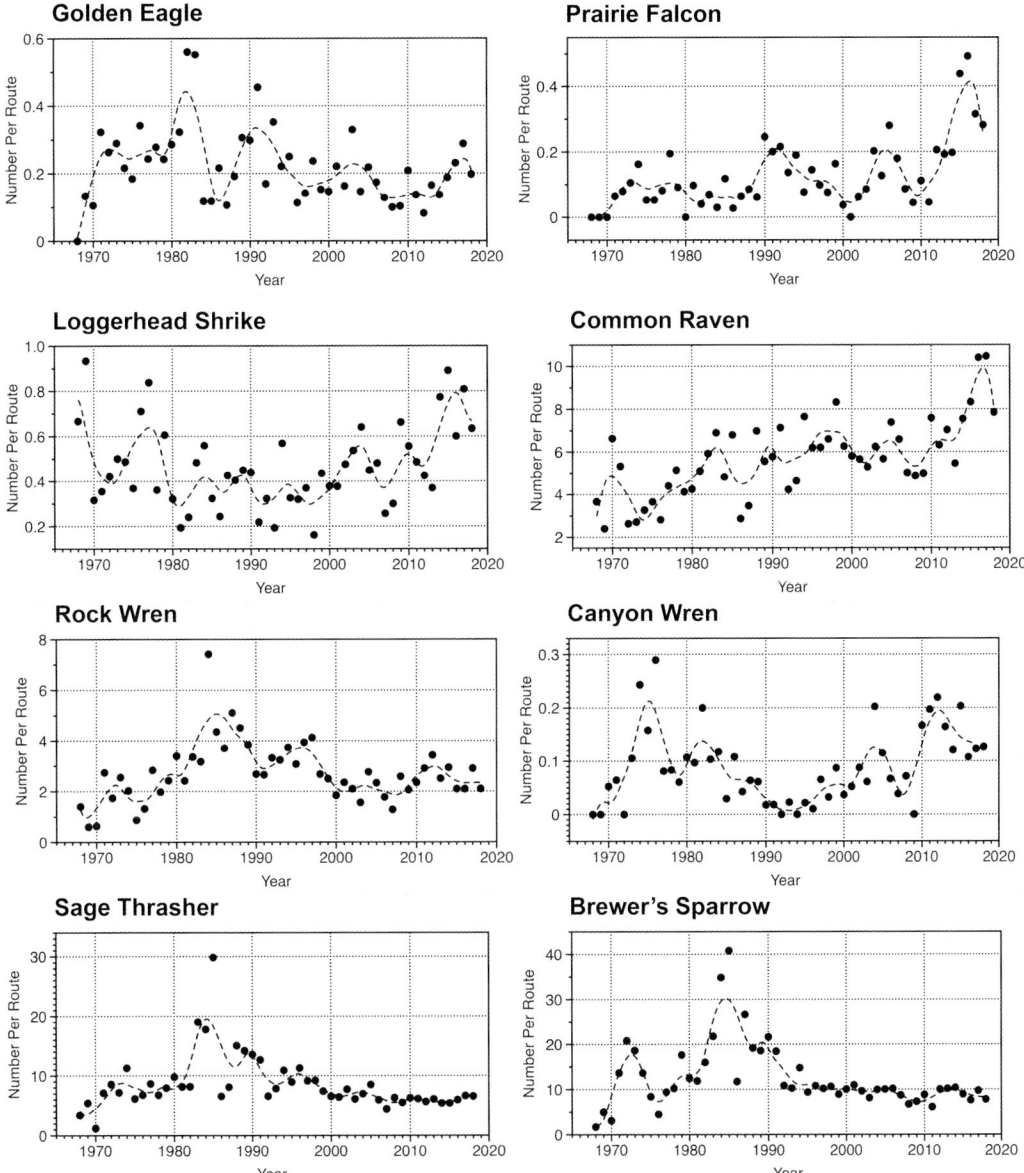

Census data of selected upland birds from around Lake Abert, based on the Oregon Breeding Bird Surveys from 1968 through 2018. The data are reported as numbers of birds counted per survey route. The numbers of routes surveyed varied from year to year. The dashed lines represent smoothed trends.

mammals that consumed concentrated amounts of the dangerous pesticide—they now suffer from another chemical: they scavenge on animals shot with lead bullets, and so they can absorb unsafe levels of lead in their blood, which weakens them and makes them vulnerable to disease and starvation. Wildlife experts are even more concerned about the impacts of energy development—especially wind development—on eagles' habitats and along their migration routes (USFWS 2016; Collopy et al. 2017). Additionally, the birds often perch on powerline poles when they hunt, putting them at risk of electrocution.

Another threat to Golden Eagles is the loss of the sagebrush habitat that is home to their prey. Sagebrush plants that sustain the eagles' prey face ongoing threats from overgrazing, the encroachment of cheatgrass, and the resulting increase in wildfires. These problems are exacerbated by the slow recovery that sagebrush makes after a wildfire dies out or after overgrazing is halted.

Eagles and other migratory birds started to receive some protection early in the twentieth century under the Migratory Bird Treaty Act, passed in 1918, but wildlife experts continued to push for more. In 1940, the federal government passed the Bald Eagle Protection Act, and then finally, in 1962, the Golden Eagle gained similar safeguards under the amended Bald and Golden Eagle Protection Act. Unfortunately, these efforts have done little to reduce the harm created by habitat loss.

Participants at a 2015 symposium on the conservation of Golden Eagles determined that the survival of these regal birds hinges on successful reproduction (Collopy et al. 2017). The eagles produce only one chick per nest each year, on average, and few of these chicks survive to become adults, so the birds face low prospects for a robust recovery. Therefore, we must continue to monitor and protect these eagles to ensure that they remain secure (Collopy et al. 2017).

Prairie Falcon

The most common raptors at the lake are not the iconic eagles but are, instead, Prairie Falcons. You can easily confuse these birds with Peregrine Falcons (*Falco peregrinus*), especially when you see them at a distance, because they are about the same size, with a wing span of forty inches, although the wing tips of Prairie Falcons are more rounded. Both birds have a dark band extending downward below the eye, called a "mustache," but the Prairie Falcon's is thinner. This common raptor's overall color is also lighter than that of the Peregrine Falcon, often cream with light brown markings. Prairie Falcons are generalist predators that find a variety of suitable prey on the slopes of Abert Rim, and their excrement can often be seen on rocks where they perch, watching for prey.

Although quite capable of capturing most shorebirds, Prairie Falcons don't seem to attack them at Lake Abert, possibly because other animals are easier to surprise and grab. Instead, the falcons feed on a variety of small mammals, especially ground squirrels, and on other types of birds. Most of the birds that they hunt are

Prairie Falcon fully spreading its wings while flying (the thin "mustache" below its eyes is visible).

on the ground or have just been flushed. This approach to hunting differs from that of Peregrine Falcons, which mostly attack birds in flight. In his book, Ryser stated that flocks of Horned Larks (*Eremophila alpestris*), which are believed to be a major food source for Prairie Falcons in the Great Basin during the winter, have been seen flying closely behind the falcons, possibly to avoid a surprise attack.

The falcons commonly nest on ledges and lay their eggs in a scrape (a shallow depression). The birds are quite vocal when anyone approaches their nest. Reliable population sizes and trends are unavailable nationwide, according to The Cornell Lab of Ornithology's *Birds of North America*. However, Oregon's breeding-bird surveys show increased numbers of Prairie Falcons in the state since 2010, although overall counts are low. The *State of North America's Birds 2016* gave the species a conservation-concern score of ten, which is moderate.

Loggerhead Shrike

Another Lake Abert regular, the Loggerhead Shrike, is unmistakable with its black mask, white throat, black wings (displaying a white bar that is visible during flight), overall gray color, and a short, black, hook-tipped bill. The shrike predominantly dwells in shrub-dominated habitats that offer excellent perches for hunting prey. This songbird has an unusual, varied diet of small mammals, lizards, and insects. Although not numerous, shrikes regularly nest in greasewood and sagebrush near the lake.

Shrikes hunt in several ways. Mostly they scan for prey from their perches and then fly out to grab it, but if they don't see anything tasty, they move to another perch. They also *hover hunt,* a technique in which they hover above their prey, trying to coax it out from cover, or they hop on the ground looking for an animal

Loggerhead Shrikes: *left*—close-up of a shrike sitting in a juniper (note the bold, contrasting, light-and-dark patterns on its face and wings, as well as its hooked bill); *right*—shrike perching after it has caught a small lizard.

or insect. Shrikes have also gained a reputation for their skill in impaling prey on sharp spines or barbed wire. Loggerhead Shrike populations in most of North America are declining, but Oregon's breeding-bird surveys from 1968 to 2018 show increases beginning in about the year 2000. The *State of North America's Birds 2016* gave the Loggerhead Shrike a conservation-concern score of eleven, which is moderate.

Common Raven

> We have wandered together through the quiet beauty of remote mountain and desert country. To me, ravens are a hardy race—an intelligent, alert, lively presence—which can ameliorate the harshness of climate and countryside.
>
> —FRED A. RYSER JR., 1985

Ravens are perhaps the most frequently seen large birds in Oregon east of the Cascades, especially along highways, where road kills provide them a never-ending banquet. They look, superficially, like crows, but they are bigger and have a much larger bill that is partially covered by feathers. At Lake Abert, pairs of ravens are often seen cavorting near the rim, where they likely nest, but they commonly fly throughout the region by themselves, in pairs, or in small groups.

Historically, the raven has been viewed as a wilderness bird, only found in remote areas far from civilization, and, in fact, it has been characterized as wary and solitary. Nonetheless, *Birds of North America* describes it as "geographically and ecologically one of the most widespread naturally occurring birds in the world": it dwells throughout much of North America, especially in the West, as well as in Europe, Asia, and North Africa. Commenting on the bird's ubiquity in the Great Basin, Ryser wrote: "Regardless of the

Common Ravens: *left*—adult looking regal and displaying its heavy bill, which is partially covered by feathers, and its shaggy throat plumes; *right*—pair of ravens flying (they likely mate for life, so this could be a mated pair).

land, whether it be arid or lush, desert or ranch, valley or mountain, hot or cold, the raven will be there."

Scientists have implicated this intelligent, adept hunter in the decline of several bird species and in that of the desert tortoise (*Gopherus agassizii*), and farmers accuse it of being an agricultural pest. We are still discovering a great deal about this intriguing bird, despite there being several thousand publications already devoted to it, including such books as *Mind of the Raven* and *Ravens in Winter* by biologist Bernd Heinrich. I never tire of watching ravens and listening to their varied conversations. I only wish that I understood their language so that I could hear their stories.

In Oregon and elsewhere, the Common Raven seems to be doing well. The breeding-bird survey data from Oregon shows an upward trend in counts that began in the early 1970s. The *State of North America's Birds 2016* gave the Common Raven a conservation-concern score of six, which is low.

Horned Lark

Lake Abert's landscape is also home to an attractive and distinctive bird, the Horned Lark. This upland bird is unmistakable, particularly the male, with his black mask, black breast band, and tiny "horns." Before I knew what these beauties were, I often saw small birds fly up from back country roads and disappear into the sagebrush and wondered about them. Finally, once I became familiar with the Horned Lark, I realized that it was a common sight, particularly during spring, summer, and fall in high-desert habitats in the northern Great Basin.

Sometimes while I hike in areas of low sagebrush during the spring, I come upon a nesting pair, and in the autumn, I see flocks of Horned Larks feeding in open areas. Kurt Kristensen and his colleagues, who studied lakeside birds in 1988 and 1989, found that

Horned Larks: *left*—male showing off its bold and distinctive facial markings and its small "horns," which make it easily identifiable; *right*—female sitting on a low sagebrush (females lack the "horns," and their facial markings are more muted than those of the males).

the larks were common nesters. At the time, they noted, "By mid-June hundreds of fledged young and adults could be seen feeding on the dense concentrations of brine flies along the northwest shore of the lake."

Oregon's population of Horned Larks seems secure, based on the high counts made during the breeding-bird surveys. However, one subspecies, the Streaked Horned Lark (*Eremophila alpestris strigata*), which dwells west of the Cascades, is listed as federally threatened. The *State of North America's Birds 2016* gave the Horned Lark a conservation-concern score of nine, which is moderate.

Rock Wren

> Even in the most barren and desolate reaches of the Great Basin, the cheerful song of this hardy wren will contradict any notion that the desert is devoid of life.
>
> —DAVID B. MARSHALL et al., 2003

Offering merry proof of life in the Great Basin's rimrock habitats, Rock Wrens are small sprites that are as much a part of Abert Rim's rock-dominated slopes as are lichens, lizards, and petroglyphs. I mostly see them at close range, when they suddenly appear on top of a boulder, bobbing up and down, and then just as quickly fly away. Robert Ridgeway, a renowned nineteenth- and early-twentieth-century ornithologist who described this little bird, noted, "Its favorite resorts are piles of rocks, where it may be observed hopping in and out among the recesses or interstices between the boulders, or perched upon the summit of a stone, usually uttering its simple, guttural notes."

Rock Wrens: *left*—adult male singing from a rocky perch; *right*—wren clinging to a lichen-covered boulder (it uses its long and sharply pointed bill to extract insects and spiders from tight crevices).

These wrens have long, sharp-pointed, slightly curved beaks adapted to pulling insects and spiders from crevices. In his seminal book, *Birds of the Great Basin,* Ryser wrote: "The Rock Wren is built for a shallow world, where the ceilings are not far removed from the floors. With flattened head and body, it creeps mouse-like on short legs and small feet, into narrow horizontal fissures to forage." The adept birds have an unusual habit of constructing pebble paths, eight to ten inches long, leading to their rock-crevice nests. Ryser also noted that sometimes the wrens pile stones so high in front of the nests that they make entry for themselves difficult, and they sometimes also build stone foundations under the nests.

In 1995, New Mexico researcher Michele Merola observed that a typical nest foundation comprised over seven hundred stones, some of which were up to two inches across, and each rock weighed one third of the wren's body weight. Merola saw only female wrens carrying the stones to the nest, at an average pace of one per minute. A researcher in Colorado found that the total weight of stones at a nest was over sixty times that of a wren's body weight. Ornithologists don't know for sure but speculate that the rock pavements and foundations may protect the nests from predators and may provide an auditory warning of a threat, as an animal noisily makes its way across the rocky path.

Whatever the purpose, the rock work involves substantial effort and therefore must be functional. Clearly the name Rock Wren is appropriate for this tiny, feathered stonemason. According to the breeding-bird survey data from Oregon, observers reported increased numbers of Rock Wrens from the late 1960s to the mid-1980s, then a decline until about 2000, and stable numbers since then. The *State of North America's Birds 2016* gives the Rock Wren a conservation-concern score of eleven, which is moderate.

Canyon Wren

> The sides of the canyon were rough and rocky, in some places very steep or even precipitous.… We saw or heard of interesting birds, but the gem of them all was the Canyon Wren. Its wild, joyous strain of sweet, silvery notes greeted us as we passed some steep cliffs: they seemed to reverberate from one cliff to another, to fill the whole canyon with delightful melody and to add a fitting charm to the wild surroundings.
>
> —ARTHUR C. BENT, 1948

The tiny Canyon Wren also dwells along Abert Rim's precipice, and you will recognize it from its low profile, long, down-curved beak, and brown back. Unlike the Rock Wren that lives amid the boulders, the Canyon Wren prefers cliffs and overhangs. Its appearance reminds me of a Brown Creeper, but one that has left the trees and adopted rocks as its home.

Weighing less than half an ounce, and only six inches long, this small bird is quite capable of climbing vertical rock cliffs. Canyon Wrens are widespread throughout the West, from British Columbia all the way into Mexico. Despite inhabiting this vast territory, they remain mostly invisible to human eyes because their primary habitats are cliffs and steep-sided canyons.

In one Colorado study, these birds appeared to be more common where there are overhangs, rather than amid vertical cliffs. The wren has adapted well to living in its rocky environment and can even climb upside down like a creeper when it searches for spiders and insects. Ryser wrote that the Canyon Wren "ranks among our foremost musicians, and its wild haunting song is a thing of beauty." In 1927, naturalist Ralph Hoffmann, writing in *Birds of the Pacific States,* described the song as "a cascade of sweet liquid notes, like the spray of a waterfall in sunshine." I regret that, due to my poor hearing, I have never heard it sing.

Adult Canyon Wren perching on a boulder and displaying its long, decurved beak, large feet and talons, pale throat, dark, spotted back, and the banding on its tail and belly.

The conservation status of the Canyon Wren seems stable, but because bird counters struggle to find it, its actual numbers remain uncertain. In fact, Cornell's *Birds of the World* says that it is one of the least studied birds in the country. Oregon's breeding-bird surveys include too few sightings from which to draw any conclusions. Nevertheless, the *State of North America's Birds 2016* gave the Canyon Wren a conservation-concern score of nine, which is only moderate, likely due to the fact that the bird's craggy habitat is secure from most types of development.

Sage Thrasher

> To appreciate fully the song of the sage thrasher, the poet of the lonesome sagebrush plain, one should visit him in his haunts in the gray of early dawn, before the chilly mists of night have lifted from the sea of gray-green billows…. As the veil lifts with the rising sun, the mists roll away…scarcely visible in the distance, a gray-brown bird mounts to the top of a tall sage and pours out a flood of glorious music, a morning hymn of joy and thanksgiving for the coming warmth of the day.
>
> —ARTHUR C. BENT, 1948

The Sage Thrasher is one of the few birds that is common in the sagebrush margins of Lake Abert in late spring and summer. It is blackbird-sized, is mostly brown on its back, has a speckled or streaked breast, and displays a long tail and legs. The Sage Thrasher is well known for its combined nonstop-singing and aerial displays during the breeding season. Ryser referred to these musical shows as "song flights." Ornithologist Ridgeway described one such display that he witnessed while traveling in Nevada in 1868: "The males, as

Sage Thrashers: *left*—thrasher sitting on a low sagebrush after it has caught an insect; *right*—close-up of an adult (it uses its long legs to forage on the ground).

they flew before us, were observed to keep up a particular tremor or fluttering of the wings, warbling as they flew, and upon alighting… raised the wings over the back, with the elbows together, quivering with joy as they sang."

The Sage Thrasher rarely finds its way outside of the sagebrush shrublands of the intermountain West and mostly makes its presence known in the Great Basin during the spring and summer. The breeding-bird survey data from Oregon suggests that this incredible singer is common in the state. Thrashers' numbers in Oregon increased between the late 1960s and the early 1980s and then gradually decreased until about 2007, a similar pattern to that of the Rock Wren. Since 2007, counts per route have remained remarkably stable, suggesting that the population may have stabilized. The *State of North America's Birds 2016* gave the Sage Thrasher a conservation-concern score of eleven, which is moderate.

Brewer's Sparrow

Another songbird that frequents sagebrush habitat around the lake, as well as in other areas of Oregon east of the Cascades and throughout the Great Basin, is the Brewer's Sparrow (*Spizella breweri*). In his description of this small inhabitant, Ryser wrote, "This tiny, nondescript sparrow is probably the most characteristic bird of our vast sagebrush country: it accompanies big sagebrush throughout the high valleys, foothills, and mountains of the Great Basin." It is a species that is perhaps best recognized because of its lack of distinctive field marks.

Brewer's Sparrows: *left*—sparrow sitting in a greasewood bush; *right*—pair of the birds perching on a dead limb.

During the spring nesting season, it can be the most common bird in this lakeside habitat, where it often sings from atop a sagebrush shrub or flies up from the ground onto a bush to keep an eye out for trespassers. Oregon's breeding-bird survey data suggests that the Brewer's Sparrow is relatively common, present in numbers similar to those of the Sage Thrasher. The data showed increasing numbers of the sparrows between 1968 and the mid-1980s, followed by a decline until about the early 1990s. Since the mid-1990s, counts have remained fairly constant, suggesting that this population has also stabilized. The *State of North America's Birds 2016* gave the Brewer's Sparrow a conservation-concern score of eleven, which is moderate.

Waterbirds and Their Natural History

Waterbirds and deserts might seem to be an unlikely combination. Each year, however, one high-desert lake in the northern Great Basin turns this assumption on its head, attracting hundreds of thousands of migrating waterbirds from as many as eighty species. What makes Lake Abert so attractive to birds is a rare combination of ecology, geography, geology, and water chemistry. At the peak of migration in midsummer, the bird numbers are so vast that volunteers find it challenging to count them. Lake Abert is ranked high as a seasonal habitat for shorebirds that include: American Avocets (*Recurvirostra americana*), Western Sandpipers (*Calidris mauri*), Red-necked Phalaropes (*Phalaropus lobatus*), and Wilson's Phalaropes. Additionally, imperiled Snowy Plovers (*Charadrius nivosus*) nest along the lakeshore. Eared Grebes (*Podiceps nigricollis*) and California and Ring-billed Gulls (*Larus californicus* and *L. delawarensis*) often flock there in great numbers too.

Multiple reasons explain why so many waterbirds fly to the lake, but perhaps the most important is an abundance of readily accessible food when they most need it. Several factors account for this rich plenty. Lake Abert has a relatively simple food web, and thus,

Flock of American Avocets resting at the edge of Lake Abert and showing off their dramatic breeding plumage.

much of the energy produced by its plants passes on to the birds, rather than to aquatic predators like insects and fish, which don't live in the highly saline waters. The dominate invertebrates—brine shrimp and alkali flies at various life stages—serve as food for the birds. Another factor helping many birds is the timing of invertebrate production, which peaks in midsummer, just as the birds arrive at the lake from nearby or distant breeding areas.

The birds also benefit because Lake Abert is extremely shallow, with much of it less than one foot deep, and this shallow water gives birds easy access to the shrimp and flies. Thus, even short-legged shorebirds like sandpipers can easily access the prey. Furthermore, birds on and around the lake have some protection from predators because they have a nearly unobstructed view in every direction and can easily detect approaching predators from a distance. Also, the soft mud bordering the lake in many places makes it difficult for terrestrial predators like coyotes to closely approach the birds. One more important factor attracting waterbirds to the lake is the presence of freshwater springs, especially the Mile Post 74 Springs, where they can drink and bathe. This is especially important for young birds that have undeveloped salt glands, the organs responsible for helping them excrete excess salt.

Based on all these factors, it is no wonder that hundreds of thousands of birds fly to Lake Abert to take advantage of its seasonally abundant banquet. Bulking up at the lake enables them to make long-distance journeys—in some cases, traveling many thousands of miles—to their overwintering sites.

Waterbirds are most plentiful at the lake in July and August during their post-breeding migration, when they stop at the lake to restore energy expended during reproduction, replace worn feathers and those molted after breeding, and prepare to migrate south. Wilson's Phalaropes are among these visitors; they first breed in the intermountain West wetlands and elsewhere and then flock to Lake Abert, sometimes numbering in the hundreds of thousands. They spend the winter at salt lakes on the east side of the Andes Mountains, such as at Lake Poopó in the Altiplano of Bolivia, and

Red-necked Phalarope sleeping on a mirror-perfect lake in midsummer. The bird has likely flown two thousand miles from the Arctic to feed at Lake Abert for one to two months. Then it will fly another three thousand to four thousand miles to the southern Pacific Ocean, spending the winter along the coastline of South America.

at other wetlands in South America as far south as southern Chile and Argentina. One half of all the world's Wilson's Phalaropes stop at just three locations in North America, prior to migrating south for the winter: the Great Salt Lake, Lake Abert, and Mono Lake (Colwell and Jehl 2020).

American Avocets nesting at Lake Abert and other wetlands in the interior West move on to winter along the coasts of California, Mexico, the US Gulf, Florida, and the southern Atlantic states. Some avocets nest along the lake's shorelines, but more of these birds first nest in freshwater marshes at Summer Lake before flying to Lake Abert to feed (Plissner et al. 1999, 2000). Gulls also raise their young in nearby areas during the breeding season before heading to Abert. For example, many California and Ring-billed Gulls nest at Summer Lake on a recently built nesting island located at Dutchy Lake, then likely fly the thirty air miles to Lake Abert to feed, remaining there until fall. In contrast, Western Sandpipers nest in the far arctic tundra of northwestern

Main breeding range (shown in light gray) and winter range (shown in dark gray) of Wilson's Phalaropes. Redrawn from a map provided by Lesterhuis and Clay (2010).

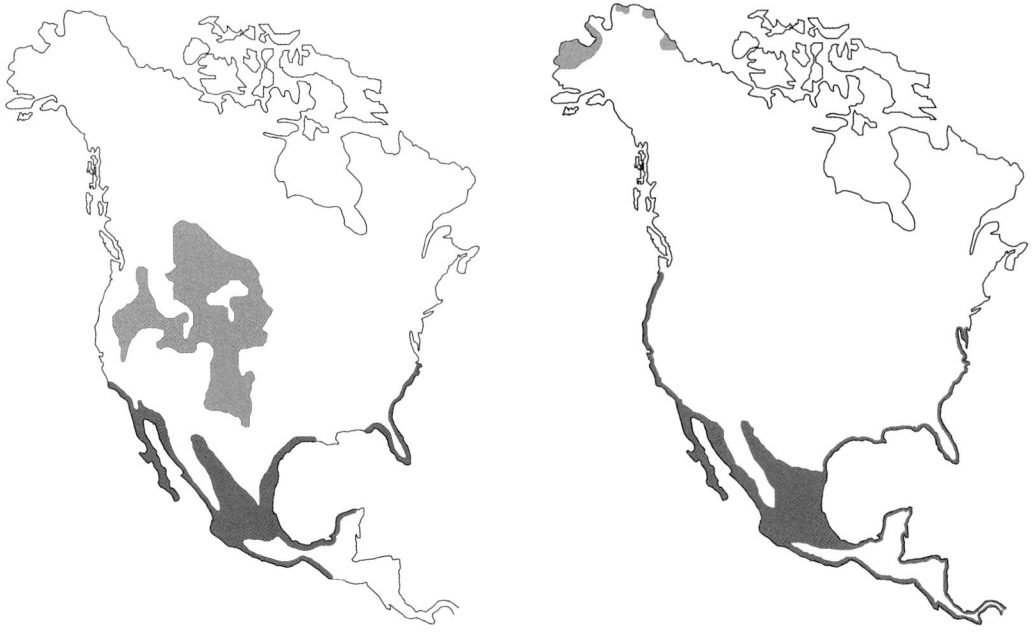

Breeding ranges (shown in light gray) and winter ranges (shown in darker gray) of American Avocets (*left*) and Western Sandpipers (*right*). Up to forty thousand avocets have been reported at Lake Abert in the summer. Up to ten thousand Western Sandpipers stop at Lake Abert to rest and feed each year. Redrawn from maps provided by Ackerman et al. (2020).

Alaska and then fly to the lake to spend time bulking up, before heading on to overwinter along Pacific and Atlantic coastlines in the United States and South America.

Waterbirds using Lake Abert benefit from nearby, large wetlands, which increase the birds' resiliency to such unpredictable environmental conditions as drought. The following large lakes, marshes, and playas are within one hundred miles of Lake Abert: Summer Lake—twenty-five miles; Warner Lakes—twenty to thirty miles; Goose Lake—thirty-five miles; Sycan Marsh—forty-five miles; Harney Lake—sixty-five miles; Klamath Marsh—seventy miles; Tule Lake—eighty miles; Upper Klamath Lake—eighty-five miles; and Lower Klamath Lake—eighty-five miles. The Klamath Basin is an especially significant waterbird habitat because it has large wetlands, about half of which are specifically managed to benefit the ecosystem.

Lake Abert has provided key nesting grounds for approximately one thousand American Avocets and one hundred pairs of Snowy Plovers (Kristensen et al. 1991), substantial portions of both species' populations in Oregon. Other lakes and wetlands in the region and throughout the Great Basin also offer essential waterbird habitats, in particular, the Great Salt Lake (Oring and Reed 1997; Warnock et al. 1998; Oring et al. 2009; Intermountain West Joint Venture 2013; Haig et al. 2019).

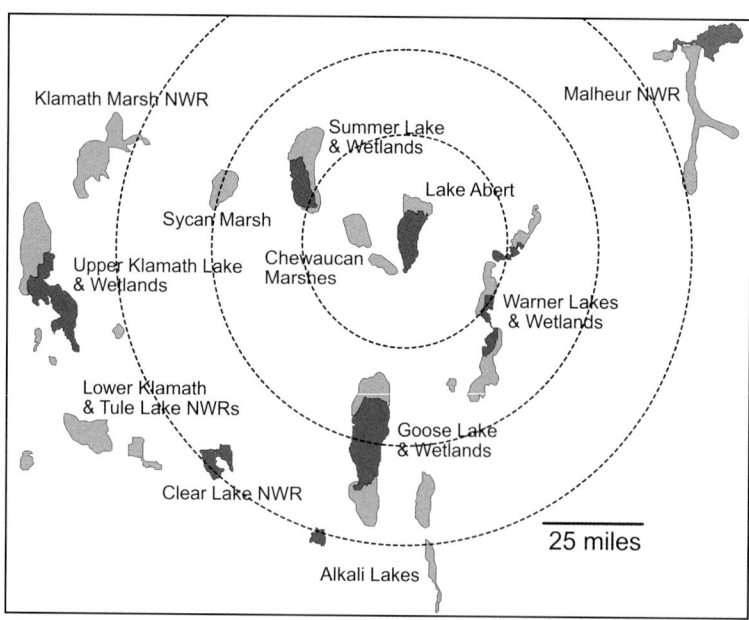

Important waterbird habitats near Lake Abert: lakes are in dark gray, marshes and playas are in light gray, and NWR stands for National Wildlife Refuge. The three dashed circles represent distances that are twenty-five, fifty, and seventy-five miles from the lake.

Living with Salt

One of the hazards for waterbirds at saline lakes and the ocean is coping with salt ingestion. All animals require some salt to remain healthy, but habitats like Lake Abert, which has salt concentrations much higher than those in the ocean, are harsh and even toxic environments for maladapted species, including birds. Waterbirds living in such environments ingest more salt than they need when they eat, and if they consume too much of the mineral, they can die. Consequently, waterbirds have evolved a variety of morphological, behavioral, and physiological ways of reducing salt's harmful effects (Rubega and Robinson 1996; Gutiérrez 2014).

These birds can minimize the adverse effects of salt by bathing in and drinking fresh water. For example, at Lake Abert, avocets, gulls, phalaropes, sandpipers, waterfowl, and Willets frequently bathe and drink at springs along the shoreline, especially at the Mile Post 74 Springs. The presence of the freshwater springs makes the environment safer and suitable for these birds.

Some of the lake's feathered visitors have developed a morphological means for reducing salt intake. For example, a phalarope's thin, needlelike bill may be such an adaptation, one that reduces the amount of salt water that the bird takes in as it feeds. Phalaropes mostly eat small invertebrates and must consume them in large numbers. Without their thin bills, they might accumulate too much salt. Waterfowl like Northern Shovelers and teal have their own adaptations, broad, spoonlike bills that can squeeze out salt water from their prey as they eat. Similarly, Eared Grebes apparently use their fleshy tongues to force salt water from the brine shrimp that they eat. In addition, ornithologists have recently proposed that shorebirds may reduce their salt intake by using the

The unmistakable American Avocet, flaunting its long and slightly upturned bill. This remarkable bird is a common presence at Lake Abert and is highly adapted to Great Basin saline-lake ecosystems.

Waterbirds taking over the eastern side of the lake at the Mile Post 74 Springs: *top*—Wilson's Phalaropes drinking and bathing; *bottom*—California and Ring-billed Gulls and other waterbirds bathing and loafing in the spring discharge.

water's surface tension to ingest prey, rather than consuming food deeper in the water.

Some salt remains inside their prey, of course, even if they manage to remove most of the external water from their food. Fortunately, invertebrates living in hypersaline environments, such as brine shrimp and alkali flies, have body fluids that are more dilute than the water in which they live because their own bodies reduce internal salt concentrations. Also, to deal with excess ingested salt, waterbirds have salt glands. These glands, however, require extra metabolic energy, and consequently, the birds have to eat even more food to meet this demand, particularly as they acclimate to increased saline levels. For example, as the shorebirds called Dunlins (*Calidris alpina*) acclimated to salt water, their metabolism increased by 20 percent (Gutiérrez 2014).

Some waterbird chicks have undeveloped salt glands, and thus they need access to water that is fresh or low in salt to survive until their glands mature (Rubega and Robinson 1996). Therefore, in some situations, the lack of access to freshwater springs could limit reproduction. However, this does not seem to be an issue at Lake Abert, where numerous and widespread distributions of springs along the shore have ensured access to enough fresh water, even during droughts.

Eared Grebes

Anyone looking for waterbirds at Lake Abert should have no trouble recognizing Eared Grebes. These small, dark-colored birds are unmistakable as breeding adults because they have ruby-red eyes, showy, golden plumes extending backward from behind their eyes, and black crests. Unfortunately, those colorful feathers led plume hunters to kill thousands of the birds early in the twentieth century, as noted by Arthur Cleveland Bent, an ornithologist who wrote an encyclopedic work about North American birds: "This species... has suffered seriously from market hunting for the millinery trade, notably in the lake regions of Oregon and California, where thousands were shot every week during the breeding season...they were much in demand for ladies' hats, capes, and muffs" (Bent 1919).

Now, however, hunters no longer chase after these birds, and the birds have recovered so well that they are characterized as the most numerous grebe species in the world. You will see them at Lake Abert most often in small flocks near the shoreline or scattered farther out in the water, and you can hear their squeaky whistles from far across the lake's expanse. In the past, Eared Grebes were one of the most numerous waterbirds at Lake Abert: observers typically report the highest numbers, thirty thousand and forty thousand, in April and September. However, recently, when the lake was nearly fully desiccated in 2014–2015 and 2021–2022, no Eared Grebes were reported using the lake. This was understandable because they are flightless for a time, the lake was too shallow for them, and there was little or no food.

An adult Eared Grebe swimming in the water and exhibiting its characteristic ruby eyes, the flame-like plumes that extend from the sides of its head, and feather tufts atop its head.

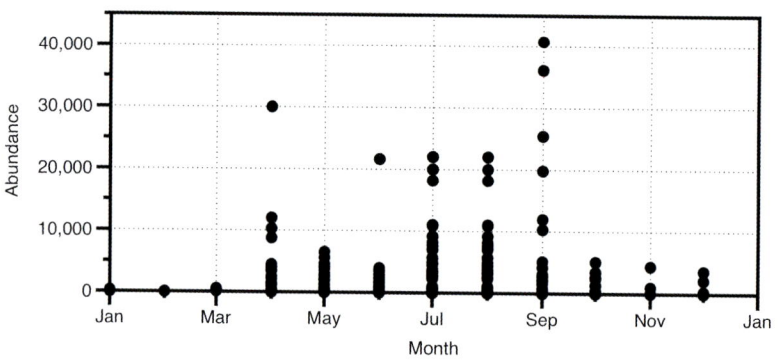

Monthly abundance of Eared Grebes at Lake Abert, January through December, based on over two hundred sightings made from 1992 through 2021. Unpublished data from BLM, ECAS, and *eBird*.

Eared Grebes have been seen at the lake in all months, at least occasionally, and they have numbered twenty thousand or more during five months of the year. Anyone tracking them at the lake will rarely see them in winter, although observers have reported seeing several thousand in December and over five hundred in January. The numbers in May and June are also relatively low because the grebes leave and fly to regional marshes, such as the large wetlands in the Klamath Basin and in the Malheur National Wildlife Refuge, where they use bulrushes, cattails, and other aquatic vegetation to build floating nests. Lake Abert does not provide enough of the marsh vegetation that they need for nesting.

In July, Eared Grebes return to the lake in large numbers to feed. They ingest their food rapidly and almost continuously, consuming tens of thousands of *Artemia* daily and increasing both the mass of their digestive organs and their fat storage. The birds are basically flightless during this period, called *hyperphagia*. Then in the

An Eared Grebe catching a midge from the water's surface. Eared Grebes subsist mostly on insects and small crustaceans.

Small flock of Eared Grebes diving in shallow water at Lake Abert. Group diving is common for this species, perhaps because the activity creates water turbulence that dislodges alkali-fly larvae and pupae from the lake bed.

fall, before migrating to wintering sites, they reverse this process, increasing the size of their flight muscles, reducing the size of their gut, and utilizing the stored fat as energy.

The number of Eared Grebes rises as food availability grows, so the birds' numbers at the lake increase through the spring, are highest in July through September, and decrease significantly from October through April, when invertebrates are scarce. Their main food sources in salt lakes are *Artemia* and the larvae and pupae of alkali flies (Jehl 1988, 2007; Jehl et al. 2003), which they peck from the water's surface, from the water column, or from the bottom, when they dive.

Ornithologists have determined that the Eared Grebes at Mono Lake migrate in the fall because their *Artemia* supply diminishes substantially (Jehl et al. 2003; Jehl 2007). Based on what we know about the seasonality of *Artemia* at Lake Abert, I suspect that the same pattern is true there. The birds aggregate in numbers of over one million at the Great Salt Lake, Mono Lake, and the Salton Sea (Roberts et al. 2013). The *State of North America's Birds 2016* gave the Eared Grebe a conservation-concern score of nine, which is moderate. However, the National Audubon Society considers the North American population of Eared Grebes to be climate endangered as a result of an estimated 100 percent loss of their summer range by 2080 (http://climate.audubon.org/birds/eargre /eared-grebe).

Waterfowl

Thirty waterfowl species, including a variety of ducks, geese, and swans, also make liberal use of Lake Abert. They come in all sizes, from Trumpeter Swans (*Cygnus buccinator*), which weigh up to twenty-three pounds, to Buffleheads (*Bucephala albeola*), which weigh less than one pound and are among our smallest North American ducks.

I never used to think of salt lakes as waterfowl habitats. After seeing the huge flocks that dwell at Lake Abert in the fall, however, I now realize that these birds are as much at home there as are Eared Grebes, shorebirds, and gulls. Unlike most waterbird counts at the lake, waterfowl numbers generally fall in the summer and increase in the spring and fall. Observers, conducting single-day counts, have tallied up over one thousand birds for each of

In the fall, tens of thousands of waterfowl, especially Northern Shovelers, migrate to Lake Abert to feed.

Waterfowl subsisting at the lake: *top*—pair of Northern Shovelers foraging at the water's surface by straining small crustaceans through their broad, spoon-like bills (shovelers are the most numerous waterfowl species at the lake, especially in late summer and fall); *bottom*—Green-winged Teal drakes foraging (the teal is one of the top ten waterfowl species counted at the lake by BLM biologists in the 1990s).

five species of waterfowl at the lake: the Northern Shoveler (*Anas clypeata*), the Green-winged Teal (*Anas crecca*), the Northern Pintail (*Anas acuta*), the Canada Goose (*Branta canadensis*), and the Ruddy Duck (*Oxyura jamaicensis*). The Northern Shoveler had the biggest presence, a one-day count of seventeen thousand birds.

Under average salinities, the chief foods for most of these Lake Abert waterfowl likely consist of *Artemia* and alkali flies' larvae and pupae. Harney Lake also attracts huge numbers of waterfowl, sometimes several hundred thousand Northern Shovelers, when biomass of small crustaceans is large enough (Littlefield 1990). Northern Shovelers feed by using their spoon-shaped bills to strain small invertebrate prey like brine shrimp from the water. They can forage by creating a vortex in the water as they paddle, bringing invertebrate prey to the water's surface. They often accomplish this by working in pairs—swimming in tandem—or by swimming in a circle with a group of three or more birds.

The Wind Birds

Writer and naturalist Peter Matthiessen thought of shorebirds as wind birds because of "their kinship with distance and swift seasons," and I agree. They are a diverse group of small-to-medium-sized wading birds that belong to the order Charadriiformes, which also includes gulls and terns. Worldwide there are approximately 220 shorebird species, with about fifty of these regularly breeding in North America (O'Brien et al. 2006). Observers have reported seeing thirty-one shorebird species at Lake Abert.

These wading birds range in size from the four-inch-tall Least Sandpiper (*Calidris minutilla*), which weighs in at less than one ounce, to the fifteen-inch-tall, two-pound Long-billed Curlew, the heaviest North American shorebird.

Shorebirds—the smallest and the heaviest: *left*—Least Sandpiper probing for food in shallow water at Lake Abert (it is North America's smallest shorebird, weighing less than one ounce, and it is mostly seen at the lake during August after breeding in the Arctic); *right*—Long-billed Curlew flying above the lake (it is the heaviest North American shorebird, weighing up to two pounds, has an eight-inch-long bill, and spends the spring and summer in salt grass and wet meadows on the northeast side of Lake Abert). Kurt Kristensen et al. (1991) found forty nesting pairs at the lake and eight nests.

According to *Birds of Oregon: A General Reference,* approximately forty shorebirds regularly spend some time in Oregon; however, some of them are marine species that mostly stay close to the ocean and rarely fly east of the Cascades. Nine species regularly nest in the western Great Basin, and they include: the American Avocet, the Black-necked Stilt (*Himantopus mexicanus*), the Wilson's Snipe (*Gallinago delicata*), the Killdeer, the Long-billed Curlew, the Snowy Plover, the Spotted Sandpiper (*Actitis macularius*), the Western Willet, and the Wilson's Phalarope. However, most Great Basin shorebirds nest in arctic Alaska or in Canada.

Fantastic Fliers—Shorebird Flocking and Flight

Nearly everyone who has watched these distinctive birds has been impressed by their swift flight and by coordinated flocking behavior that is so precise, it seems to defy what is possible. Even at a substantial distance, you can readily see their characteristic "flashing" as they twist and turn in marvelous unison and precision, as if following well-practiced choreography. Your eyes see this flashing effect when the birds fly together, rapidly alternating between exposing their dark backs and white bellies. This effect really stands out if the sun is behind you and the birds are in front of a darker background. E. H. Forbush, a New England ornithologist in the late-nineteenth and early-twentieth centuries, poetically described this flocking behavior:

> The little birds approach over the water in a dense column of perfect order.... Changing pattern, direction, color and formation with every turn, each individual yet keeps the same distance from its neighbor, the same momentum and same angle of the body, as though pulled hither and thither with lightning rapidity from the ends of an infinite number of invisible and equidistant threads, all radiating from a common point. Thus, they cut one design after another out of the fabric of space—3,000 leaderless birds, executing intricate movements with the single cohesion of one body, supported upon one pair of wings.

Another interesting aspect of shorebird flight is what appears to be erratic flight behavior, or "crazy flight": when a bird or a flock flies straight ahead but then spontaneously makes unpredictable turns and twists, as if being chased by a falcon. Dennis Paulson, in his 1993 book *Shorebirds of the Pacific Northwest,* described their "occasional erratic flight performances, which have been termed 'crazy flight.' A bird will fly at top speed, zigging and zagging in a manner that is breathtaking to watch, with ninety-degree course

Flying above the lake, this Western Sandpiper flock makes it difficult for predators to single out individual birds: the sandpipers appear to be part of a confusing, undifferentiated mass.

changes vertically as well as horizontally, and this may go on for several seconds and a hundred yards or more."

Scientists have long puzzled over how the birds manage to fly in a coordinated way. Now, however, with improved, high-speed video, powerful computer analyses, and excellent simulations, they are getting closer to understanding how the flocks operate. Studies on the amazing, coordinated movements of a flock, or murmuration, of European Starlings (*Sturnus vulgaris*) have determined that the birds constantly evaluate the behavior of nearby birds and instantaneously adjust their own movements accordingly. This behavior is similar in some ways to what is required of athletes in fast-paced team sports like soccer. However, we still don't understand how the birds can constantly monitor their neighbors and make those split-second corrections in speed and direction, or how the flocks perform this feat with no apparent leader.

We can assume, however, that shorebirds' elusive flight patterns are an adaptation intended to help them escape from predators, particularly because they live without cover, in the open and exposed. Small birds otherwise have few defenses and are vulnerable to predators. To see birds' adaptive, self-protective behavior in action, you can watch how sparrows approach a bird feeder in winter, mostly remaining close to vegetation and only darting out briefly into the open to feed. Shorebirds' erratic flight behavior likely benefits them and is meant to show that they are not easy prey.

Flocking also helps protect shorebirds from predators by giving them collective vigilance, because of many watchful eyes. Living in an open habitat has certain advantages, as well: the birds have an unobstructed view that allows them to detect predators early (Van den Hout et al. 2010). Nevertheless, for open space to be advantageous requires the birds to quickly identify a threat and then to

Peregrine Falcon keeping watch: it is one of the few avian predators swift enough to regularly prey on shorebirds.

react fast. These responses are especially critical when a Peregrine Falcon is on the attack, flying toward the flock at several hundred feet per second and giving the birds just a fraction of a second to identify the threat, burst into the air, and undertake elusive flight.

For their size, shorebirds are fast fliers, reaching speeds of thirty to fifty miles per hour, but they are no match for Peregrines that swoop in at speeds of over two hundred miles per hour. Thus, shorebirds' survival depends on hypervigilance, coordinated flocking, and evasive flight behaviors (Van den Hout et al. 2010).

Aerial escape behavior is highly developed in shorebirds and is best known in Dunlins, a shorebird that sometimes migrates to Lake Abert in the spring but which more commonly visits the coast. Dunlins often form compact flocks of several hundred birds, and their movements are so coordinated that the birds appear to maneuver

Dunlins surviving together: sizable, dense flocks such as this one show how intricately coordinated the Dunlins' flying is and how flocking helps protect each individual bird from predators.

as a single unit. Ornithologists refer to this behavior as a "socially coordinated aerial-escape tactic." When one of these shorebirds is threatened, it tries to avoid being caught by flying upward, by diving into the water, or by hiding in vegetation (Lima 1993). Dunlins also use several other evasive flock behaviors, such as engaging in "rippling" or "columnar" flight patterns when they try to elude falcons (Buchanan et al. 1988). Rippling flight involves waves of movement that start at one side of the flock and sweep through it, either horizontally or vertically. In columnar flight, Dunlins form a tornado-like, vertical column that undulates throughout its length. These flocking behaviors likely confuse predators and reduce exposure of individual birds to predation. Surprisingly, despite the great numbers of shorebirds at Lake Abert during the summer, they never seem to be attacked by falcons. Falcons do routinely pursue shorebirds at other lakes (De Deckker 1983).

Killdeers—The Adaptable Shorebird

Another shorebird at Lake Abert, the Killdeer, is a regular visitor but never flocks there in large numbers. Despite its shorebird pedigree, it easily adapts to manmade habitats, including roadsides,

Killdeers resting on the shore: Adult Killdeers have distinctive bands on their necks and foreheads. Like other plovers—their avian relations—these shorebirds forage by watching for insects and other invertebrates and then darting out to secure them.

Killdeers at Lake Abert: *left*—four conical Killdeer eggs resting in a nest (a simple scrape made of short sticks and pebbles); *right*—downy chick making its way to the water (the precocial chicks have long legs, which make them independently mobile, and they can feed themselves soon after hatching).

parking lots, and the gravel-covered, flat tops of buildings, where it sometimes nests.

Killdeers generally nest in a variety of habitats at the northwest corner of the lake, but they prefer areas dominated by salt grass near freshwater springs (Kristensen et al. 1991). They also nest on the gravel shoreline along the east side of the lake and visit scattered locations along the lakeshore, in small numbers, from spring through fall. Because Killdeers' numbers are so limited, they are frequently overlooked. The *State of North America's Birds 2016* gave the Killdeer a conservation-concern score of eleven, which is moderate.

The Snowy Plover—Living on Alkali

> What intrigues me about these tiny white birds with brown bands across their breasts is how they manage their lives in such a forbidding landscape. The only shade on the salt flats is the shadow they cast. There is little fresh water, if any. And their diet consists of insects indigenous to alkaline habitats—brine flies and beetles.
>
> —TERRY T. WILLIAMS, 1991

Snowy Plovers are small, pale-colored shorebirds that nest on coastal beaches and interior playas and margins of large salt lakes, including those at Lake Abert. They range widely along the Pacific coast (from Canada to Mexico), along the Gulf of Mexico's coast, in the central United States, in the Great Basin, and in North-Central Mexico.

A male Snowy Plover looks smart during the breeding season, in his contrasting white, gray, and black outfit. His belly, breast, throat, and face are snow-white; his brow has a coal-black band; and a similar band extends from his eyes to the back of his neck. Over his shoulder is a delicate black line that tapers toward his back, like the line a skilled calligrapher might make using a thin, flat brush. The back of his head and his upper wings are pale brown, and his wing tips are edged in jet. Each of his large, elliptical eyes is so dark that it's like looking into a cave. His bill is black, and his legs are a light pinkish-gray, with white feathers covering the upper parts. Female Snowy plovers are similar but lighter in color.

These shorebirds have adapted to foraging and nesting on ocean beaches and on alkali-encrusted playas of interior salt lakes, where there is no vegetative cover and where temperatures are in the nineties by midsummer. Snowy Plovers have good vision and run quickly over the dry, salty crust to grasp insects, including adult alkali flies. At Summer Lake, I have seen them trying to capture

Male Snowy Plover, displaying his breeding plumage and resting on an efflorescent crust, formed when alkali migrates to the surface, at the Summer Lake Wildlife Area (his coloration provides good camouflage, helping him blend in with his surroundings).

Female Snowy Plover adult, keeping watch (females lack the bold markings of the males but have the large eyes and short beaks typical of all plovers).

fast-moving Oregon tiger beetles. The birds, like other plovers (including Killdeer), tend to forage by standing still and scanning the ground around them for insects and other small invertebrates. Once they see their prey, they rush forward and use their short, stout bills to grab it.

One June, while I was at the Summer Lake Wildlife Area, I got a close-up view of a Snowy Plover foraging. As it moved along the shore of Dutchy Lake, it stopped and scanned the area in front and to the sides and then dashed out and poked at something in the mud near the water's edge. Although I watched carefully from a short distance, I couldn't determine what it was after, but I believe it was hunting for fly larvae, which are small and difficult to see on the muddy substrate. During the time that I watched the bird, I saw it frequently tilt its head, directing one eye skyward, apparently watching for aerial predators like harriers or falcons.

Because Snowy Plovers often forage alone, they have to be more vigilant than flocking shorebirds like phalaropes and sand-pipers, which watch out for predators as a group. Snowy Plovers are

Snowy Plover female apparently signaling her concern over the presence of an avian predator overhead, looking skyward and making herself less conspicuous by sitting on the ground.

difficult to see because they are often solitary or forage in a small group of just a few, scattered birds. Their light coloration also helps to protect them from predators in the open habitats where they live. I have also noticed that when they are threatened, they crouch down, making themselves even less visible.

Because of their cryptic coloration and tendency to forage alone, Snowy Plovers are also difficult to count. However, beginning in 1980, ODFW staff and volunteers began conducting periodic surveys, counting adult Snowy Plovers at up to 20 sites in Oregon's interior, including at Lake Abert (Simon Wray, ODFW pers. com. 2015). During the 21 surveys that followed, the ODFW team counted over 11,000 birds, and they observed nearly one-third of them at Lake Abert. The total number of Snowy Plovers per survey ranged from 7 to 345, with an average of 211 birds per count. In ODFW's last two survey years—2014 and 2021—Snowy Plover numbers at Lake Abert fell to their lowest levels since the project began, in large part due to serious drought and to related high salinity, which reduced the invertebrate population that the birds feed on (Larson et al. 2016).

Ecologist Mark Stern and his colleagues at the Nature Conservancy conducted some earlier research on Snowy Plovers at Lake Abert during the springs and summers of 1988 and 1990 (Stern et al. 1991). Their work focused on the ecology and post-breeding movements of birds nesting on the playa at the north end of the lake. In the first two years of their study, they found over 120 nests and determined that nesting begins in late April, peaks from mid-May through late June, and lasts until early July. They also discovered that at least one egg hatched from nearly 70 percent of the clutches that they monitored. They attributed any nest failures to a variety of causes, including predation, wind, sandstorms, flooding, and nest abandonment. They also saw a peak number of nearly 260 adults.

To assess the birds' movements from the lake, Stern's team banded nearly four hundred plovers. During the winters that

Snowy Plover survey counts made at Lake Abert from 1980 through 2021 by ODFW staff and volunteers. Kaly Adkins and Simon Wray, ODFW.

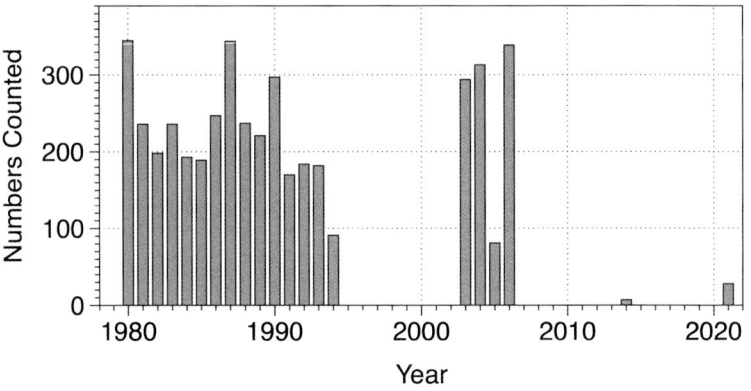

followed, observers reported seeing the tagged birds along the California coast as far south as San Diego and on the west coast of Baja California in Mexico. Volunteer counters also saw twelve banded birds on the California coast during the breeding season at Lake Abert and at Oregon's other interior sites. During the summer of 1990, the Nature Conservancy team found that nearly 50 percent of banded adults and 15 percent of banded young birds returned to the lake. This study showed that the lake is an important breeding site, that the coastal and interior Snowy Plover populations join up during the winter months, and that the birds maintain fidelity to their breeding sites, generally returning to them.

In the summer of 1990, Stern's team, with the assistance of ODFW staff, extensively surveyed Oregon's interior sites for Snowy Plovers. One of their findings was that drought conditions that year had altered the plovers' range and had eliminated their nesting habitats at some smaller playa lakes in northern Harney County that were dry. However, in lowering water levels at Harney Lake, the drought also created shoreline habitat that the birds had not previously used. These results showed that, although the birds are loyal to their breeding sites, they will move to new locations regionally as water conditions fluctuate.

Nevertheless, Snowy Plovers are considered imperiled because of declining numbers, a small overall population size, negative effects of predators and invasive plants, and impacts from human recreation and coastal development (Thomas et al. 2012). Consequently, those birds nesting along the US Pacific coast are listed as threatened under the federal Endangered Species Act. In 2007 and 2008, ornithologist Susan Thomas and her colleagues conducted their own surveys and determined that the total North American population size was only twenty-six thousand birds (Thomas et al. 2012). During that period, only twelve of fifty shorebirds regularly breeding in the United States and Canada had smaller populations than that of the Snowy Plover (Andres et al. 2012). The National Audubon Society has raised its own alarm bells, declaring the

Snowy Plover possibly hunting for an insect hidden in the sand.

Snowy Plover to be climate endangered and estimating that 56 percent of their winter range will be gone by 2080, but the report does not assess any impact to the birds' summer range (http://climate.audubon.org/birds/snoplo5/snowy-plover). The *State of North America's Birds 2016* gave the Snowy Plover a conservation-concern score of fifteen, which is high, and put it on the watch list.

Thomas has acknowledged that her count of Snowy Plovers at Lake Abert represented just over 1 percent of the birds' estimated total for all of North America, but she has recommended that we confer management priority to any site with at least 1 percent of the country's total population of plovers and work toward the goal of maintaining or increasing their numbers. Indeed, from a conservation standpoint, we should more broadly consider Lake Abert as just a part of the northern Great Basin habitat complex. After all, within seventy miles of the lake—perhaps a two-hour flight for a Snowy Plover—Thomas and her team counted over two thousand Snowy Plovers, equal to nearly one-tenth of the total North American population (Thomas et al. 2012).

Obviously, the northern Great Basin deserves special attention as a key region for Snowy Plover conservation. Although more Snowy Plovers nest at Lake Abert and at other Great Basin lakes than at anywhere along the coast, their population in the interior is far from secure. As the drought conditions that impacted Lake Abert and other regional playa lakes in 2014–2015 and in 2021–2022 showed, Great Basin birds face threats from a highly variable and warming climate, pressures that are further exacerbated by our water diversions (Larson et al. 2016; Moore 2016; Williams 2016; Wurtsbaugh et al. 2017). The perils faced by the plovers and by other Great Basin birds will likely get more severe in the future in our ever-drier climate.

American Avocet—A Bird at Home at Great Basin Saline Lakes

Its favorite resorts seem to be the shallow, muddy borders of alkaline lakes, wide open spaces of extensive marshes, where scant vegetation gives little concealment, or broad wet meadows splashed with shallow pools. If muddy pools are covered with reeking scum, attracting myriads of flies, so much the better for feeding purposes.

—ARTHUR C. BENT, 1927

Perhaps no other waterbird is more symbolic of Great Basin alkali wetlands than the American Avocet, and in fact, this species has its peak nesting population in the Great Basin (Ryser 1985). Avocets are usually the first shorebirds to arrive at Lake Abert in the spring,

Photo action sequence of an American Avocet landing.

with some stray birds appearing as early as March. They are also among the last shorebirds to leave in the fall, with a few hardy ones staying on even into December.

Avocets are also the most common shorebirds at the smaller playa lakes that are widespread and abundant in the northwestern Great Basin. Their robust numbers at Lake Abert increase steadily through the spring and peak in August and September. Observers have seen high numbers of at least thirty-five thousand in August, September, and October. Avocet monthly averages are much lower but still exceed five thousand birds. The highest count was forty thousand reported in September 1993 (*eBird* website data accessed on February 14, 2019).

When avocets arrive at the lake, they come already strikingly dressed in their breeding plumage: reddish-brown-feathered heads and necks, black and white wings, gray-blue legs, and glossy-black bills. After they breed, the birds lose their colorful head and neck feathers, which are replaced by white ones, and then they look much like the plainer juvenile avocets.

Avocets have bill shapes that identify their sexes, a characteristic that is unique among shorebirds. A female's bill curves upward more than a male's bill does, although the shape can vary (Hamilton 1975). This difference may relate to their different feeding habits, a variation also documented in the group of hummingbirds called hermits. A female hermit's bill also has a greater curvature than a male's bill, which helps the female to get nectar from particular kinds of flowers (Temeles et al. 2010). Ornithologists have not studied this feature in avocets, but the more-pronounced bill curvature may enable the females to feed more effectively from a muddy surface by *scything,* the avocets' method of foraging on alkali fly larvae and pupae from the sediment, while the straighter bill of the male might be an adaptation for pulling prey from the water column. If this were true, it would reduce competition between the

Lake Abert waterbirds, *from top to bottom, left to right*: Eared Grebe; American Avocet; Western Willet; Black-necked Stilt; Lesser Yellowlegs; Least Sandpiper; Western Sandpiper; Red-necked Phalarope; and male and female Wilson's Phalarope.

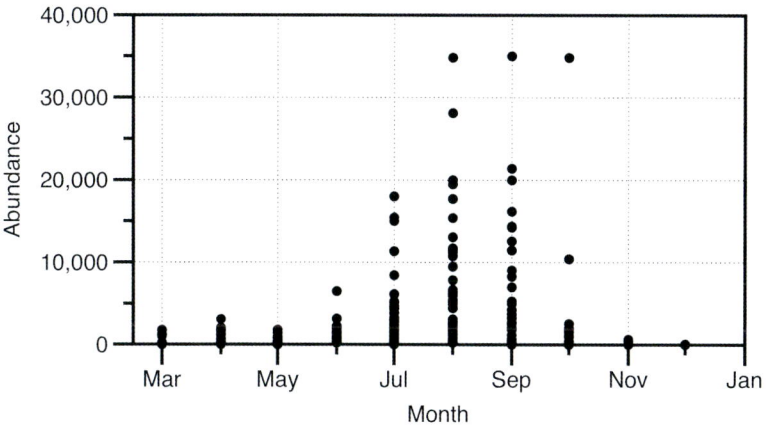

Monthly abundance data for American Avocets, March through December, at Lake Abert, based on sightings made on more than 230 census dates (1992–2021). Unpublished data from BLM, ECAS, and *eBird*.

Trio of breeding American Avocets foraging in shallow water. The leader (the bird on the left) is a male, based on his nearly straight bill, and the other two are probably females, based on the greater curvature of their bills.

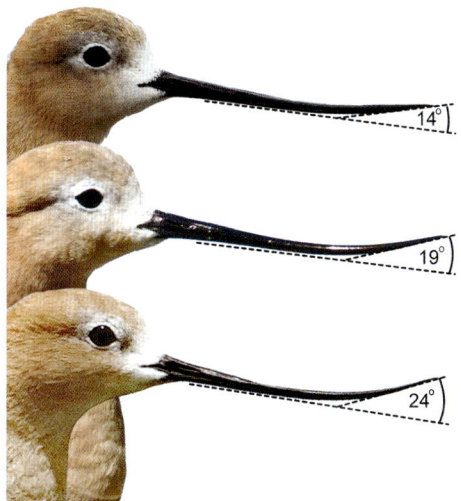

Shape variability of American Avocets' bills, with approximate angles shown. The top angle is typical of a male bird, and the bottom angle is typical of a female bird. The middle angle is also common for a female bird, but the intermediate curvature shows just how variable the shape can be.

A pair of avocets swimming in the lake. Although avocets usually forage while wading in the water or walking on the mud, they occasionally do swim and peck at prey near the water's surface, in the water column, or from the lake bottom.

sexes for food because their adaptations would allow them to feed in slightly different habitats.

Avocets forage visually, tactilely, or by tasting food. They not only hunt for alkali fly larvae and pupae, but they also hunt for *Artemia* that they peck or scythe from the water column. Females seem to use their extra-curved bills to forage in very shallow, spring-fed, muddy pools, like those that remain behind when Lake Abert's water recedes during a drought. In 2014 and 2021, when the lake was nearly dry, these muddy pools were some of the only habitats containing invertebrates, mostly alkali fly larvae.

The birds also likely go after water boatmen (aquatic insects in the family Corixidae) and amphipods in shallow pools near springs and in the lake when salinities are low (Boula 1986). They also forage while swimming, pecking prey from the water column or from the sediment as they go, something they have in common with phalaropes, but an uncommon behavior among shorebirds.

In 1986, Oregon State University graduate student Kathryn Boula reported that brine shrimp do not factor importantly in avocet diets; however, she likely underestimated their significance because she did not account for how quickly *Artemia* are digested. Also, the relative proportion of invertebrates in bird's diets may well change from year to year as the lake's salinity changes, due to the salt level's impact on invertebrate populations.

Avocets nest in small groups, and they build their nests as simple cups on the mud or in higher shoreline areas of sand or gravel. They usually lay their khaki-colored, dark-flecked eggs in groups of four and in various sizes. The population of avocets breeding at the lake in 1989 reached a high-water mark of one thousand—the largest number ever reported for any waterbird there—and mostly occurred at the northwest corner of the lake (Kristensen et al. 1991).

American Avocets brooding: *left*—an avocet, probably a male, sitting on eggs at Summer Lake; *right*—nest made of dried vegetation, resting on the Lake Abert shore (the clutch of four eggs is typical for shorebirds).

Avocets on the move: *left*—a female avocet flying above me, close overhead, and calling, apparently alarmed by my presence (she probably had a nest or young nearby); *right*—a young avocet, displaying its still-downy plumage.

Avocets sometimes nest along the eastern shore of the lake; however, I have never seen them raising their hatchlings there, so they may routinely lose those nests to predators.

Most avocets arrive at the lake after breeding and therefore are most numerous in August and September. Wildlife biologists from BLM and ODFW reported peak counts of 35,000 and 40,000 in September 1993. This undoubtedly is a record for the state. Other important sites for American Avocet populations, based on highest counts, include: the Great Salt Lake—250,000; the Lahontan Valley wetlands—64,000; the Salton Sea—15,000; and Mono Lake—9,000 (Neel and Henry 1996; Paul and Manning 2002; Strauss et al. 2002; Patton et al. 2003). Based on these data, it is evident that Lake Abert is one of the key sites for avocets in the West.

Experts do not know where the avocets that breed and forage at Lake Abert spend their winters. Researchers have, however, observed that birds banded at Honey Lake, California, and at the Great Salt Lake migrate together to wintering sites in western Mexico and coastal California (Robinson and Oring 1996). More recently, John Cavitt, an ornithologist from Weber State University, told me that he has used small satellite transmitters to track avocets from the Great Salt Lake to numerous coastal sites in western Mexico. Thus, it is likely that Lake Abert's birds also fly to that area or to coastal California.

Although the current conservation status of the American Avocet appears secure, because of its large population size of nearly 500,000 (Andres et al. 2012), it may not be. The *State of North America's Birds 2016* gave the American Avocet a conservation-concern score of twelve, which is moderate although not quite serious enough to put it on the watch list. Some researchers consider the birds' breeding, wintering, and migration habitats to be at an increased risk from climate change (Galbraith et al. 2014), and the National Audubon Society has estimated that the shorebird will lose 96 percent of its summer range by 2080 and therefore has assessed the American Avocet as climate endangered (http://climate.audubon.org/birds /ameavo/american-avocet). Unfortunately, due to a drying climate, we may already be seeing evidence of the loss of summer range in recent reductions of waterbird habitat along much of the Great Basin. The number of avocets at Lake Abert in 2021 fell to only 1,400, the lowest recorded total in the past decade.

Black-Necked Stilt—All Dressed Up

> Stilts are essential waders; for they are highly specialized, and here they show to best advantage. At times they seem a bit wobbly on their absurdly long and slender legs, notably when trembling with excitement over the invasion of their breeding grounds. But really, they are expert in the use of these well-adapted limbs, and one cannot help admiring the skill and graceful way in which they wade about in water breast deep.... The legs are much bent at each step, the foot is carefully raised and gently but firmly planted again at each long stride.
>
> —ARTHUR C. BENT, 1927

Avocets have dapper relatives: the Black-necked Stilts. These shorebirds are always formally dressed in black and white attire, and their long, pink legs make them real standouts. They belong to the Recurvirostridae family, which actually gets its name from the curved bill of their cousin, the avocet. Stilts are slenderer than avocets, have straight, black bills, and both sexes have contrasting black and white plumage and pink legs. These shorebirds feed by wading in shallow water, where they search for swimming invertebrates and quickly grab them with their needlelike bills. They can be noisy during the

Black-necked Stilts flying. No other Great Basin waterbirds share their bold, black-and-white pattern and long, pink legs.

Black-necked Stilt preparing to grab prey from the water column with its long, pointed bill.

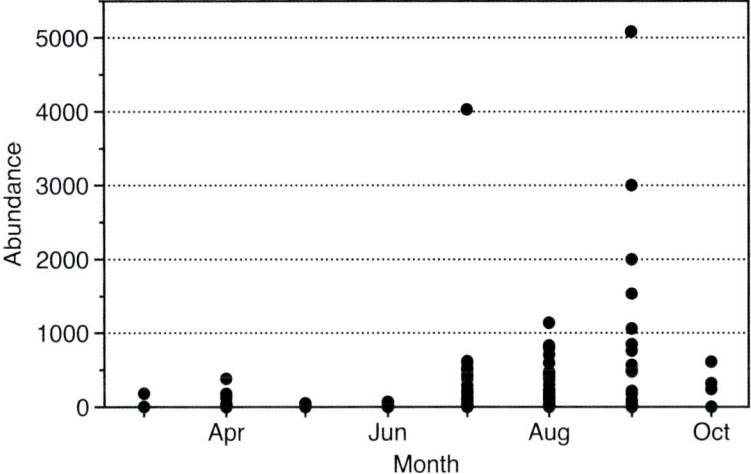

Monthly abundance data for Black-necked Stilts, March–October, at Lake Abert, based on over 170 census dates (1992–2021). Unpublished data from BLM, ECAS, and *eBird*.

breeding season and routinely alert nearby birds to the presence of trespassers.

Stilts arrive at Lake Abert as early as March and April, prior to breeding. Then, most of the birds leave to nest elsewhere in freshwater and brackish marshes across the region, so their numbers at the lake decrease in May and June. By July, their numbers rise again and continue to do so until September. Most of these shorebirds have left the lake by October.

Breeding stilts are widespread in South Central Oregon marshes and adjacent areas of the northern Great Basin, especially in the Chewaucan, Klamath, and Malheur wetlands (Marshall et al. 2003). They nest on the mud by building a platform of marsh vegetation. No one has observed the birds breeding at Lake Abert; however, some researchers once reported seeing stilts during the normal

Black-necked Stilt sitting on a marsh-grass nest nestled in the mud at Summer Lake. Because their legs are so long, stilts fold their legs backward when they sit. Relative to their body size, stilts have the longest legs of any North American bird.

breeding season at a small, vegetated pond at the northwest corner of the lake (Kristensen et al. 1988). That particular May, the birds were engaging in distraction behaviors, such as aerial mobbing, broken-wing displays, and mock incubations, which suggests that the birds were guarding nearby nests. Personally, I have never seen stilts nesting at the lake, in all my visits there.

Stilts are active, visual feeders that select individual prey, and they even try capturing adult alkali flies by snatching them out of the air. They probably peck alkali fly larvae and pupae, as well as *Artemia* and water boatmen, from the sediment and water column.

The highest count of stilts at the lake was over five thousand on September 19, 2012, which is a record for the state, based on data in *Birds of Oregon* and from the *eBird* website. Over the past several years, some researchers have estimated that the North American population of Black-necked Stilts is between 150,000 and 200,000 birds, and they have not reported any obviously shifting trends in these numbers (Andres et al. 2012). The *State of North America's Birds 2016* gave the stilt a conservation-concern score of eight, which is low. However, experts have warned that this species faces increased risks from climate change, including a reduction in its

Black-necked Stilt trying to capture adult alkali flies by running after them and grabbing them from the air.

breeding habitat (Galbraith et al. 2014). We need to monitor stilts in the Great Basin to ensure that they do not decline as a result of our drying climate.

Western Sandpiper

Sandpipers add even more variety and vitality to Lake Abert's water-bird community. The Western Sandpiper, sometimes just called a western or a peep, is the most numerous of the lake's small shore-birds. I think westerns are the cutest of the shorebirds because they are tame and sociable, they rarely quarrel, and they usually gather together in small flocks, feeding, preening, or sleeping with one eye partially open and their bill tucked under a wing.

Westerns are the best-dressed sandpipers during the breeding season, showing off snow-white undersides with scattered, jet-black flecks. Their wings and backs display beautiful, cinnamon-brown feathers tipped with white fringes, and with jet-black patches between the bases and tips. Their heads have streaks of cinnamon, jet, and white, and their stout bills and legs are an elegant, glossy black.

These peeps stop at the lake only briefly and irregularly in late April and early May on their way north to the Arctic, where they breed at sites along Alaska's Arctic Ocean coast and on western Alaska's Bering Sea coast, and many of them nest at the Cooper River Delta. However, when they do stop at the lake, they can be numerous. In May 1989, approximately fifteen thousand sandpipers—mostly westerns—visited the lake (Kristensen et al. 1991), and BLM biologists counted twelve thousand in late April 1994. You will not find greater numbers of them in Oregon, east of the Cascades, except at Malheur National Wildlife Refuge (Nehls 1994).

Small flock of post-breeding Western Sandpipers loafing and preening at Lake Abert.

A nonbreeding Western Sandpiper foraging along Lake Abert's shore, where receding water levels had exposed an accumulation of alkali-fly pupae.

A Western Sandpiper, decked out in its colorful breeding plumage, foraging for invertebrates in shallow water along the Lake Abert shoreline.

Western Sandpipers mostly appear at the lake in July and stay until mid-September. When they arrive, they still have some colorful breeding feathers on their wings, atop their crowns, and around their eyes. They slowly lose these feathers as gray and pale-gray ones grow in as replacements during their stay at the lake.

These little shorebirds have a short and stout bill adapted for probing mud and sand in search of invertebrate prey; there is often a slight droop at the bill's tip. When they feed, they walk along and repeatedly probe the sediment in shallow water, using their bills in a rapid, sewing-machinelike way. They have some webbing at the base of their toes, which probably helps them traverse the soft mud, but they rarely swim.

As they forage, they most likely hunt for larval alkali flies, amphipods, or worms. Westerns have special organs located near the tip of the bill, called *sensory pits,* that help them find prey, so their probing behavior may be a way to discover dense concentrations of food. They also probe around the base of rocks in the water to find alkali fly pupae, and they search through accumulations of pupal cases that have washed ashore. However, I have never seen them show any interest in catching adult alkali flies.

Western Sandpipers are the most numerous peeps in North America, with an estimated population of about three and a half million, and they are especially plentiful on the West Coast. Lake Abert's westerns winter along the Pacific, Gulf, and Atlantic coasts of the United States and also migrate farther south, along the coasts of Mexico and Central America. The *State of North America's Birds 2016* gave the Western Sandpiper a conservation-concern score of thirteen, which is moderate. Nevertheless, some experts consider this species to be highly vulnerable to climate change, due to ongoing losses of breeding, migration, and wintering habitats, so its status warrants our attention (Galbraith et al. 2014). Few westerns and other sandpipers migrated to the lake from 2017 to 2021, but we are not sure why.

Western Willet

If you see a Western Willet (*Tringa semipalmata inornata*) in or near Lake Abert's waters, you may consider it to be drab in comparison to some other shorebirds. These birds are medium-sized and mostly dull in appearance, mottled and covered in brown and white bars. When they lift off and fly, however, their wings reveal a bold and distinctive black-and-white pattern. They also have a loud and characteristic alarm call, for which they are named. Willets have

Photo action sequence of a Willet taking off. The bird's contrasting black-and-white wing pattern, which is so visible when it flies, is unmistakable. The Willet also has a loud and distinctive alarm call—*pill-will-willet*.

Western Willet foraging in shallow water along Lake Abert's shore.

medium-length legs and bills that enable them to forage in deeper water, probing more deeply into sediment than smaller sandpipers can. You can distinguish them from the eastern Willet subspecies by observing minor differences, including lighter-colored plumage.

During their breeding season, Willets migrate to lakes across the western half of Canada and the United States. They winter as far south as Chile in South America. They are frequent visitors to Lake Abert, where they forage along the shore in shallow water, but they are never present in high numbers. Unlike most shorebirds, they prefer nesting in the uplands.

One May, I accidentally discovered a nesting Willet on the lower slope of Abert Rim, nearly a quarter mile from the lake. She was well camouflaged, looking like a lichen-covered rock, and she remained very still, so I could have missed her very easily. The nest was well hidden among rocks and grasses. I don't know who was more surprised, but she continued sitting on the nest without making the slightest movement as I watched her. However, finally after several minutes, she burst off the deep, grassy, cup-shaped nest, revealing four mottled eggs, colored in earth tones of blue, brown, gray, and black. Elsewhere in the West, Willets have been found nesting nearly two miles from water, which makes for quite a long hike for young Willet chicks, even though they are precocial—quickly independent—and capable of running within a few days of hatching.

Willets are never very numerous at the lake: the highest recorded count was 250 birds on July 27, 1994. Observers have seen these shorebirds there throughout the year, except during December and February. However, bird counters have seen the biggest overall numbers—150 or more—from April through July, with the highest average counts of nearly 90 birds in May and June. No one has

Western Willet nesting: *left*—female Willet sitting on a nest on the lower slope of Abert Rim; *right*—she laid her clutch of four eggs in a simple, grass-lined depression hidden among grasses and rocks.

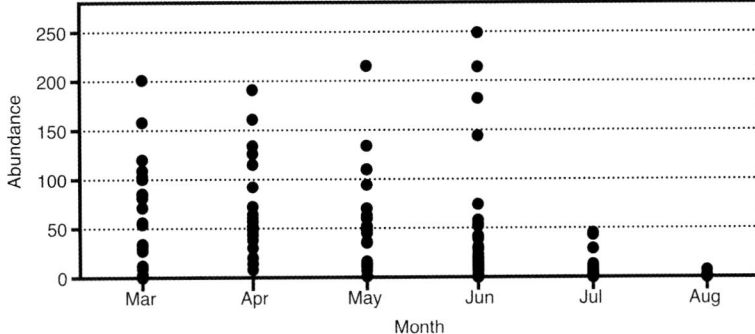

Monthly abundance data for western Willets at Lake Abert, March through August, based on sightings made on nearly 200 census dates (1992–2021). Data for January, February, September, and October were not plotted because people tracking the birds saw only one or two of them in a single sighting in each of those months. Unpublished data from BLM, ECAS, and *eBird*.

figured out why there is so much variation in the Willets' numbers or why the average total in July is lower. Lake Abert has the second highest number of Willets in the state, only lagging behind Goose Lake, based on data in *Birds of Oregon* (Marshall et al. 2003).

Most Willets in South Central Oregon depart for their coastal wintering sites in late June, soon after nesting (Haig et al. 2002); however, at Lake Abert, their numbers often remain quite high, even through July in some years, suggesting that some of the birds stay longer to feed. Their main foods are probably alkali fly larvae and pupae, or perhaps amphipods and worms, when the birds feed near springs. I have never seen Willets trying to catch adult flies or brine shrimp.

The *State of North America's Birds 2016* gave the Willet a conservation-concern score of fourteen, which is high enough to put it on the watch list (the western and eastern subspecies were not separated in the analysis). The National Audubon Society lists the Willet as climate endangered, estimating that it will lose 70 percent of its winter range by 2080 (http://climate.audubon.org/birds/willet1/willet). Other researchers consider the Willet's migration habitat to be at an increased risk from climate change (Galbraith et al. 2014). In light of these concerns, we should regularly monitor and work to protect the Willet populations in the Great Basin.

Phalaropes—The Spin

Only three species of the medium-sized shorebirds called phalaropes live on our planet, and two of them frequent Lake Abert each summer—the Red-necked Phalarope and the Wilson's Phalarope. The third species, the Red Phalarope (*Phalaropus fulicarius*), is a rare visitor east of the Cascades, according to *Birds of*

Female Wilson's Phalarope, displaying her breeding plumage as she moves through the water. The phalarope has adaptations for foraging on small prey near the water's surface: a long, needlelike bill and forward-looking eyes that provide binocular vision.

Oregon. However, observers have reported catching sight of one at the lake on two occasions. Red Phalaropes normally migrate offshore over the Pacific Ocean, flying between the Arctic and the equatorial Pacific. In some years, many of these birds die during their long migration, apparently due to starvation, and are found washed ashore along the Oregon and Washington coasts (https://sites.uw.edu/coasst/2016/11/22/watchoutforphalaropes/).

Female phalaropes are unusual, although not unique, for being larger and showier than the males. For most shorebirds, the sexes are similar both in size and coloration. The females are also the pursuers: they mate with several males, and then the males incubate the eggs and care for the young until the juveniles are old enough to be independent.

Phalaropes have evolved a specialized and nearly unique foraging technique called *vortex* feeding, which Northern Shovelers also use. Their behavior, when foraging in open water, involves creating a whirlpool by continuously swimming in circles, using their partially lobed toes (Obst et al. 1996). The birds are very buoyant, they do not dive, and their primary habitat is the open ocean or salt lakes, and so they need a way to reach small, invertebrate prey that are too deep in the water for easy access. The whirlpools that they create bring prey right up to the surface. In a 2015 article in *Audubon Magazine,* writer Nicholas Lund says this about vortex feeding: "It's peculiar behavior when seen in a single bird, but when a group of phalaropes is spinning together it's downright dizzying, like some kind of intensely competitive avian dance contest."

A phalarope appears to be spinning when it feeds in this way, but it is actually swimming in a tight circle that has a diameter about equal to the bird's length. This action forms a vortex: water is pushed outward at the surface by rearward propulsion created by the bird's feet and also by the bird's breast pushing the water ahead. As water at the surface is pushed outward, water at the center of the circle is also drawn upward, bringing weakly swimming or

Vortex-feeding behavior of the Wilson's Phalarope: *left*—the bird's movements in the water bring prey up to the surface (a circular swimming motion forces water outward, which forms the vortex and pushes the prey upward; modified diagram based on the work of Obst et al. [1996]); *right*—female Wilson's Phalarope vortex feeding (she swims clockwise and turns her head inward to the right and toward the center of the vortex, which likely enables her to see any prey that the vortex brings up).

suspended prey, such as *Artemia* and alkali fly larvae, closer to the surface, where the bird can grab them.

When vortex feeding, the birds complete a circle approximately once every second or even faster, and they can turn either clockwise or counterclockwise. In my observations of a small sample of seventy birds, I saw most of the phalaropes turning clockwise. Brian Obst, who has studied the behavior of captive phalaropes, has found that the phalaropes consistently swim in one direction, rather like people favoring their left or right hand.

The center of the vortex is not directly under the birds but is instead near the center of the circle that the birds make when they swim, because phalaropes swim in circles rather than spinning in place. The birds watch for prey moving up to the surface by turning their heads toward the center of the vortex, either turning to the left when they swim counterclockwise or to the right when they swim clockwise.

Versatile foragers, phalaropes can even capture prey that are too deep to reach from the surface by "tipping up," or rather, by immersing their heads and chests under water. The birds also quickly walk in circles in shallow water, apparently to dislodge prey living in the sediment. Typically, they feed at one location for several minutes, and, apparently when they can no longer see additional prey, they move to another site to repeat their foraging behavior.

These clever shorebirds also have a needlelike bill and binocular vision, which enable them to capture their prey precisely. They

Phalaropes vortex feeding at Lake Abert in August 2012, scattered uniformly over most of the water's surface. Bird counters estimated that there were more than one hundred thousand phalaropes—mostly the Wilson's variety—at the lake that summer.

can sneak up on their prey with little warning because their narrow bill barely disturbs the water. Additionally, the thin shape of the bill lessens the amount of salt water that they swallow while feeding.

Interestingly, phalaropes' bills are also adapted to take advantage of the water's surface tension and capillary action, both of which help them to transport small prey up to their mouths. Recent studies using high-speed photography show phalaropes using a technique called *capillary ratcheting:* the birds rapidly open and close their bill slightly, several times, and with each bill movement, the water droplet containing the prey moves upward to their mouth (Prakash et al. 2008).

I have seen just how effective the vortex-feeding technique is in open water, when I have watched the phalaropes feeding. In my observations, I have seen the birds pecking at rates of ten to thirty times per minute, when *Artemia* have been numerous. Phalaropes, however, apparently only use this feeding method when prey is abundant but too deep to reach in any other way, because the technique involves expending a great deal of energy (DiGiacomo et al. 2002).

In early August 2012 and 2013, when Wilson's Phalaropes numbered one hundred thousand or more at the lake, they were scattered over much of the surface, vortex feeding, with approximately one bird per several square yards. They were likely eating *Artemia,* which were plentiful at the time. Generally, these shorebirds feed by themselves, but I have also seen pairs of Wilson's Phalaropes vortex feeding within one or two body lengths of each other. In these instances, I have wondered whether they might be cooperating, as Northern Shovelers do, to bring food to the surface through teamwork.

Red-Necked Phalaropes

Red-necked Phalaropes are annual visitors to the lake, migrating there seasonally in great numbers. In 2016, a small flock of these shorebirds arrived in late May and remained there for about a month. They were in breeding plumage, with a curved stripe or patch—actually chestnut or reddish-orange, rather than a true red—extending from behind the eye and spreading down the side of the neck, where it broadened as it approached the back.

These birds were likely headed north to breeding sites across arctic Alaska and northern Canada. During the winter, Red-necked Phalaropes are pelagic—living in open waters—and they swim and feed offshore along the southeastern Pacific Ocean from southern

A flock of adult Red-necked Phalaropes flying, exhibiting their breeding plumage, including reddish necks, white throats, and conspicuous, white wing stripes.

Red-necked Phalaropes gliding through the lake, *clockwise from top left*: three birds exhibiting their breeding plumage; a post-breeding bird beginning to molt; and a dark-plumed, black-capped juvenile making its way.

Mexico to South America, as well as elsewhere in warm oceans worldwide. They gather in the water in areas called *fronts,* where ocean currents bring zooplankton to the surface.

In most years, Red-necked Phalaropes arrive at the lake in early August after breeding, and they remain there through September, remaining after most Wilson's Phalaropes have left. By the time Red-necked Phalaropes have migrated to the lake, most of them have partially molted into their nonbreeding plumage, and their only remaining colorful courting feathers consist of just a few reddish-brown plumes on their heads and wings. By September, they have finally finished molting, and their colors are similar for both sexes: gray mixed with white on their wings and tail, a dark cap, and a dark, curved patch extending from in front of their eyes to the back of their head. Juveniles are noticeably darker than adults.

You may find it challenging to differentiate between Wilson's and Red-necked Phalaropes at a distance, especially in late summer, when they look so similar. However, the former is somewhat larger, has a light gray back and a white belly, and has a mostly white face; and the latter is smaller, has darker streaking on its back, and sports a dark cap and mask.

Red-necked Phalaropes live in the open ocean for most of the year, feeding on zooplankton, and when they visit the lake, they also spend most of their time offshore foraging on the water's surface. Unlike Wilson's Phalaropes, Red-necked Phalaropes only infrequently wade in the shallows or come ashore. Their feet have larger webs than those of Wilson's Phalaropes, an adaptation to a more aquatic life.

Some researchers determined that Red-necked Phalaropes at Mono Lake were unable to survive on a diet of *Artemia* (Rubega and Inouye 1994). This might not be the case at Lake Abert, however, because when the birds forage in open water at the lake, brine shrimp are the most abundant prey, especially when salinities are higher. Also, the birds make frequent pecks while they forage, which suggests that they are feeding on *Artemia.* They undoubtedly also feed on all life stages of the alkali fly, when they can, because the flies are a higher-energy source.

The peak count of Red-necked Phalaropes at the lake was nearly forty-three thousand in August 1995 (Sullivan 1996), which is likely a record for the state. The global population of Red-necked Phalaropes is estimated at approximately four million birds, with about half living in North America. Although their numbers are declining in some regions, these shorebirds still maintain a large population worldwide, which means that they are currently secure. The *State of North America's Birds 2016* gave the species a conservation-concern score of twelve, which is moderate. Nevertheless, researchers do warn that climate change and rising sea levels in the Arctic will likely eliminate some of this shorebird's breeding habitat (Galbraith et al. 2014), so we need to continue monitoring the bird's status.

Wilson's Phalarope

The most numerous waterbirds at the lake are Wilson's Phalaropes, and they are one of my favorites. They first appear at the lake in small numbers in April, May, or June. Then, they leave to nest in regional wetlands and don't return to the lake until July, so their numbers peak in late July and early August. By mid-September, most have left the lake and have migrated to South America (Jehl 1988, 1999).

Females fly to the lake after they breed, followed by the males, and the juveniles arrive last. This sequence results from the males staying with their nestlings until they fledge, which allows the females to migrate to feeding areas after they breed and nest, so that they can replenish their greatly depleted fat reserves in preparation for their southward migration in the fall. They nest in small groups within shallow wetlands, primarily in the interior West, the prairie states, and the Canadian prairie provinces (Colwell and Jehl 2020).

No one knows where the Wilson's Phalaropes nest before journeying to Lake Abert. However, based on the large population at the lake, they likely come from a wide area, perhaps even from wetlands hundreds of miles away. Most ornithologists believe that the Oregon breeding sites for this species lie in wetlands in the South Central and southeastern parts of the state, including in Klamath, Lake, Harney, and Malheur Counties (Marshall et al. 2003). I have personally seen Wilson's Phalaropes breeding in many of these areas, but their numbers are far too small to account for the hundreds of

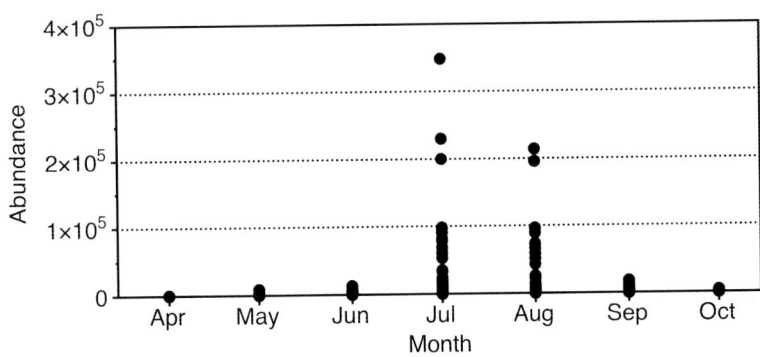

Monthly abundance data for Wilson's and Red-necked Phalaropes, April through October, at Lake Abert, based on approximately 195 census dates (1992–2021). The abundances on the vertical axis are in scientific notation, so 3×10^5 equals 300,000. Unpublished data from BLM, ECAS, and *eBird*.

A nonbreeding Wilson's Phalarope foraging on Lake Abert's muddy shore for alkali-fly larvae and pupae.

thousands that come to the lake each summer. Thus, it appears that the birds are coming from many places and from much farther away.

When female Wilson's Phalaropes first arrive at the lake in late June, they still have their colorful breeding plumage, which consists of a broad, black stripe along the side of the neck that extends through the eye to the base of the bill; a gray cap; a white throat; light, reddish-brown feathers on the lower front of the neck; and a rich, chestnut pattern extending along the side of the neck onto the back and the adjacent part of the wing. During their time at the lake, they lose their breeding plumage, which is replaced by pale gray feathers on their backs and white plumes on their bellies, similar coloration to that of the juveniles.

At Lake Abert and at other post-breeding, feeding, and staging sites, Wilson's Phalaropes undergo hyperphagia, like Eared Grebes, feeding persistently, both day and night. This process enables them to increase their weight so much that they can become temporarily flightless (Jehl 1997). They need this extra fat to provide energy for their three-thousand-mile to five-thousand-mile, nonstop flights to wintering sites in South America (Jehl 1997). They produce the water required to stay hydrated during these long flights as a

A flock of post-breeding Wilson's Phalaropes flying along the Lake Abert shore. This species flies more than ten thousand miles each year between Lake Abert and the phalaropes' wintering areas in South America.

Wilson's Phalaropes trying to catch adult alkali flies: *top*—a bird singling out an isolated fly, located just in front of its bill, and about to grab it; *bottom*—another bird trying to grab a fly from a swarm, a nearly impossible feat.

byproduct of metabolizing their fat. Their wings, which are relatively long and pointed, are a helpful adaptation that lowers air resistance—while still creating sufficient lift to keep the birds aloft—during their long migration.

Of all the waterbirds at the lake, Wilson's Phalaropes may be the most interesting to watch because they constantly show new behaviors. This is especially true when they forage for adult alkali flies, because they keep switching their fly-catching tactics. Consistently catching these flies requires skill because the insects move in large swarms and have keen vision: as soon as one fly detects a predator and manages to escape, the others, alerted to the threat, leave too. Stealth is mandatory for the birds to succeed.

One method they use, as they swim on the surface or wade in shallow water, is to approach the flies as covertly as possible by stretching their heads far forward and lowering their bills to nearly touch the water's surface. Another behavior that I have seen involves a small group of the phalaropes (I once saw a cluster of five) walking together as a pack along the shore. This teamwork may help them better detect and capture their prey.

Wilson's Phalaropes also find prey in shallow water by *puddling* or *foot-paddling*. This method involves quickly walking in a tight circle within shallow water, a technique that apparently stirs up prey that might otherwise be invisible. Foot-paddling behavior is

"Wolf pack" of Wilson's Phalaropes hunting along the shoreline of Lake Abert.

perhaps most associated with gulls, but when I watch gulls doing this, they stand in one place and do not move in circles.

In addition, the phalaropes *scythe-feed* or *sweep-feed* by swinging their bills back and forth in shallow water, in a manner similar to that of avocets. These latter techniques may be useful when they forage for fly larvae that live in the watery surface layer of the mud and which are difficult to see. The shorebirds also use other feeding methods, such as following behind avocets in shallow water, perhaps relying on the other birds to dislodge prey from the sediment.

Wilson's Phalaropes are sometimes quarrelsome, flying up and toward each other and engaging in aggressive fights. What motivates this behavior is unknown. I also once watched a Red-necked Phalarope trying to grab prey that was stirred up in a vortex created by a Wilson's Phalarope, at which point the Wilson's phalarope threatened its red-necked counterpart by lowering its head, extending its bill forward, and chasing after it.

Aggressive behavior of Wilson's Phalaropes: *left*—interspecific aggression involving two birds flying toward one another and vocalizing loudly; *right*—bird demonstrating its threat posture (lowering its head, pointing its bill forward, and rapidly swimming toward another phalarope).

When they are not feeding, Wilson's Phalaropes often gather in flocks to rest and preen. Usually, these flocks are relatively small, totaling up to 100 birds at most. However, sometimes much larger groups of birds, as many as 100,000, gather in what could be described as a super flock. During such times, the flock will extend along the lakeshore for a quarter mile or more. In late July 2013, an East Cascades Audubon Society (ECAS) volunteer named Chris Hinkle was counting birds at the lake and photographed a Wilson's Phalarope super flock containing an estimated 330,000 birds (*eBird* data for July 24, 2013). Other Great Basin areas where large flocks of Wilson's Phalaropes gather, when conditions allow, include: Mono Lake and Honey Lake in northeastern California; Boca, Harney, Malheur, Stinking, and Summer Lakes in Oregon; and Washoe Lake and the Lahontan Valley Wetlands, including Stillwater National Wildlife Refuge, in Nevada.

Once Wilson's Phalaropes have molted and have fattened up enough, rebuilding their energy reserves, they migrate south. They spend the austral summer (the warm season in the Southern Hemisphere) feeding at shallow, saline lakes in the eastern Andes of northern Argentina, Chile, Bolivia, and southern Peru, as well as in lowlands extending from Peru diagonally south to Uruguay and Patagonia (Colwell and Jehl 2020; Lesterhuis and Clay 2010). One of these sites, the four-thousand-square-mile Mar Chiquita saline wetland in central Argentina, has concentrations of Wilson's Phalaropes estimated to reach up to a half million birds (Lesterhuis and Clay 2010). Interestingly, in salt lakes like Lake Poopó in the high Andean Plateau of Bolivia, Wilson's Phalaropes sometimes capture prey that Chilean flamingos (*Phoenicopterus chilensis*) first stir up (Hurlbert et al. 1984). This makes me wonder whether, during the Pleistocene, Wilson's Phalaropes also greedily followed after the now-extinct Cope's Flamingo (*Phoenicopterus copei*), which then lived in saline lakes located in the region we now call South Central Oregon (Grayson 2016).

Lake Abert is listed as one of thirty globally important Wilson's Phalarope sites in the Western Hemisphere (Lesterhuis and Clay 2010). The only other such site in Oregon is Harney Lake, but the birds' numbers there—the highest count was eighteen thousand—are just a small fraction of those typical for Lake Abert, which can have populations of several hundred thousand. The largest concentrations of Wilson's Phalaropes in North America are at the Great Salt Lake, the Lahontan Valley wetlands, Mono Lake, and Lake Abert (Neel and Henry 1996; Jehl 1988, 1999; Lesterhuis and Clay 2010).

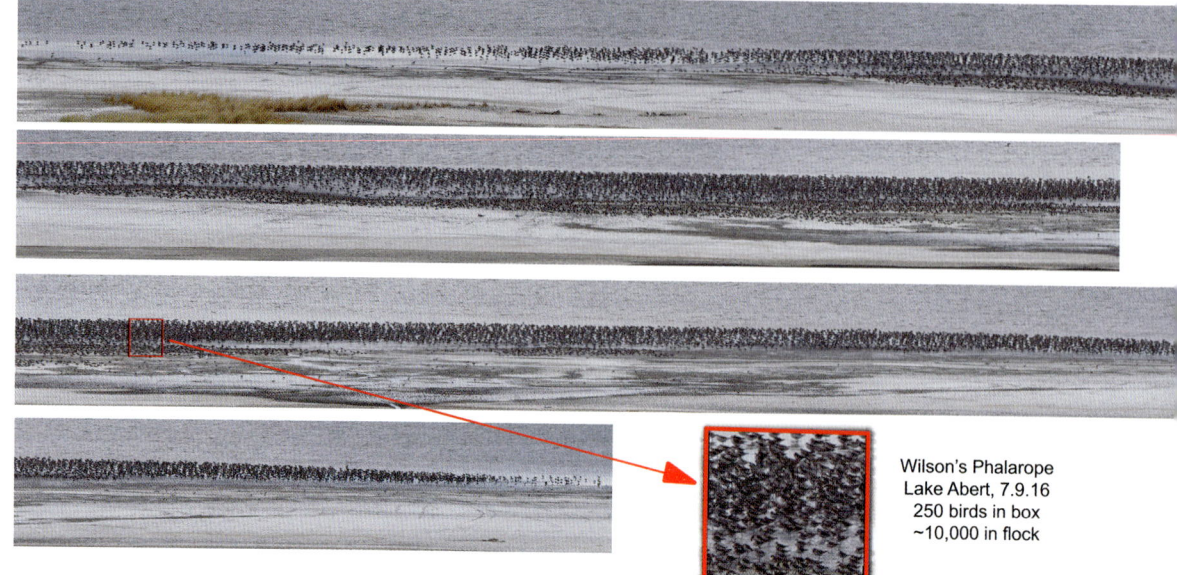

Wilson's Phalarope
Lake Abert, 7.9.16
250 birds in box
~10,000 in flock

Wilson's Phalarope super flock, which I photographed at Lake Abert in July 2016. The flock comprised about 10,000 birds that spread out over about one quarter mile of the shore. To figure out this number, I combined ten photographs to create a panorama, and then I cut the panorama into the four pieces shown here. I selected a small area of the flock in one of the cutouts (shown in the red box) and enlarged it so that I could count this subset of birds: about 250. From this portion, I extrapolated the total number. Bird counters can find it challenging to get an accurate tally from a large, dense flock, so they find a technique like this to be helpful.

A Wilson's Phalarope super flock photographed at Lake Abert in June 2012, with tens of thousands of birds. Similar abundances have not been seen in recent years.

The current global status of Wilson's Phalaropes seems secure, based on an estimated world population of 1.5 million (Lester-huis and Clay 2010). Nevertheless, the fact that so few Great Basin saline lakes serve as major feeding sites for the birds after they breed is worrisome, especially in view of the threats posed by climate change. Of major concern for this species is loss of habitat at the Great Salt Lake. In 2016, the Great Salt Lake reached its lowest recorded elevation, and scientists warned that continued water diversions for cites, agriculture, and other development, coupled with climate change, were putting the globally important Great Salt Lake ecosystem at serious risk (Wurtsbaugh et al. 2017).

The *State of North America's Birds 2016* gave the species a conservation-concern score of twelve, which is moderate. Ecologists rate the Wilson's Phalarope as highly vulnerable to effects of climate change, due to the likely loss of breeding, wintering, and migration habitats (Galbraith et al. 2014). Additionally, the 2015 National Audubon Society's climate-change report assessed this species as climate endangered, estimating that Wilson's Phalaropes will lose 98 percent of their summer range by 2080 (http://climate.audubon.org/birds/wilpha/wilsons-phalarope).

Gulls—The Most Adaptable Waterbirds

> Although we had an excellent chance to study gull life from our blind, yet we found little pleasure in it at the time. The sun was pelting hot and there was not the faintest movement in the sultry atmosphere. We had to breathe the foulest kind of air on account of the dead birds and decaying fish scattered about, and we were standing in muck that was continually miring deeper. Swarms of mosquitos harassed us constantly, while the perspiration kept dripping from our bodies…we were compelled to quit for the day. But for all we suffered there was a fascination in watching these wild birds going and coming fearlessly almost within arm's reach.
>
> —WILLIAM FINLEY, 1907

Gulls are common and conspicuous members of the waterbird fauna at Lake Abert, with six different species spending time there. The most numerous of these are California Gulls (*Larus californicus*) and Ring-billed Gulls (*Larus delawarensis*). Franklin's Gulls (*Leucophaeus pipixcan*) also make regular use of the lake, although in small numbers. Three additional species include Bonaparte's (*Chroicocephalus philadelphia*), Sabine's (*Xema sabini*), and Herring Gulls (*Larus argentatus*). You can find California Gulls and Ring-billed Gulls at the lake in any month of the year, but they are

A California Gull (*middle*) and two Ring-billed Gulls loafing on a boulder in Lake Abert.

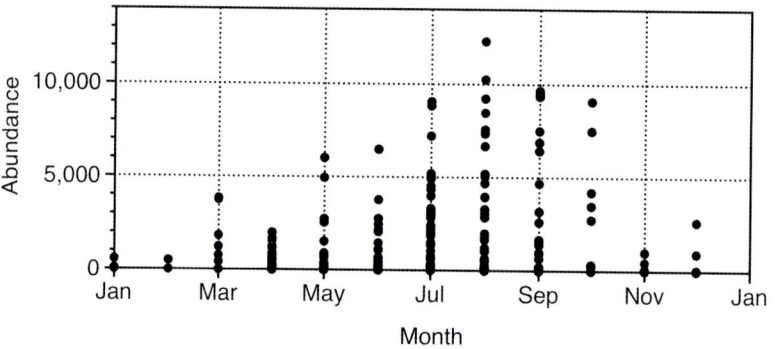

Monthly abundance data for California and Ring-billed Gulls at Lake Abert, January through October, based on observations made on over 240 census dates (1992–2021). Unpublished data from BLM, ECAS, and *eBird*.

A Bonaparte's Gull soaring through the sky and showing off its first winter plumage. This gull is the smallest of the common gulls, all of which are agile fliers. Gulls' long and narrow wings minimize drag and are adapted to both soaring and long-distance flight.

Lake Abert's varied gulls, *from top to bottom, left to right*: an adult Ring-billed Gull exhibiting a characteristic broad, black band near its bill tip; an adult California Gull displaying a narrow band close to its bill tip and a red spot on its lower bill; an adult Franklin Gull looking dapper, with its dark head, partial white eye ring, and red bill; and an adult Bonaparte Gull exhibiting its dark head, partial white eye ring, and a black bill.

most plentiful in the summer months, especially in August, when their combined total often reaches over ten thousand birds.

Ring-billed Gulls at the lake have been seen more often than any other waterbirds. On many different days, observers have counted over five thousand of these gulls, especially in the summer, and they tallied a high of ten thousand birds in September 2009. Volunteers have also regularly seen California Gulls, although less commonly. They have counted over five thousand of the birds on several dates and once tallied eight thousand in July 1995.

Many of the gulls using the lake are likely ones that first nest at Summer Lake. In 2009, wildlife managers built two artificial islands at the Summer Lake Wildlife Area, constructing them specifically for Caspian Terns (*Hydroprogne caspia*), but gulls have used one of them extensively (Bird Research Northwest, no date).

Most gulls are highly opportunistic and forage for food in diverse environments, including farms and cities. Their reputation, however, sometimes greatly exceeds that of a scavenger: Utah honors the California Gull as its state bird, crediting the gull with saving the Mormon pioneers in 1848, when their crops were threatened by a plague of Rocky Mountain locusts (*Melanoplus spretus*), an extinct

Bonaparte's Gull puddling, shaking its left foot to agitate the water and sediment so that the adept bird can expose burrowing invertebrates.

species of grasshopper. According to legend, the gulls descended on the Mormon's fields and ate the pests, saving the crops.

If you visit Lake Abert, you will often see gulls scattered offshore, where they peck at the water's surface, likely feeding on brine shrimp and alkali fly pupae and adults. Gulls also often forage along the shore and may well eat all life stages of alkali flies. I have watched Bonaparte's Gulls puddle as they forage for food in shallow water at the Mile Post 74 Springs. When gulls puddle, they shake one foot to agitate the sediment, which exposes burrowing invertebrates like aquatic worms and amphipods that live in the spring.

The National Audubon Society lists the California Gull as climate endangered and estimates that it will lose 91 percent of its summer range and 45 percent of its winter range by 2080 (http://climate.audubon.org/birds/calgul/california-gull). The same assessment also lists the Ring-billed Gull as climate endangered and estimates that this gull will lose 43 percent of its summer range and 32 percent of its winter range by 2080 (http://climate.audubon.org/birds/ribgul/ring-billed-gull). The *State of North America's Birds 2016* gave the California, Franklin's, and Ring-billed Gulls conservation-concern scores of eleven, eleven, and six, respectively, which are low to moderate. Because these gulls (especially Ring-billed gulls) are numerous and adaptable, their status is unlikely to become a concern anytime soon. The large number of gulls willing to use the artificial island at the Summer Lake Wildlife Area suggests that we could build similar ones to increase the gulls' breeding success in the future, if needed.

Waterbirds as a Gauge of Ecosystem Health at Lake Abert

Birds are often our best indicators of ecosystem health: they are truly our "canaries in the coal mine." We know that many birds are highly sensitive to environmental changes, and we can also often easily identify and count them, even those of us who are amateur birders, so they are a useful marker of the environment's robustness.

Fortunately, over one hundred people have participated in bird counts at Lake Abert. Most of these people have been members of ECAS, but many other volunteers have also contributed data to the *eBird* website at https://ebird.org. Gathering data on the seasonal and annual populations of birds is critical, not only for their management but also to help us understand how inflows, water levels, salinity, and climate change affect the productivity of the lake.

Graduate student Kathryn Boula made the first seasonal waterbird counts at the lake in the summers of 1982 and 1983. A decade later, BLM biologist Walter Devaurs made near-weekly counts from 1992 to 1994 (Bureau of Land Management 1995). Since 2011, volunteers from ECAS in Bend and other volunteers have conducted census counts of waterbirds twice a month during the peak-use summer period. Recent data are posted on the *eBird* website.

Boula's seasonal data were the first to show that large numbers of waterbirds use the lake, that a few species are especially numerous, and that the populations for some species, such as gulls and Wilson's Phalaropes, vary considerably, while populations for others, such as avocets, and shovelers, remain fairly constant.

Lake Abert Waterbird Populations from July to Mid-October, 1982 and 1983.

WATERBIRD SPECIES	PEAK ABUNDANCE 1982	PEAK ABUNDANCE 1983
American Avocet	5,000, early August	5,000, early July
Eared Grebe	4,000, early August	8,000, early August
Gulls	200, late August	7,000, early and late July
Northern Shoveler	8,000, mid-October	6,000, early September
Red-necked Phalarope	2,000, early August	5,000, mid-August
Wilson's Phalarope	50,000, early August	7,000, early August

Source: Data from Boula (1986).

The more comprehensive 1992–1994, BLM census data provided a valuable baseline for waterbird populations and remains the only systematic, year-round, bird data for the lake. During the three years that BLM biologists counted the birds regularly and frequently, they identified a total of eighty waterbird species, including some rarer ones like the Red Phalarope, the Red-throated Loon (*Gavia stellata*), the Sabine's Gull, the Stilt Sandpiper (*Calidris himantopus*), and the Wandering Tattler (*Tringa incanus*). Observers have occasionally seen these birds east of the Cascades, but only rarely (Marshall et al. 2003).

The BLM biologists most often saw Ring-billed Gulls, American Avocets, Mallards (*Anas platyrhynchos*), Northern Shovelers, Eared Grebes, Green-winged Teals, Killdeer, and Canada Geese, and they

Northern Pintail drake swimming at Lake Abert. This species was one of the ten most-abundant waterbirds at the lake **1988–1998**, according to BLM data.

saw all these birds on at least one-quarter of the survey dates. Interestingly, about half the species that they counted during the three-year period represented less than 1 percent of the total number of birds, which indicates that many waterbird species were rare. These data may also suggest that Great Basin waterbirds often need to search widely for suitable habitats, which makes sense, because wetlands can vary considerably in size and quality over time.

The BLM data showed considerable year-to-year variability in bird numbers, with the maximum number of waterbirds ranging from twenty-eight thousand in 1992 to eighty-one thousand in 1993. The daily population size changed seasonally and was highest between April and October and lowest in the winter months. Also, the seasonal populations differed among the waterbird groups, a fact best seen in the 1992 data, which showed that shorebirds were most numerous during the summer and peaked in late July.

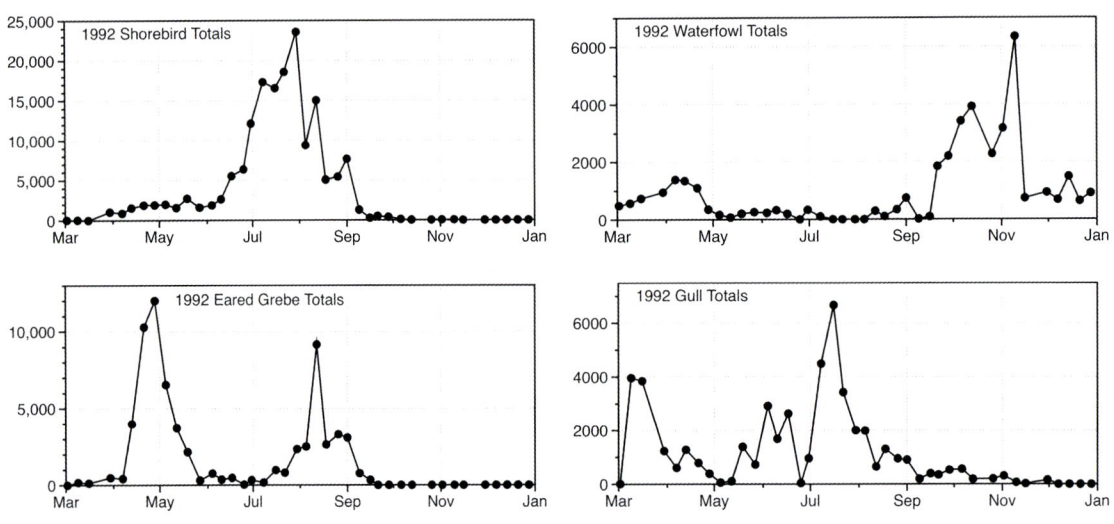

Seasonal abundance data for major waterbird groups counted by BLM in 1992 at Lake Abert. Note that the abundance scale, shown on the vertical axis, is different for each graph.

Seasonal and Peak Abundance Data for Common Waterbirds at Lake Abert

WATERBIRD SPECIES	SEASONAL OCCURRENCE	SEASON OF PEAK ABUNDANCE	PEAK NUMBER AND DATE
American Avocet	March–December	Summer	35,000 on 9/8/1993
Eared Grebe	Year-round	Spring	30,000 on 4/28/1994
Wilson's Phalarope	Spring and Summer	Summer	20,000 on 7/30/1992 and 7/12/1998
California Gull and Ring-billed Gull	Year-round	Summer	18,400 on 8/10/1993
Northern Shoveler	Year-round	Fall Through Spring	17,000 on 4/1/1997
Western Sandpiper	Spring and Summer	Summer	12,000 on 4/28/1994
Red-necked Phalarope	Spring and Summer	Summer	11,400 on 9/1/1994
Least Sandpiper	Spring and Summer	Summer	6,000 on 4/27/1998
Green-winged Teal	Year-round	Fall	4,000 on 11/2/1992
Northern Pintail	Year-round	Winter	2,900 on 3/1/1993
Canada Goose	Year-round	Winter	1,500 on 2/16/1993
Ruddy Duck	Year-round	Fall Through Spring	1,500 on 4/8/1994
Mallard	Year-round	Winter	745 on 3/9/1994
Black-necked Stilt	Spring and Summer	Summer	600 on 8/8/1994

Source: Based on 1988–1998 BLM census data and information reported by Kristensen et al. (1991).

That year, waterfowl numbers peaked in the fall, and Eared Grebes and gulls had two seasonal peaks, one in the spring and one in the summer.

Two factors probably explain these differences: (1) some of the birds using the lake left because they needed to nest and to molt in freshwater wetlands; and (2) their populations were affected by changes in food availability.

During this 1992–1994 period, the BLM data showed that eight species had peak populations of greater than ten thousand birds on a given date, and that American Avocets and Eared Grebes were the two most abundant birds, having peak counts of thirty-five thousand and thirty thousand.

Waterbird Populations from 2011 to 2022

Starting in 2011, volunteers—mostly members of ECAS—conducted annual waterbird surveys at the lake from the spring through the fall. Their goals were to gather bird population data for wildlife managers and to help assess the health of the ecosystem. This invaluable information on the movements and population changes of Lake Abert birds cost the state and federal governments nothing. We refer to this type of data collecting, handled by experienced, amateur birders rather than by professional ornithologists, as citizen science, volunteer science, or community science.

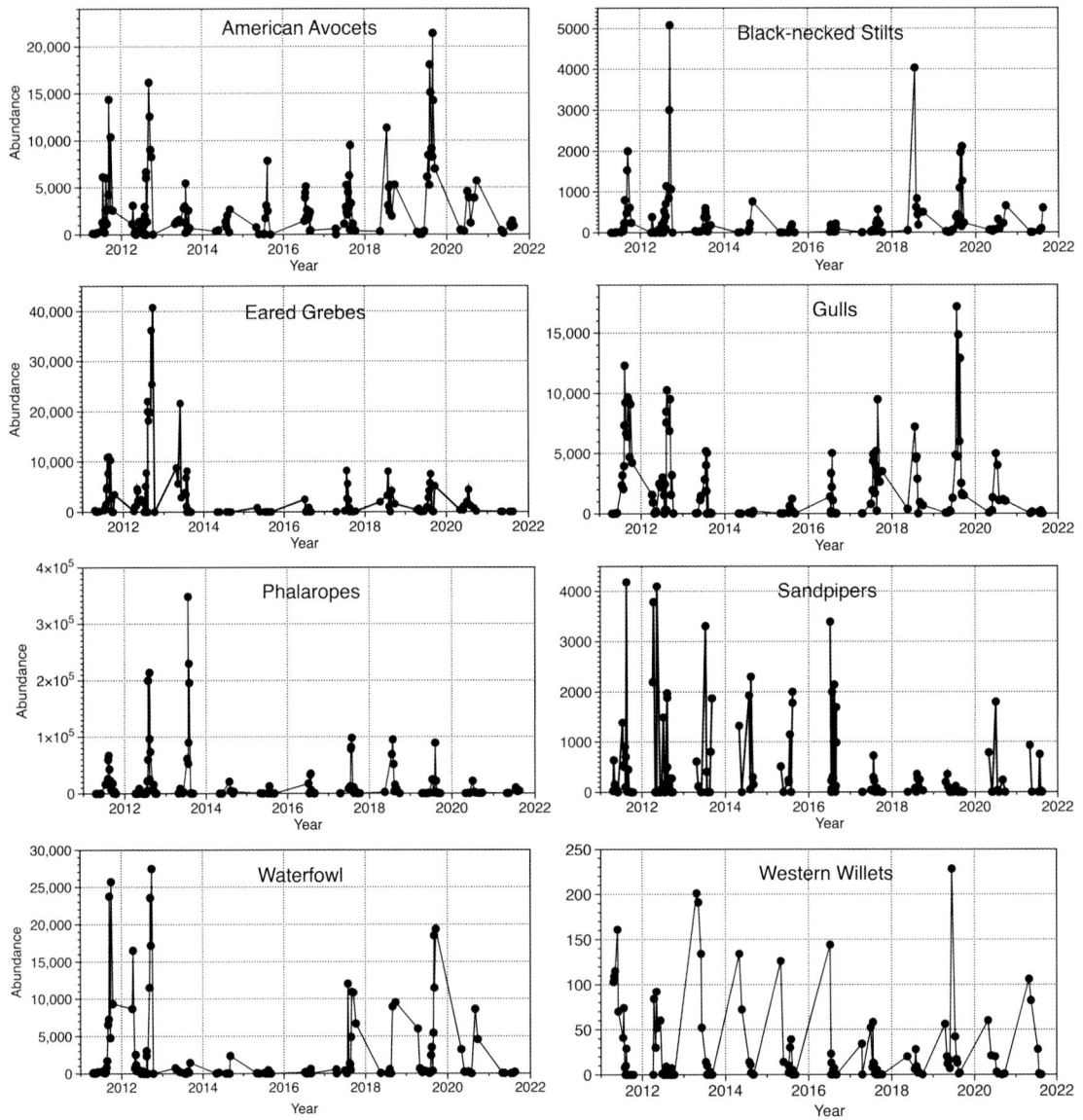

Abundance data for selected waterbirds at Lake Abert, 2011–2021, counted by the ECAS and volunteers who reported sightings on the *eBird* website. Note that the vertical axis scale varies in magnitude for each graph. Because phalaropes were so abundant, the vertical axis scale is in scientific notation, so 2×10^5 equals 200,000, 3×10^5 equals 300,000, and so on.

Although not as intensive as the BLM waterbird census effort, the ECAS surveys were perhaps even more valuable because they provided a relatively long data set during a period in which the lake was experiencing dramatic changes in water levels, salinities, invertebrate prey populations, and bird abundances.

The volunteers' data clearly showed that the numbers of birds varied considerably over the study period. For most of the birds, the counts were: (1) relatively high in the early part of the study period (2011–2012); (2) lower in the middle part (2014–2016); (3) somewhat higher again for a time (2018–2019); and (4) very low at the end of the study period (2021–2022). This suggests that most of the birds, except for the sandpipers and the Willets, were impacted by the low water levels and high salinities that existed from 2014 to 2016 and again in 2021–2022. Eared Grebes were especially impacted and were not seen at the lake during the two periods of very low water levels. These data also showed that the populations of some of the birds, for instance, avocets, gulls, stilts, and waterfowl, decreased or increased, depending on the lake's level, while other populations of birds, like Eared Grebes and phalaropes, never recovered in abundance once they decreased.

We can't attribute these changes solely to the lake's high salinities, although the increasing salt levels very likely impacted the waterbirds, by reducing their food supply. The causes of population changes in migratory birds are likely complex and involve multiple factors related to their breeding, migration, and wintering habitats. The information gathered at Lake Abert, however, does seem to show that some birds respond strongly to shifting habitat conditions—including food availability—at the lake (Larson et al. 2016; Senner et al. 2018). For example, it seems likely that brine shrimp almost vanished in 2014, that they increased from 2015 through 2018, and that they were likely scarce again during 2020-2022, although we don't have scientific verification. My observations indicate that the alkali fly population also was very low in 2014 but increased in 2018 and 2019, based on the presence of conspicuous fly mats along the shore, and that it was low again in both 2021 and 2022. This variability shows how difficult it can be for birds seeking a consistently plentiful food supply at the lake.

Food availability impacts all waterbirds at the lake, but regional habitat conditions are also important factors. Great Basin waterbirds depend on regional habitats and fly between them as conditions change (Haig et al. 1998; Plissner et al. 1999 and 2000; Haig et al. 2002), particularly as water conditions shift. I saw this in action one day when I was hiking on Lynches Rim in Lake County, Oregon, and noticed a procession of American White Pelicans (*Pelicanus erythrorhynchos*) using thermals—masses of rising, warm air—to gain enough elevation to fly west from the Warner Lakes and over the 6,500-foot-high escarpment. At the time, I thought that the pelicans' movements were unusual, but I later learned that the birds were likely flying west from their nests at Pelican Lake in the Warner Basin to get to feeding sites elsewhere, perhaps including

Avocets resting on one leg at sunset.

Goose Lake and the Klamath Basin lakes, thirty to one hundred miles away.

We can therefore infer that some of the many shorebirds, including phalaropes, that visited Lake Abert in 2013 may have come from other lakes in the region where habitat conditions were perhaps less favorable. We don't know where the birds that have departed the lake have gone. However, we do know that one species—the Wilson's Phalarope—needs a high biomass of invertebrates that only live in salt lakes, which means that they may have flown hundreds of miles to reach the Great Salt Lake or Mono Lake, or even much farther away to reach the Salton Sea in Southern California. It is also unclear what the survival rate is of birds that are forced to leave one feeding area and search for another, but doing so is likely risky.

The dynamic nature of Great Basin wetland habitats affects waterbird movements and likely determines what we are seeing at Lake Abert. Although Lake Abert's waterbird data probably surpasses that of any other lake in the northwestern Great Basin, it doesn't explain everything. To fully understand how the lake's bird populations are affected by environmental changes, we need to look across the entire region, because for waterbirds, "no lake is an island."

Unfortunately, up until recently there was no systematic and coordinated monitoring of shorebirds in the Great Basin. However, Oikonos, a nonprofit that monitors a variety of waterbirds from Alaska to Chile, has taken on the job of coordinating surveys, sharing data, and writing reports—with an emphasis on the Wilson's Phalarope—through the International Phalarope Working Group, led by Ryan Carle of Oikonos. This effort includes Lake Abert, the Great Salt Lake, Mono Lake, and other North American lakes, as well as South American lakes where Wilson's Phalaropes overwinter. Eventually Oikonos plans to track the Wilson's Phalarope remotely across its entire seasonal range (https://oikonos.org /species/wilsons-phalarope).

Comparing Waterbird Populations at
Lake Abert with Those at Other Salt Lakes

To really understand the regional and global significance of Lake Abert as a waterbird habitat, we must compare it with other saline lakes around the world. We can perhaps best relate it, in terms of waterbird diversity and populations, to the many shallow salt lakes in Australia, where researchers have taken great interest in water-bird ecology (e.g., Halse et al. 1998; Timms 2009). Like the Great Basin, much of Australia is arid, and the country has numerous basins containing shallow salt lakes that attract waterbirds. Just like the shallow lakes in the Great Basin, Australia's salt lakes have vari-able levels and different amounts of flooding, due to annual fluctu-ations in precipitation.

For example, Lake Gregory, located in northwestern Australia, varies in size from 150 to 500 square miles, depending on yearly pre-cipitation, and it has a salinity ranging from 1 percent to 8 percent (Halse et al. 1998). It has some of the highest waterbird diversity in the world and one of the greatest populations of these birds for an interior lake on the continent. Observers have reported seeing nearly 70 waterbird species there and have counted waterbird popu-lations exceeding 100,000 in three of six years, with a high count of 650,000 birds.

Researchers have speculated that the high numbers of birds at the lake were due to a continental-scale drought in 1998 that forced the birds to seek a new habitat and food (Halse et al. 1998). We can

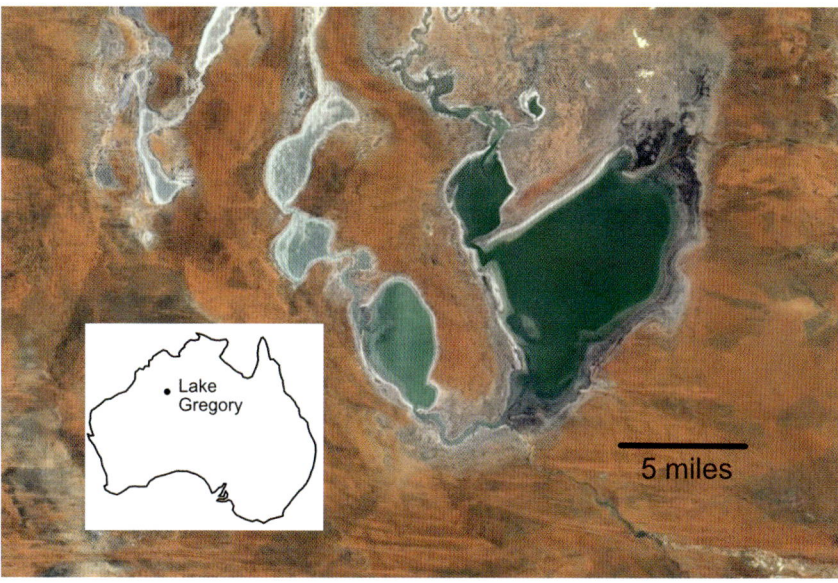

Northwestern Australia's Lake Gregory, a salt lake located in the Kimberly District and notable for its high diversity and abundance of waterbirds. Google Earth image.

compare both the species diversity and the high population of birds at Lake Gregory with those at Lake Abert during peak periods; however, Lake Gregory has had a higher maximum population, and Lake Abert has had higher populations per unit area because of its smaller size.

Another important waterbird habitat in Australia is the Paroo River, a catchment that is part of the Murray-Darling Basin and which actually consists of a series of waterholes, lakes, and wetlands that are located in both Queensland and New South Wales. The Paroo River includes 200 small salt lakes that support large numbers of migratory waterbirds when water conditions are suitable (Timms 2009). These lakes have a total surface area of 60 square miles and a salinity that ranges from 0.3 percent to greater than 5 percent. The Paroo lakes have supported nearly 60 species of waterbirds. The maximum population of waterbirds seen on a one-day count was 280,000 and a maximum number of birds observed per square mile was approximately 4,700.

The level of salinity in these lakes drives the types of birds that migrate to them and impacts what food is available for the birds that come (Timms 2009). When the lakes are hypersaline, they mainly support crustaceans, such as brine shrimp, and the saltier water attracts avocets, stilts, and a variety of other shorebirds, as Lake Abert does. Australian research indicates that waterbirds move around regionally—even continentally, in the case of Lake Gregory's birds—to find suitable habitats and plentiful food (Halse et al. 1998; Timms 2009), which is the same pattern that seems to be occurring at Lake Abert.

Back in the United States, we can also make interesting comparisons between Lake Abert and two other Great Basin lakes—Mono Lake and the Great Salt Lake—because they also draw large numbers of waterbirds. The Great Salt Lake is particularly remarkable: it has had the highest concentrations of Wilson's Phalaropes in the world; is one of the two largest staging areas for Eared Grebes in North America; and attracts the world's largest breeding populations of California Gulls and of White-faced Ibis (*Plegadis chihi*; Paul and Manning 2002).

Mono Lake, approximately 75 square miles in surface area, also attracts a wide variety of waterbirds, including Eared Grebes, which number in the hundreds of thousands seasonally, and numerous shorebirds, especially Wilson's Phalaropes (Boyd and Jehl 1998; Strauss et al. 2002). Researchers tracking shorebird populations at Mono Lake in the early 1990s found that the total number varied from 20,000 to 34,000 between 1989 and 1993 (Strauss et al. 2002), or approximately 250 and 450 shorebirds per square mile.

A White-faced Ibis displaying its impressive profile in flight. Ibis feed by probing the sediment with their long, sensitive bills. Normally, very few ibis migrate to Lake Abert, but for unknown reasons, an unusually large number of them spent the summer of 2019 at the lake, climbing to a peak of more than two hundred that August.

Ornithologists do not know how these numbers compare with longer-term maximums. However, Lake Abert's maximum shorebird population of 230,000 birds in 2013, or approximately 7,000 shorebirds per square mile, far exceeds Mono Lake's 1990s values, and numbers of birds at Abert in 2013 were even greater. It is clear that, factoring in the size differences of all three lakes, Lake Abert has comparably high waterbird densities.

Southern California's Salton Sea, at 350 square miles in surface area, is also a major saline lake habitat for waterbirds (Shuford et al. 2002; Patton et al. 2003). Estimates of maximum total waterbird populations at the Salton Sea range from 430,000 to 580,000, or approximately 6,000 to 8,000 birds per square mile, which are, again, comparable to Lake Abert's 2013 maximum density. One additional habitat that is important for Great Basin waterbirds is western Nevada's Lahontan Valley (Neel and Henry 1996).

Importance of Lake Abert for Waterbird Conservation

Lake Abert's waterbird population clearly compares favorably with populations at various lakes around the world on a per unit-area basis, even with that of the Great Salt Lake, which might support the most waterbirds anywhere. So, how much emphasis should we place on Lake Abert as a conservation priority?

We know that the conservation status of many shorebirds is either uncertain or declining. In fact, the *State of the Birds 2022* indicates that shorebirds declined by 33 percent between 1970 and 2019 (https://www.stateofthebirds.org/2022/). Because of this reality, several organizations, including the Ramsar Convention for Wetlands of International Importance, the Important Bird Areas Program of the National Audubon Society, and the Western Hemisphere Shorebird Reserve Network consider a wetland to be of global importance if it supports at least 1 percent of a species' North American population (Andres et al. 2012). Lake Abert supports over 1 percent of the North American Snowy Plover population, and the

Shorebird Species Whose Maximum Numbers at Lake Abert Represent at Least 1% of the North American Population

SPECIES	ESTIMATED NORTH AMERICAN POPULATION SIZE	MAXIMUM POPULATION AT LAKE ABERT	DATE OF COUNT	PERCENT OF NORTH AMERICAN POPULATION[4]
American Avocet	450,000	35,000[*1]	September 1993	8
Snowy Plover	25,800	345[*2]	June 1987	1.3
Black-necked Stilt	175,000	5,000[*3]	September 2012	2.9
Red-necked Phalarope	2,500,000	52,000[*1]	August 1995	2.1
Wilson's Phalarope	1,500,000	300,000[*3]	July 2013	20

Data sources: [*1]=BLM, [*2]=ODFW, [*3]=C. Hinkle, *eBird*; [*4]=Reported by Andres et al. (2012).

lake is also designated as a site of international conservation importance for American Avocets, Black-necked Stilts, Red-necked Phalaropes, and Wilson's Phalaropes.

Based on these populations, Lake Abert should be included in the Western Hemisphere Shorebird Reserve Network. Additionally, the lake provides a seasonal habitat for Killdeer, Pectoral Sandpipers (*Calidris melanotos*), and Western Sandpipers, all of which show consistent range-wide declines (Andres et al. 2012). Within Oregon, Lake Abert is ranked either first or second in terms of peak populations for the following eleven waterbirds: American Avocets, Black-necked Stilts, California Gulls, Eared Grebes, Least Sandpipers, Northern Shovelers, Red-necked Phalaropes, Ring-billed Gulls, Western Sandpipers, Western Willets, and Wilson's Phalaropes. More broadly, the lake is an integral part of a mosaic of regional Great Basin wetlands that includes most of South Central

Pectoral Sandpipers are Arctic breeders that occasionally stop at Lake Abert on their southward migration. This species shows consistent, range-wide declines and thus merits special conservation monitoring.

Oregon and adjacent lakes and wetlands in California and Nevada. Clearly, the lake merits high priority for concerted and ongoing conservation efforts.

Birding at Lake Abert

Lake Abert is one of the most accessible of the Great Basin's waterbird habitats because Highway 395 passes along the entire eastern shore of the lake. Nevertheless, I am amazed at how few people stop to appreciate the lake and the abundance and diversity of birds that spend time there. However, it has certainly become a must-see stop for birders in the know, and some of them come long distances to experience the lake and its vast numbers of shorebirds.

During the peak midsummer season, almost innumerable waterbirds busily forage for food along parts of the shore. The springs and mudflat at the north end of the lake are often the best places to see these birds. Gulls and shorebirds regularly bathe and drink in the shallow water coming from the Mile Post 74 Springs at the northeast side of the lake. You can easily do some bird watching from the highway, using either binoculars or a spotting scope, but be sure to pull off on a wide turnout to avoid the large trucks that barrel fast down the highway. Also, don't forget to listen for Soras, the small and rarely-seen rails, that repeatedly call "koEE, koEE, koEE" from the marshes along the lake's eastern shore in May and June.

You can access beaches along the lake's eastern shore from trails off nearly every highway pullout, so you have many ideal spots for watching the birds. Be cautious about approaching the birds too closely, however, because you can easily disturb them. I have found that the best strategy for observing the birds at close range is to

Sora, a type of rail, foraging for insects and seeds among bulrush stems.

sit on the shore and wait for the birds to come to you. I often take a lawn chair, and an umbrella if it's sunny and hot. I sit near the water, quietly waiting for the birds to come along, and my patience usually works out well. Plus, honestly, what could be better than spending some time along a beautiful lake in the shadow of the impressive Abert Rim?

In the uplands in May and June, you will invariably see Logger-head Shrikes, Sage Thrashers, and Brewer's Sparrows, which perch on greasewood and sagebrush around the lake's perimeter. Some lucky birders have even reported seeing and hearing the uncommon Black-throated Sparrow along the slope of Abert Rim. You can also observe neotropical migrants—those coming from the tropical region south, east, and west of Mexico's central plateau—including various warblers, vireos, and Western Tanagers (*Piranga ludoviciana*) along the riparian habitats along Poison Creek. Look for Rock Wrens and Chukars (*Alectoris chukar*) among the boulders at the base of the rim. You might also get a glimpse of Canyon Wrens, which seek out areas of large talus and which visit the upper part of the rim and its numerous overhangs and crevices.

Don't forget to check out the habitats at the north end of the lake along Lake County Road #309 (also called the XL Ranch Road), because nearly 120 bird species have been seen around the XL Ranch house and springs. Some rarely seen birds spotted there have included: the Eastern Kingbird (*Tyrannus tyrannus*), the Palm Warbler (*Setophaga palmarum*), the Northern Water-thrush (*Parkesia noveboracensis*), the Yellow-breasted Chat (*Icteria virens*), and the Black-throated Sparrow. If you visit the ranch house location, please respect the private property there. Burrowing Owls (*Athene cunicularia*) have nested in the gravel pit near the highway and County Road #309. You can download a checklist of birds that observers have reported seeing around the lake from the *eBird* website (https://ebird.org/home).

If you are planning a trip to the lake, you should also consider stopping at the Summer Lake Wildlife Area, an excellent location for both birding and camping. There is lodging at Summer Lake and additional camping and lodging near Paisley at Summer Lake Hot Springs. You may also want to stop near Paisley at the Chewaucan River, which provides additional opportunities to see riparian birds, such as warblers and Western Tanagers, and which offers trout-fishing opportunities and its own camping sites. You can also camp near Lake Abert at Valley Falls.

Consider traveling farther east and stopping at the Warner Wetlands and nearby Hart Mountain National Wildlife Refuge, where you will find camping sites and a hot spring. If you have more time, continue even farther east to Malheur National Wildlife Refuge and the nearly 10,000-foot-high Steens Mountain nearby; both provide excellent seasonal birding opportunities. One final option is to go southwest and visit the Klamath Basin National Wildlife Refuges and Wood River Wetland, where birding is exceptional. You can easily reach all these areas from Lakeview, Klamath Falls, or Bend.

Finally, some useful references to consult prior to or during your visit include: *Birds of Malheur National Wildlife Refuge, Oregon* by Carroll Littlefield; *The Birders Guide to Oregon* by Joseph Evanich Jr.; *A Birders Guide to the Klamath Basin* by Steven Summer; and *Oregon Desert Guide* by Andy Kerr, as well as such websites as Oregon Birding Trails (http://www.oregonbirdingtrails.org).

CHAPTER 7

The Lakeshore People

The Ancient Chewaucanians

In the old days they used to dig food all summer—until it was gone. They gathered seeds and roots and buried them in the ground. In winter they stayed until the buried food was gone and then moved to the next place. They hunted every day, all year. In those days there were many sage hens, ducks, geese, swans, jackrabbits, cottontails, deer, and antelope.

—Northern Paiute Indian informant,
interviewed by Isabel Kelly, 1932

The Paiute's world was and is bleak, open, inhumanly spacious at first encounter. And their stories evoke this great emptiness unforgettably—a blank space, with here a sagebrush, there a rock (both capable of talking), and over there on the horizon, a butte. Time is a spring that has gone dry. Then someone appears, traveling through, coyote perhaps, and he happens to meet another wanderer...these Paiute stories seem to be premised on the fact of sheer desert space...it is a viable Indian way of imagining the local reality, a way of making "Home" into good stories, ones to live by.

—JAROLD RAMSEY, 1977

Three thousand years ago, South Central Oregon was a much different place. Shallow lakes and marshes were much more numerous than they are today, and Lake Abert's water was probably fresh enough to support fish. Perhaps even more surprising, many people called the lake and its marshes home.

We know that Native Americans lived in the region then because they left abundant evidence of their villages on both sides of the lake, upstream in the marshes, and along the Chewaucan River. The first archaeologist to extensively study these people, University of Oregon researcher Richard Pettigrew, referred to them as the Ancient Chewaucanians (Pettigrew 1980, 1985). But who were they?

Rare lizard pictograph displayed on a large boulder at the base of Abert Rim, created using iron oxide as a pigment. Although we don't know who made these images, or when they were created, we believe that the likely artists were Northern Paiute Native Americans.

What did they call themselves? How many of them were there? How long did they stay, why did they leave, and what happened to them?

We may understand only bits and pieces of their lives, but we can rightly imagine that they had their own joys and fears and that they likely experienced real deprivations as they struggled to feed and provide shelter for themselves and their children in a changing climate and an isolated landscape. Their Great Basin home was probably often harsh and unforgiving. They certainly experienced bitterly cold winters, and they may well have exhausted their food stores in difficult years. Deep snow would have made gathering food and firewood difficult and even dangerous. Nonetheless, the northern Great Basin was also seasonally bountiful, offering them an abundance of fish, game, and edible plants. These hardy people also tempered the rigors of the environment with finely honed survival skills passed down through many generations.

The northern Great Basin became a principal focus for archaeologists when Luther Cressman began his pioneering work in the 1930s and 1940s and found numerous artifacts in South Central Oregon's rock shelters and caves. Today, archaeologists continue the work, unearthing new objects and locating new sites that have greatly increased our knowledge of early humans in the Americas.

The Chewaucan Basin has gradually revealed a long and rich history of Native American occupation, and we increasingly understand that the basin's varied and productive habitats, especially its wetlands, supported people for thousands of years, perhaps beginning as early as 14,500 years ago.

Archaeological surveys near the lake and around the Chewaucan Marshes have uncovered diverse artifacts, including six hundred pit-house depressions; over seventy rock-ring structures that may have been houses; nearly one hundred examples of rock art; a large number of wood, bone, and stone tools, such as obsidian projectile points; and even rarely preserved, woven basketry and leather moccasins, all indicating a well-developed human occupation (Cannon 1977; Pettigrew 1980, 1985; Oetting 1989, 1990; Aikens et al. 2011; Connolly et al. 2016). In fact, in 1995, the BLM stated, "There is virtually no portion of the immediate shoreline of the lake where some form of cultural resource cannot be found.... [The shoreline]...is literally one continuous [archaeological] site." In recognition of the lake's cultural abundance, a three-hundred-acre area of federal and state land was designated as the East Lake Abert Archaeological District, and in 1978, it was listed on the National Register of Historic Places.

The BLM's assessment of the lake's shoreline may well apply to nearly all the Chewaucan Basin. Paisley Caves, which became a key focus for Cressman, have proven to be a particularly rich cultural resource within the basin. The caves lie along what was once the northwestern shore of Paleolake Chewaucan. There, wave action eroded the rock, creating the series of shallow caves during a high-water period in the Pleistocene, more than fourteen thousand years ago.

In the 1930s, Cressman decided to explore this remarkable part of the basin's landscape. He headed the University of Oregon's new Anthropology Department and would go on to become a preeminent archaeologist because of his work in the region. He unearthed numerous, remarkably well-preserved artifacts, including many pairs of ten-thousand-year-old sagebrush sandals buried in several caves in the Fort Rock Basin, located in northern Lake County, Oregon.

In 1937, Cressman journeyed to the caves with Ernst Antevs, a geologist from the Carnegie Institution for Science in Washington, DC. Antevs was particularly interested in how lake sediments could provide evidence of past climate conditions. Cressman hoped that the geologist could help him clarify the relationship between lake deposits and the precontact sites that the archaeologist had happened upon on earlier trips. A third man, a retired Forest Service employee named Walter Perry, urged the two scientists to explore the caves because he believed that the geological formations were archaeologically significant. What Cressman and Antevs saw, once they got there, was beyond anything they could have hoped for. Looking back on the experience in 1988, Cressman wrote:

Walt now took us to something that would be absolutely new in our experience if our examination corroborated his inspection. His cave was at Five-Mile Point outside of Paisley on the high terrace on the eastside of Summer Lake. Uncovering the small test hole he had dug, perhaps fifteen inches across, he cleaned it out with his hands and then cleaned the sides of all loose material. Kneeling, he looked up at us and said, "There it is." It was clear there was a level of occupation fill, then clean undisturbed pumice resting on another bed of cave fill with bits of pieces of artifacts clearly in situ in the wall. Ernst examined the test pit with his usual care and said, "If this stratigraphy, for that is what you have here, holds when you excavate, you will have one of the most important sites for Early Man in the whole West."

In fact, Antevs's declaration was a gross understatement, because archaeologists now herald the Paisley Caves as one of the most significant sites in the entire Western Hemisphere. They have yielded the oldest human DNA ever found in the Americas, extracted from *coprolite* (fossilized excrement). Scientists used high-resolution radiocarbon dating to determine that the remains go back 14,500 years (Gilbert et al. 2008; Jenkins et al. 2012).

Another remarkable set of artifacts from the caves are stone projectile points, or spearheads, that are characteristic of the Western Stemmed Tradition. Early hunters used these points by hafting their characteristic narrow-stemmed bases onto wooden shafts, which they threw at their prey. These stemmed points are contemporary with, or even older than, those of the Clovis Tradition, which have been found at archaeological sites in the US Southwest, the Midwest, and the East (although rarely in Oregon). Archaeologists once believed Clovis points to be the oldest stone projectiles in the Americas and have dated them to a period between 12,800 and 13,300 years ago. That belief gave rise to the Clovis First hypothesis, which argued that the Clovis people who used these tools were the first humans to populate the Americas, that they traveled over the Bering Land Bridge that connected northeastern Asia with North America, and that they then journeyed south through an ice-free corridor in Alaska's interior that was present near the end of the Pleistocene. These people and their spear points are named for Clovis, New Mexico, the town where the points were first discovered.

The Clovis points, which range in size from about two to four inches in length, are distinctive: they have a pair of rear-facing flutes and a concave area between them for hafting the spearhead onto a wooden handle. Hunters probably threw the spear by using

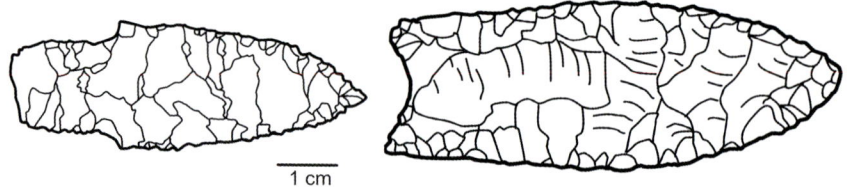

Drawings of stone projectile points: *left*—western stemmed point from Parman, Nevada; *right*—Clovis point from Naco, Arizona. Both redrawn after Grayson (2016).

an *atlatl,* a rod that aided them in lobbing the weapon farther and with greater penetration. However, the discovery of 14,500-year-old human DNA and western stemmed points in the Paisley Caves means that the Clovis First hypothesis may not hold up. Archaeologists continue to find additional relics at other sites and to prove that these items are older than Clovis artifacts, using more accurate and thorough dating techniques (Adovasio and Pedler 2016).

Living with Giants

The archaeologists who uncovered the Clovis points found them near fossilized bones of many now extinct *megafauna* (particularly large animals). They theorized that "Ice Age man" used the stone-tipped spears to obtain the food that he needed, killing mammoths, mastodons, and other large mammals. Although this is partially true, the late-Pleistocene indigenous people were more likely to be generalized hunter-gatherers who hunted and foraged for a broad range of animals and plants, rather than subsisting on megafauna (Aikens et al. 2011; Adovasio and Pedler 2016). Their tool kits included not just projectile points but also stone knives, scrapers, and hand stones for pounding and grinding, and they very likely used plant fibers to make nets and snares to secure fish, birds, and small game, although these less-durable objects have mostly disappeared. They used still other tools to make clothing and footwear, waterproof shelters, and baskets to transport and store their food (Adovasio and Pedler 2016). Bow and arrow technology that incorporated smaller arrowheads was a much later development, appearing about two thousand years ago (Aikens et al. 2011).

The fact that humans lived in the Paisley Caves 14,500 years ago means that they likely shared the area with a variety of large mammals, many (although not all) of which are now extinct in North America. Donald Grayson's 2016 book—*Giant Sloths and Saber-tooth Cats: Extinct Mammals and the Archaeology of the Ice Age Great Basin*—provides a fascinating account of the huge creatures that lived in North America in the Pleistocene. Ten such mammals roamed around the northern Great Basin with humans around

Large Mammals in Oregon's Great Basin, 14,000 to 11,700 Years Ago

COMMON NAME	SCIENTIFIC NAME	APPROXIMATE WEIGHT IN POUNDS
American lion	*Panthera leo atrox*	500
American mastodon	*Mammut americanum*	10,000
Ancient bison	*Bison antiquus*	3,500
Columbian mammoth	*Mammuthus columbi*	17,000
Dire wolf	*Canis dirus*	150
Giant bear	*Arctodus simus*	1,500
Helmeted muskox	*Bootherium bombifrons*	1,300
Horse	*Equus* spp.	600
Sabertooth cat	*Smilodon fatalis*	600
Yesterday's camel	*Camelops hesternus*	2,400

Source: Data from Grayson (2016).

12,000 years ago, including a sabertooth cat with an appropriate name—*Smilodon fatalis*. It had formidable, 6-inch long, scimitar-shaped canine teeth.

Additional megafauna that lived in this region at the time included grizzly bears (*Ursus arctos*) and gray wolves (*Canis lupus*). Humans, however, wiped all of them out in this part of the Great Basin during the last several hundred years (Bailey 1936; Verts and Carraway 1998), and the American bison (*Bison bison*) was also killed off in the area (Verts and Carraway 1998). Prior to their extinction in the region, bison likely coexisted with the Paleo-Native Americans. In fact, forty miles northwest of the lake, you can find an ancient petroglyph of one of these great animals (Loring and Loring 1996).

Scientists examining the Paisley Cave coprolites have proved that the cave dwellers ate a broad range of animals, including bison, camels, deer, horses, bighorn sheep, pronghorn, sage grouse, and even such carnivores as coyotes, dogs, foxes, and wolves (Aikens

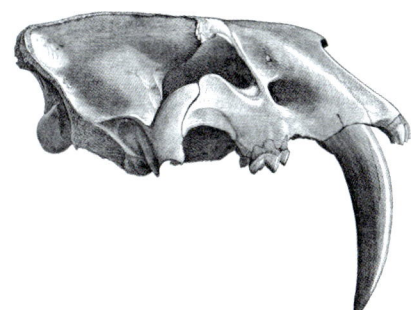

Drawing of the skull of a sabertooth cat (*Smilodon fatalis*) from *On the Extinct Cats of America* by Edward Drinker Cope (1880). Sabertooth cats weighed six hundred pounds and were dangerous predators: they had six-inch-long, curved, canine teeth and jaws capable of opening much wider than the jaws of any large cats today.

et al. 2011). They have also recovered either mammoth or mastodon DNA from a stone tool collected at the site. Protein residues from the coprolites indicate that these people ate a wide variety of plants too.

Archaeologists explain that a rich diversity of foods and the use of stone tools clearly show *ecological adaptation*: these early people knew how and where to harvest locally available foods and how to gather useful tool-making materials, such as obsidian. Early Native Americans did not rapidly move through an area, solely hunting for big game before moving on to another location. Instead, they spent enough time in the region to gather the knowledge necessary for their subsistence. However, the evidence coming from the Paisley Caves indicates that these people used only temporary shelters. More permanent settlements in the Chewaucan Basin would come much later.

Evidence of Native American Settlements at Lake Abert

In 1937, Cressman published the first archaeological information about the lake in a report titled *Petroglyphs of Oregon,* which included his sketches of lizard, rattlesnake, and geometric rock carvings. He searched for petroglyphs around the state, and he sent letters to rural postmasters, asking if they knew where the carvings were located. Cressman teamed up with Howard Stafford, a University of Oregon graduate student in geology, and they traveled thousands of miles around Oregon in a Model-A Ford on the hunt for petroglyphs.

Much later, in the mid-1970s, Pettigrew began surveying sites along Highway 395, seeking evidence of ancient cultures on the

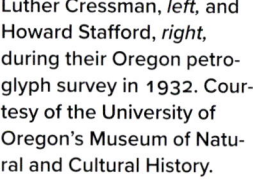

Luther Cressman, *left,* and Howard Stafford, *right,* during their Oregon petroglyph survey in 1932. Courtesy of the University of Oregon's Museum of Natural and Cultural History.

eastern shore of Lake Abert, and in the 1980s he produced several publications and a detailed report on what he had found (Pettigrew 1980, 1985). This is how Pettigrew introduced his 1985 report on the archaeology of Lake Abert:

> Excitement is an emotion that archaeologists hope often to achieve in the course of their work, but perhaps only occasionally experience. It was my luck to feel this emotion on the east shore of Lake Abert in 1976 when I found there astonishingly clear traces of a substantial prehistoric society. Having been accustomed to recognizing archaeological sites on the basis of subtle clues such as lithic debitage and soil coloration, I was suddenly met by a staggering array of circular depressions, art motifs, concentrations of all kinds of lithic tools, and even stone-walled ruins of apparent houses. As I clambered over the rocks, finding yet another such site around each bend, my normal restraint slipped away and I found myself shouting and gasping with incredulity. To this day a measure of that feeling returns to me whenever I think of that time, whenever I return to look again at the sites on the lakeshore, and whenever I summarize information about the place for a report to others.

Clearly, Pettigrew really believed that he and his team had stumbled upon a place of great archaeological significance. Their search led them to rock art comprising 350 design elements, including many geometric shapes and drawings of lizards, snakes, bighorn sheep, deer, unidentified animals, and humans. What struck me when I saw some of these sites were the numerous lizard petroglyphs, and, in fact, Pettigrew found more lizard petroglyphs—more than

Lizard petroglyphs, each about six inches long, from the lower slope of Abert Rim. The petroglyph on the left appears to be a desert horned lizard, while the other two are probably western fence lizards. The petroglyphs were scratched into the brownish desert varnish that formed on the surface of the rock.

forty—than any other designs. I was not surprised by that because western fence lizards continue to sun and display, doing push-ups, sometimes even on the very same rocks that contain these petroglyphs.

One detail particularly impressed me as I surveyed the sites: the presence of several lizard petroglyphs on rocks barely a foot above the ground. Knowing that the carvers had often created their designs near or even inside their villages, it occurred to me that kids who were fascinated by the lizards may have made these petro-glyphs. I even imagined a child gazing back and forth between their model, a lizard sunning itself on a nearby rock, and their "canvas," the boulder on which they carved.

In 2011, Melvin Aikens, a University of Oregon archaeologist, asserted that the early Native Americans' rock art was more every-day than mysterious to its creators. He said that the designs are "contrary to common speculation that such images reflect secre-tive shamanistic practices or conjuring of hunting magic at remote camps. Though certainly having spiritual meaning, rock art occurs here generally in a living room or community context." When I saw the numerous petroglyphs, rock rings, and stone kitchen tools and looked out over the lake, I felt a sense of connection with the lake-shore people, who must have seen a similar scene of blue water stretching far into the distance and the fawn-colored hills of Coglan Buttes visible on the western shore, of course without our modern highway and traffic noise.

As Pettigrew searched for early Native American artifacts on the eastern side of the lake, he located dozens of possible village sites with likely dwellings, evidenced by pit-house depressions and rock

Stone kitchen tools made from boulders located on the lower slope of Abert Rim at Archaeological Site #534: *left*—shallow depression called a *metate,* where seeds were ground using a rounded, handheld, stone tool called a *mano* (the gray, circular area was worn smooth from grinding); *right*—rock mortar (six inches across) filled with water, where seeds, bones, and other hard, tough, or fibrous foods were ground or pounded using a rock or wooden pestle. Both boulders are approximately three feet across.

rings; forty cultural locales associated with petroglyphs; numerous food-processing artifacts, such as manos, metates, mortars, and pestles; stone hunting blinds and fences; and many examples of smaller objects, such as obsidian points and shell ornaments.

These varied cultural artifacts, including those used for food preparation and shelter, clearly show that people settled at Lake Abert, at least seasonally. Pettigrew believed that these objects reflect a unique Chewaucan culture. Other archaeologists, however, went on to locate many similar sites in basins throughout South Central Oregon, including those at Catlow, Fort Rock, Goose Lake, Guano, Malheur, and Warner (Aikens et al. 2011). All of these settlement sites combined suggest that the region supported a large population prior to Euro-American immigration. When conditions were favorable, early people in the region may well have outnumbered today's human inhabitants.

Scientists have radiocarbon dated the Lake Abert archaeological sites, estimating that the remains come from 3,500 to 500 years

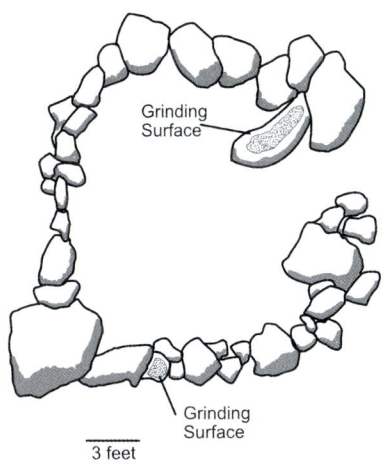

Grinding
Surface

Grinding
Surface

3 feet

Evidence of a Native American settlement, which I have nicknamed "Cove Village," along the Lake Abert shoreline, *clockwise from top left*: drawing of a rock-ring "house" based on an original sketch by Pettigrew (1985); a photograph of the same rock ring; and a metate or grinding stone located at the Cove Village site and formed from a basaltic boulder. Villagers used the smaller, handheld mano (found elsewhere) by moving it back and forth across the surface of the metate while pushing down on it to grind seeds into a flour.

ago, so they believe that the so-called Chewaucan culture lasted for approximately 3,000 years. Pettigrew thought that the villagers settled near the shoreline to have easy access to the water, which was key to their survival. During that period, the lake was probably much higher than it had been earlier in its history, and the water was likely fresh enough to support fish, mussels, and other edible aquatic species. Also, extensive marshes probably surrounded the lake, giving the indigenous people easy access to waterbirds and their eggs.

Incidentally, the engineers who designed Highway 395 routed it through many of the same beaches where the Chewaucanians built their villages because these flat areas made road building easier. Unfortunately, because no one conducted archaeological surveys prior to constructing the highway in the 1930s, the builders likely destroyed or damaged some artifacts (Loring and Loring 1996; Beauchamp 2013).

Indigenous people clearly found it advantageous to live close to the shore. The proximity gave them easy access to water for transportation, drinking, food preparation, and washing. Pettigrew discovered that the people had built their lakeshore villages at varying elevations, and he reasoned that they had done so because the lake level gradually decreased. Assuming that the sites he located were near the shore 3,500 years ago, we can infer that Lake Abert was about 40 feet higher than it is today (Pettigrew 1985). However, by about 2,000 years ago, the lake had evidently fallen to an elevation of about 30 feet above contemporary levels, based on the different location of village sites from that period. By about 500 years ago, the water had apparently fallen so low that the people had abandoned their villages altogether. The lake likely rose and fell numerous times in response to more temporary climate changes too, just as it does

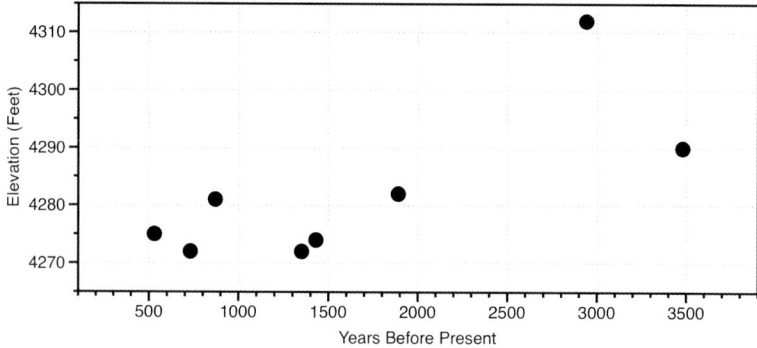

Respective elevations of radiocarbon-dated charcoal and tufa obtained from archaeological sites along the eastern shore of Lake Abert. The recent average lake level is approximately 4,250 feet, so these elevations are 20–50 or more feet above the lake's current elevation. Data from Pettigrew (1985).

Dried tui chubs (approximately eight inches long) on exhibit at the University of Oregon's Museum of Natural and Cultural History in Eugene. Tui chubs were likely an essential source of protein for the Chewaucanians, who likely ate both fresh and dried fish.

today. In fact, at one of the sites, archaeologists found beach gravel covering cultural artifacts, so they knew that the lake had risen and buried the site after the area had been occupied.

Although we don't know for certain why the villagers ultimately abandoned the lakeshore settlements, we can guess that lower lake levels, and the resulting higher salinities, made the water unpotable and also reduced the availability of lake- and marsh-derived food, especially protein sources, which were often scarce in the Great Basin. Pettigrew argued that the Chewaucan culture could only have prospered if the lake contained fish. Many other available foods, including plant seeds and tubers, as well as waterbirds and their eggs, were seasonal. Large game like bighorn sheep, deer, and pronghorn would have become scarce fairly quickly, as a result of hunting (Pettigrew 1985). So, villagers may well have relied on fish, particularly on tui chubs.

These little minnows don't live in the lake now, but observers have spotted them in XL Spring at the lake's north end and in the Chewaucan River. Scientists have found widespread evidence that Great Basin Native Americans did depend on tui chubs as an important food (Raymond and Sobel 1990; Fowler 1992; Butler 1996; Stevenson and Butler 2016; Aikens et al. 2011). Villagers caught the minnows using nets. Because the fish were small, they were easy to dry and store for later use with little preparation (Fowler and Bath 1981; Fowler 1989). In fact, archaeologists have unearthed the remains of many dried tui chubs in Great Basin caves (Raymond and Sobel 1990).

Some scientists believe that fish constituted about half the diet of some Native Americans who lived around Great Basin lakes and

Northern Great Basin plants that the Klamath and Northern Paiute Native Americans ate, turned into fiber-based goods, and used for other purposes, *top to bottom and left to right*: bitterroot (*Lewisia rediviva*); taper-tip onion (*Allium acuminatum*); camas (*Camassia quamash*); Indian hemp (*Apocynum* sp.); blazing star (*Mentzelia laevicaulis*); biscuit root (*Lomatium* sp.); arrowleaf balsamroot (*Balsamorhiza sagitta*); yampah (*Perideridia oregana*); Delphinium (*Delphinium nuttallianum*); Klamath plum (*Prunus subcordata*); yellowbell (*Fritillaria pudica*); and chokecherry (*Prunus virginiana*).

rivers (Fowler 1986). Dried fish would have been a good source of fat and protein for people who otherwise relied a lot on plant-based carbohydrates. Pettigrew found numerous plant-processing tools in his search of the area, so he surmised that plants were another important part of the lakeshore people's diet. Plant foods and fibers available regionally might well have included: balsamroot, basin wild rye, biscuit root, bitterroot, lilies, camas, cattails, chokecherries, elderberries, Indian hemp, Indian rice grass, Klamath plums, service berries, wax currants, wild onions, and others. In fact, these plants still grow around the lake, especially Klamath plums, which flourish in thickets near Poison Creek.

The search for food was undoubtedly a major occupation for Great Basin indigenous people throughout much of the precontact period, especially as the region became drier, and this search became much more difficult when they lost their access to a broader territory. As University of Colorado anthropologist Omer Stewart put it, in writing about the Northern Paiutes, "The poverty of the landscape is reflected in the small size of each band and the sparse population relative to the large area occupied by this tribe." The barrenness of that landscape became increasingly problematic. For example, the Yahooskins, who had lived in an estimated five-thousand-square-mile territory around Silver Lake and the Chewaucan Basin, totaled only about one hundred people in 1867, when they were forced into the confines of the Klamath Reservation (Stewart 1939). For Northern Paiutes, food's important role was reflected in the different names they gave to their bands, often basing these appellations on locally available food: Deer Eaters, Tule Eaters, Grass-Seed Eaters, Marmot Eaters, and others (Stewart 1939; Fowler and Liljeblad 1986).

Pettigrew focused his hunt for archaeological sites on the eastern side of the lake, but other researchers surveyed the western side, including William Cannon, a BLM archaeologist based in Lakeview, who found twenty circular depressions and thirteen rock rings there (Cannon 1977). Another team recently described a number of interesting artifacts that they found in Rattlesnake Cave, a tufa-rock shelter also located on Lake Abert's western side (Connolly et al. 2016). The objects include bone, wood, and stone tools; leather moccasins; and woven items made from tule (also called hard-stem bulrush, *Schoenoplectus actus*) and bundled grasses. These woven items consist of cordage, basketry, matting, and a 1,500-year-old sandal.

Indigenous people living in the Great Basin preferred woven-fiber sandals to moccasins, and for obvious reasons, they made them out of readily accessible materials like tule and sagebrush. The sandals, which served as their footwear of choice for nearly

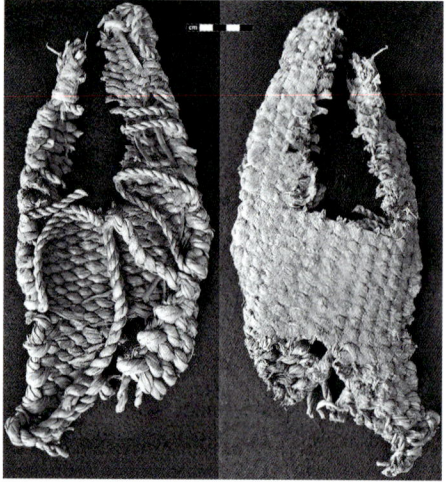

Rattlesnake Cave and artifacts: *left*—the cave is located under a small tufa mound on the west side of Lake Abert, is approximately 4 feet high, 12 feet wide, and 12 feet deep, and has a rock ring at the opening, which was perhaps part of a wood and tule-mat structure that provided additional protection from the weather; *right*—this worn-out, 12-inch-long, 1,500-year-old sandal from Rattlesnake Cave was woven from tule stems using a technique called "multiple warps." The loops along the edges were used to secure the sandal to the foot. Both images courtesy of Tom Connolly, the University of Oregon's Museum of Natural and Cultural History.

10,000 years, were easy to repair too. (Connolly and Baker 2008; Connolly et al. 2011). Cressman found over 100 sagebrush sandals, some even tied in pairs, when he excavated Fort Rock Cave in the Fort Rock Basin, which is located 50 miles northwest of Lake Abert. The sandals were buried under a blanket of 7,600-year-old Mount Mazama ash. In 1962, Cressman described the dusty work he had undertaken, writing:

> We excavated one of these caves near the village of Fort Rock in 1938. Down below, the old lake bed was white with alkali in the blazing August sun. Dust devils swirled across miles of shimmering space in the dancing heat mirage. The accumulated refuse of centuries, rodent droppings, dust, ash, all had a characteristic stench that clung to our nostrils as did the dust to our sweaty bodies. As we dug, we went through a bed of volcanic ash from an ancient eruption and suddenly under this came a sandal. It was made of rope of twisted sage-brush bark, and unlike any we had ever seen.

Northern Great Basin sandal, 9,600 years old, from Fort Rock Cave and made from shredded sagebrush bark. Native Americans in the region made this Fort Rock-style sandal until about 9,000 years ago. It is on exhibit at the University of Oregon's Museum of Natural and Cultural History.

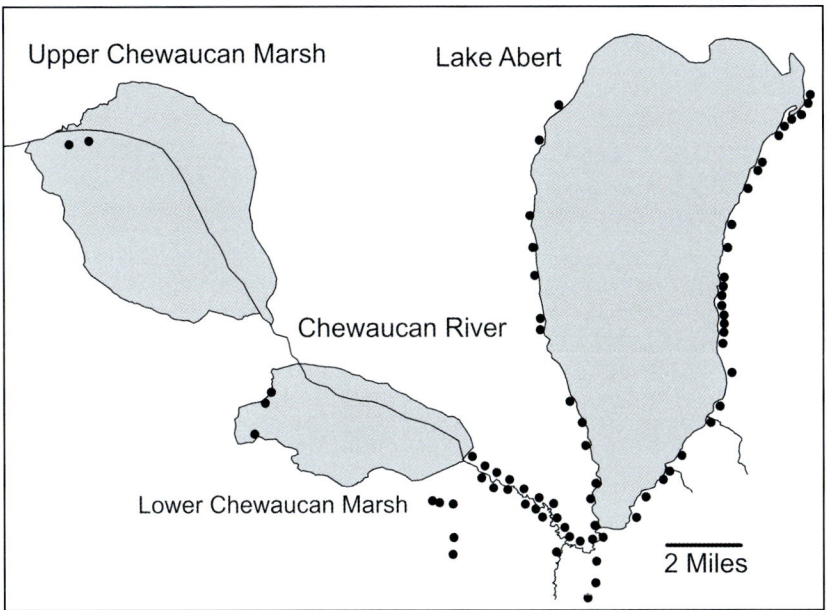

Approximate village locations that researchers have identified around Lake Abert, in the Chewaucan Marshes, and along the lower Chewaucan River. Additional unidentified sites are likely present in the basin, but large portions of the Upper and Lower Chewaucan Marshes remain unsurveyed. Redrawn from Oetting (1989).

Another archaeologist, Chip Oetting of Heritage Research Associates in Eugene, searched for artifacts near the lake and found over 40 village sites and nearly 180 pit-house depressions and rock-ring structures along the lower Chewaucan River, upstream of the lake (Oetting 1990). The artifacts that he encountered on the riverbank, like those that Pettigrew had located along the eastern shore of the lake, were also dated from 500 to 3,500 years ago, which indicates that they, too, were relics from this Chewaucan cultural period.

The abundance of archaeological sites around the lake, the lower river, and possibly in adjacent, unsurveyed areas of the Chewaucan Marshes, add up to a major settlement. Based on the remains of many houses, we can surmise that hundreds of people of all ages lived there, at least seasonally. They likely traveled long distances to hunt, gather plants, obtain obsidian, and to trade with members

A tule decoy resembling a Canvasback Duck, approximately ten inches long, made by Ivan Jackson of the Klamath Tribes. Decoys like this would probably have been covered with a duck skin or feathers to make them more lifelike.

Extensive tule marshes like this one at Upper Klamath Lake have likely flourished in the Chewaucan Basin during the past three thousand years when sufficient water has been available. These marshes once provided a wealth of food and fiber resources to the indigenous people.

of other communities. The river, lake, and nearby marshes would have provided fish in season, as well as a wide variety of birds, eggs, and valuable plants, including tules that could be woven into various items like mats, hut coverings, clothing, sandals, duck decoys, and even small boats (Fowler and Liljeblad 1986).

The Paisley Cave discoveries show that people also lived along the Chewaucan Basin in the late Pleistocene epoch, 14,000 years ago, long before the Chewaucan culture arrived. Who were they? That remains a mystery because archaeologists have found so little evidence of their presence. However, what we have found suggests that they likely belonged to small, relatively mobile bands of hunter-gatherers who had no fixed settlements and who relied on whatever nature provided. Gradually, over thousands of years, people occupied the northern Great Basin in greater numbers and left considerable evidence of their presence in places such as the Fort Rock Basin (Jenkins et al. 2004), but evidence of people during that period in the Chewaucan Basin is scarce. However, starting about 3,500 years ago, the Chewaucanians had made the region their home, and they stayed there over the next 3,000 years. They were also hunters and gatherers, but they settled, at least seasonally, in small villages.

They were technologically advanced, especially in their methods of gathering and using a wide variety of animal and plant resources from the wetlands for food, shelter, and clothing. In addition, they made sophisticated use of plants that they obtained from the uplands. Chewaucanians' wetland adaptations and lifestyle suggest that they were similar in many ways to the Klamath and Modoc

Native Americans who lived near the Klamath Basin wetlands well into the period of contact with Euro-Americans. In fact, these similarities make many archaeologists think that these people were related (e.g., Pettigrew 1985; Oetting 1989; Aikens et al 2011; Connolly et al. 2016).

Up to the contact period with Euro-Americans, the Klamath and Modoc Indians lived in permanent winter villages consisting of pit houses with circular depressions similar to those found in the Chewaucan Basin (Gatschet 1890; Spier 1930; Cressman 1956). Chewaucanian artifacts found near Lake Abert also resemble those left by the early Klamath and Modoc peoples (Pettigrew 1985; Oetting 1989; Oetting 1994b; Connolly et al. 2016). Additionally, archaeological sites in the Warner, Harney, and Surprise Valley Basins, which are similar in age to those near Lake Abert, show such cultural affinities (Cressman 1942; Weide 1968; Eiselt 1997; Aikens et al. 2011). Thus, there is ample evidence suggesting that these three cultures were related, if not actually the same.

Nineteenth- and early-twentieth-century ethnographers extensively studied the Klamath and Modoc Native Americans and showed that they had developed a complex technology based on extensive use of wetland resources (Gatschet 1890; Coville 1897; Spier 1930). It seems probable, then, that the Chewaucanians followed the same developmental path. We also know that the people who adapted to living in the marshes of what would become South Central Oregon, as well as in adjacent parts of what would become California, traded with and likely adopted new technologies from the Northern Paiutes when they moved into the region in the late Holocene (Eiselt 1997).

If the Chewaucanian, Klamath, and Modoc peoples were related, who were their ancestors? We know that the first humans to reach North America came from Asia, but there is considerable uncertainty over how and when this happened (Adovasio and Pedler 2016). For some time, archaeologists supported the Clovis First hypothesis. More recently, however, experts have come to believe that humans came to our hemisphere not only by journeying across the Bering Land Bridge but also by traveling along the coast, on the "Kelp Highway" (Eshleman et al. 2004; Erlandson et al. 2007). If maritime-adapted people migrated from Asia to North America along the coast, they would have already acquired technologies and the ecological know-how to enable them to use resources from a range of aquatic environments, both marine and freshwater. Thus, it seems logical that maritime-adapted people were the common ancestors of Chewaucanian, Klamath, and Modoc peoples. Also, the Klamath and Modoc language is part of a group of

Penutian languages that include those used by coastal Native Americans from along the Columbia River and elsewhere, further evidence of a common origin.

Chewaucanians appear to have left their settlements in the basin approximately four hundred to five hundred years ago; we have found no evidence of their long-time presence there in later years (Pettigrew 1985; Oetting 1989, 1990). We don't know why they left, but we can surmise that increasing climate variability—especially more frequent or severe droughts and resulting declines in wetland habitat, food, and fiber resources—may have driven them away. Scientists have reported that analyses of tree rings and charcoal in the Great Basin indicate that brief, but severe, droughts and related forest fires slammed the environment between one thousand and five hundred years ago (Miller and Wigland 1994; Weppner et al. 2013). Another study has pointed to major droughts drying up what would become South Central Oregon sometime between the late 1420s and the early 1430s, and again in the mid-1630s (Nebert 1985). These conclusions come from reconstructed Goose Lake water levels over the past five hundred years, which show evidence of droughts based on changes in the widths of tree rings.

Even short-term droughts can strongly impact the lake because of its high rate of evaporation, more than three feet per year. Lower water levels and increased salinities caused by drought could have eliminated tui chubs and other fish in the lake, could have killed the salt-intolerant tule marshes, and therefore could have reduced, or perhaps even eliminated, the associated wetland biota that supported the Chewaucanians. Evidence suggesting an increasingly difficult and less predictable environment in the northern Great Basin, requiring the people to travel greater distances to find food and other resources, comes from the study of human remains (Hemphill 1999; please see my note in the Preface regarding the study of these remains).

Although people abandoned the shoreline villages at Lake Abert and along the nearby river about five hundred years ago, some returned, at least seasonally, to gather resources, and we have evidence of this pattern. One example is an archaeological site called the Chewaucan Cave, which is located upstream along the river and near The Narrows, west of the lake. Chewaucan Cave has yielded relatively recent artifacts similar to those made by Klamath Native Americans, which suggests that people who had adapted to living in the marshes were still utilizing the basin's resources into the more recent precontact period. The cave contained an amazing assortment of highly perishable and rarely recovered objects dating to approximately 350 years ago, collectively known as the Chewaucan

Cave Cache because the artifacts were likely stored there for later use, but the owner never returned.

Items from the cache include a stitched grass bag filled with textiles; twined baskets; fiber nets for possible use in catching waterfowl; a leather bag; and even a pouch made from a badger head (Kallenbach 2013). University of Oregon archaeologist Elizabeth Kallenbach, who studied the artifacts, said this about them: "The cache represents a single event in time in that the larger basket contained nets and other hunting tools and supplies, perhaps stored for future use. Many of the items show significant wear; the nets and leather bag have been repaired many times suggesting the bag and nets were highly valued. When I think about the care and preservation that were given to these valuables by their owner, I am reminded of what a rare and wonderful opportunity it is to continue to care for these artifacts for future generations."

In the last five hundred years, coincident with the disappearance of most of the northern Great Basin's marsh-adapted people, a new group of desert-adapted people arrived. They were the Northern Paiutes, who likely came from drier areas farther south and who spoke one of the Numic languages. Thus, ethnographers refer to the arrival of these people as the Numic expansion (Bettinger and Baumhoff 1982). Instead of having permanent, multiple-family, winter settlements like those of the Chewaucanians, the Northern Paiutes lived in small, highly mobile family groups who gathered resources from large areas and accessed many different foods as they became available (Kelly 1932; Wheat 1967; Couture et al. 1986; Eiselt 1997; Aikens and Couture 2007; Soucie 2007; Aikens et al. 2011). Northern Paiutes were less dependent on the wetlands, and so they may have been pre-adapted to thriving in the drier and

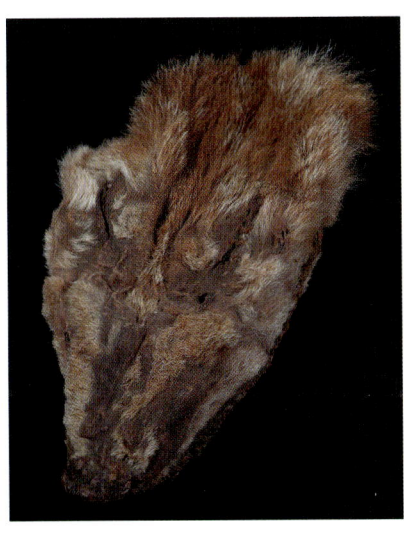

A rare badger-head pouch recovered from the Chewaucan Cave Cache. Courtesy of the University of Oregon's Museum of Natural and Cultural History.

Carved petroglyphs displayed on large, basaltic boulders resting on the edge of Abert Rim near Poison Creek.

more variable climate that was becoming more prevalent in the northern Great Basin.

Captain John Frémont, in 1843, noted that he had seen recent evidence of Native Americans at the lake. He likely observed signs of the Northern Paiutes who occupied the region between Lake Abert and Silver Lake, to the east in the Warner and Harney Basins, to the south in Surprise Valley, and elsewhere in the Great Basin (Kelly 1932). Isabel Kelly, a late-nineteenth- and early-twentieth-century ethnographer who interviewed the Surprise Valley Paiutes, said that her informants had described a Northern Paiute winter camp near Paisley called Soho. Kelly, along with ethnographers Albert Gatschet and Frederick Coville, interviewed surviving members of the Great Basin Native American tribes in an effort to document their cultures before they were lost.

We can make some pretty good guesses about what happened to the Northern Paiutes who lived in the Chewaucan Basin. Gatschet, who studied the ethnography of the Klamath peoples, stated that over one hundred Yahooskin (also called Yahuskin) and Walpapi Paiutes lived on the Klamath Reservation in 1888. He also said that prior to signing a treaty in 1864, the Northern Paiutes had "hunted the shores of Goose Lake, Silver Lake, Warner Lake, Lake Harney, and temporarily stayed in Surprise Valley, on Chewaukan [sic] and Saikan [Sycan] Marshes, and gathered wokash [wocus, a yellow water lily, *Nuphar lutea*] on Klamath Marsh." Faced with growing numbers of Euro-American settlers, and the likelihood of having to fight them, the surviving Yahooskin and Walpapi people had little choice but to move onto the reservation.

Thus, less than three decades after John Frémont and his men had explored the Chewaucan Basin, all the indigenous peoples who had lived there for hundreds, if not thousands, of generations, disappeared from the region. They died in battle against the Euro-Americans, died from the settlers' diseases (against which they had no natural immunity), or moved to reservations to avoid being harmed. According to Warren D'Azevedo, who founded the Anthropology Department at the University of Nevada, Reno, "The impact upon the way of life of the native people, who had at first cautiously welcomed the intruders and later attempted sporadic resistance, was devastating. Starvation and diseases brought by Whites decimated large numbers. Those that survived were forced onto reservations containing lands least desirable to the new settlers, and their access to the wide range of resources that had sustained traditional economy and society was denied them. Others became dependents and laborers on ranches and on the fringes of White communities. So desperate was their condition that White observers in the latter part of the nineteenth century predicted their imminent extinction, and some even welcomed the possibility as a solution to the Indian problem."

Indigenous people lived, at least periodically, in the Great Basin for over fourteen thousand years. Few, if any, areas in the Americas provide comparable evidence of their history or of their rich and complex lives. What we have learned speaks to the productivity of the basin's ecosystems, especially to the periodic abundance of the lakes, rivers, and marshes. Even more importantly, we have come to understand the incredible tenacity of the people who once made their homes in this environment and who passed their amazing ingenuity and resourcefulness down through many generations.

CHAPTER 8

The Future of Lake Abert and Salt Lakes Worldwide

Lake Abert and all salt lakes worldwide are important natural resources and are clearly worth protecting. They serve as critical ecosystems, especially for migratory waterbirds, and they also provide considerable aesthetic benefits, as anyone who has ever visited Lake Abert can appreciate.

However, climate change, water diversions, pollution, and development are placing these lakes—including those in the Great Basin—under extreme pressure. Of particular concern are water management practices in the arid West that continue to focus almost solely on maximizing agricultural profits, with almost no regard for adverse environmental impacts. However, these harms are not inevitable. We can sustain both agriculture and our natural world if we focus on best practices and on working toward sensible compromises that help our farmers while we protect our waters.

The Status of Salt Lakes Worldwide

In many parts of the globe, fresh water is in limited supply, and this constraint threatens humanity and the environment as a whole (Hassan et al. 2005; Gleick and Cooley 2021). Most salt lakes are in arid regions with high evaporation rates, so they are especially sensitive to reduced inflows and, consequently, face perils from a warmer and drier climate (Melack et al. 2001; Williams 2002; Wurtsbaugh et al. 2017). In fact, as William Williams, formerly of the University of Adelaide, wrote two decades ago, "by 2025, the natural character of most of the world's salt lakes will have changed...[and] most permanent salt lakes will have become smaller and more saline, with extensive if not complete exposure of their beds to the atmosphere." If that happens, the lakes' ecosystems will suffer harms that will be particularly detrimental to waterbirds, and human health will be put at risk because of increased alkali dust. Unfortunately, Williams's prediction seems to be on track to becoming a reality because of climate change, but we still can minimize the adverse effects, if we make the effort to do so.

Lake Abert on a calm, early-spring morning, reflecting fresh-fallen snow on the slope of Abert Rim.

Recent Changes to Salt Lakes Globally

LAKE	CHANGES	TIME PERIOD	CAUSES	REFERENCES
Urmia, Iran	70% decrease in area	Last 14 years	Drought and water diversions	Heydari and Jabbari 2012; Stone 2015; NASA 2016a
Walker, United States	90% loss of volume, salinity increasing from 0.4 to 2.2%	Since 1880	Water diversions	Umek et al. 2008; Walker Basin Conservancy (https://www.walkerbasin.org)
Mar Chiquita, Argentina	30-foot increase in elevation	1970–2000	Increased rainfall	Bucher and Curto 2008
Dead Sea, Jordan and Israel	100-foot decrease in elevation, with recent increased rate to over 3 feet/year	1975–2008	Water diversions and mineral extraction	Oren et al. 2009
Aral Sea, Kazakhstan and Uzbekistan	90% decrease in surface area and 10x increase in salinity in South Aral Sea	1960–2021	Water diversions	Micklin 1988; Jellison et al. 2004; Heydari and Jabbari 2012
Corangamite, Australia	10-foot decrease in water levels and salinity increase from 3.5 to 5%	1960–1990	Water diversions	Williams 1995
Great Salt Lake, United States	50% decrease in surface area by 2016	1988–2016	Drought and water diversions	Wurtsbaugh et al. 2017
Ebinur, China	60% decrease in surface area	1955–2010	Drought and water diversions	Heydari and Jabbari 2012; Ma et al. 2014
Poopó, Bolivia	Near total desiccation	1986–1994, 2013–2015, and 2017	Drought and water diversions	Perreault 2019

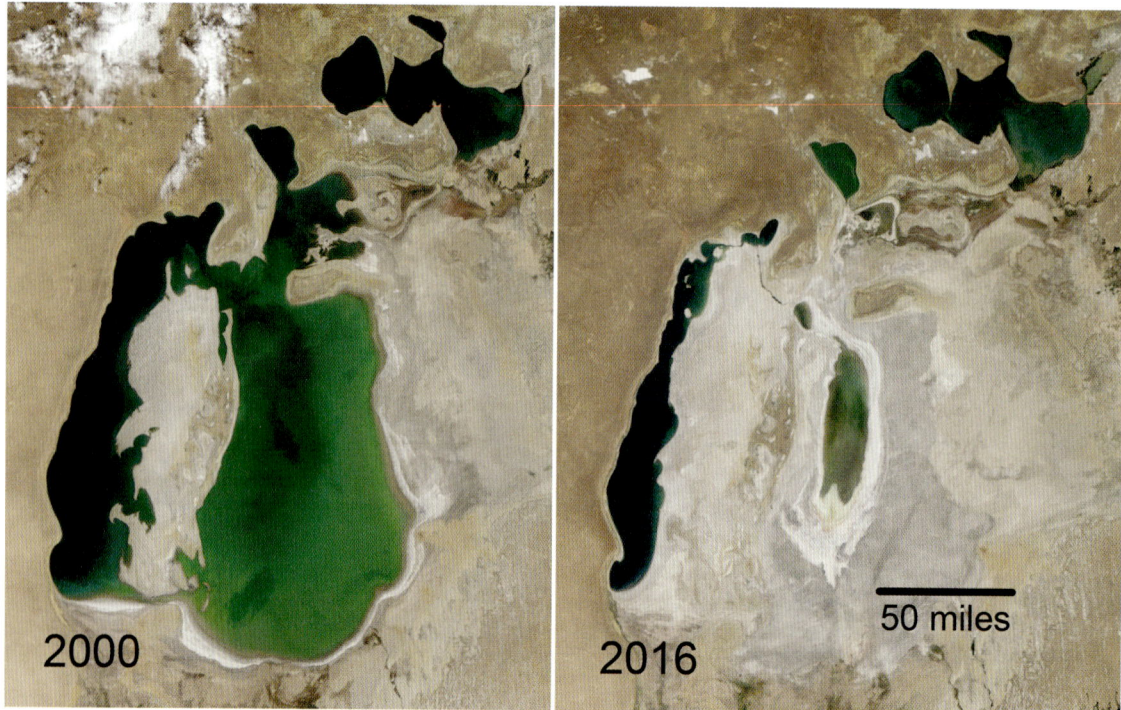

Changes in the surface area of the Aral Sea over sixteen years. Images from NASA Earth Observatory, "Shrinking Aral Sea," August 21, 2016.

Our awareness of the number of salt lakes at risk from climate change, water diversions, and other causes seems to be limited only by the availability of data. Most salt lakes have lost both surface area and water volume, with only rare exceptions, such as Mar Chiquita in Argentina, where water levels have recently risen.

The Aral Sea is perhaps the foremost example of an ecological disaster caused by water diversions. Historically the Aral Sea covered 26,000 square miles and was the world's fourth largest lake. However, in the 1960s, Soviet engineers built 80 reservoirs and 20,000 miles of canals on its main tributaries to increase agriculture upstream of the lake. Consequently, they reduced the Aral Sea to 6,500 square miles by 2003, and by 2016, their diversion efforts had made it even smaller (Jellison et al. 2004; Aladin et al. 2009; NASA 2016b).

Scientists, as they assessed this damage, were able to determine that one lobe of the Aral Sea had not been dry for at least six hundred years, so they concluded that the lake had not fallen so low for at least that long. Kazakhstan was desperate to save some of the Aral Sea, and so the country's engineers built a dam in 2005 across the northern part of the lake to try to retain some water (NASA 2016b). Unfortunately, the Aral Sea's loss of water has led to a cascade of economic, cultural, and ecological problems, including: harm to

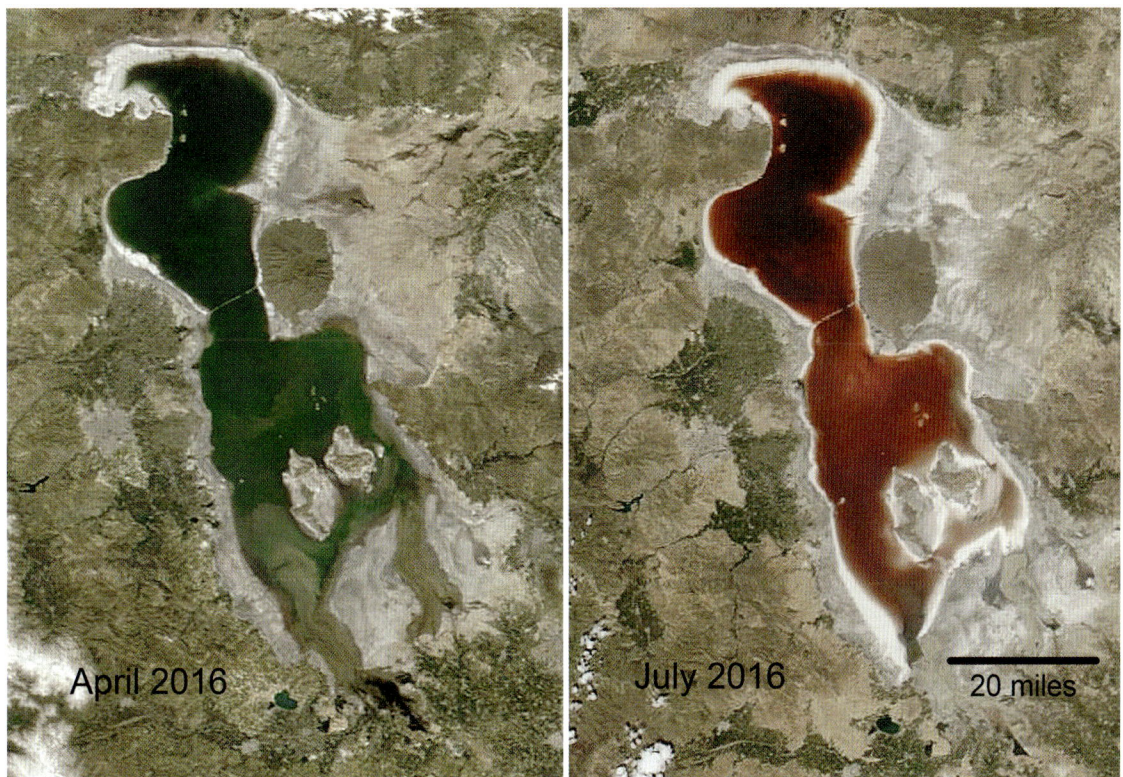

Iran's Lake Urmia losing surface area and hosting an expansive population of red, hypersaline microbiota in 2016. Images from NASA Earth Observatory, "Red Lake Urmia," July 26, 2016.

fisheries; hazardous, wind-blown dust; and a change in local climates, due to the loss of the moderating effects normally provided by a large water body (NASA 2016b).

Another environmental catastrophe has occurred at Iran's Lake Urmia, a national park and a UNESCO Biosphere Reserve that is 2,300 square miles in area (Asem et al. 2014). The lake used to be the largest in the Middle East and was the seventh largest saline lake in the world (see chapter 3), even larger than the Great Salt Lake. Recently, however, agricultural water diversions and drought have caused it to nearly disappear (Asem et al. 2014; Stone 2015; NASA 2016a). Historically, Lake Urmia has had high biological diversity with numerous endemic species. Until recently, it supported large populations of rare waterbirds, including the Great White Pelican (*Pelecanus onocrotalus*), the Eurasian Flamingo (*Phoenicopterus ruber roseus*), the Eurasian Spoonbill (*Platalea leucorodia*), and the Glossy Ibis (*Plegadis falcinellus*). The birds there fed on fish, on a native species of brine shrimp (*Artemia urmiana*), and on a native brine fly (*Ephydra urmiana;* Asem et al. 2014). In July 2016, Lake Urmia turned red, as Lake Abert periodically does, as a consequence of halobacteria and *Dunaliella salina* algal blooms that

Changes in the surface area of Walker Lake, Nevada, since 1984. Google Earth images.

began spreading as salt concentrations increased in the water (Stone 2015; NASA 2016a).

Here in the western United States, water diversions have severely impacted several Great Basin terminal lakes. Owens Lake in eastern California was the first to suffer serious consequences from agricultural diversions, starting in the mid-to-late 1800s. Then in 1913, engineers diverted the lake's inflows to Los Angeles, and by 1926, they had completely desiccated it (Robinson 2018).

The next unfortunate target was Nevada's Winnemucca Lake (formerly one hundred square miles in area), which was also parched by water development, and by 1930, little if any water flowed into it (Pratt 1997; Eilers and Walker 2014). Another salt lake in Nevada, Walker Lake, has also faced an ever-increasing loss of inflows, due to upstream water diversions that have appropriated nearly all surface water in the basin. By 2016, its volume had declined from ten million acre-feet to less than two million acre-feet, and its salinity had increased significantly, which caused substantial adverse ecological effects (Umek et al. 2008; Herbst et al.

2013; USFWS 2013). The surface area of the lake fell from fifty-nine square miles in 1984 to forty-three square miles in 2016.

Walker Lake's increasing salinity and rising pH are harming the aquatic ecosystem, including native fishes, and as a result, most fish have disappeared, possibly all of them. The lake used to be one of just a few that naturally supported the threatened Lahontan cut-throat trout. Unfortunately, no one has reported seeing a single trout there since 2011. Tui chubs may still live in the lake, due to their higher salt tolerance, but if the salinity continues to climb, they may also vanish.

The Great Salt Lake, which is a critical habitat globally for many migratory waterbirds—including avocets, stilts, Eared Grebes, Snowy Plovers, and Wilson's Phalaropes—also faces significant harm. This famous body of water has supported a multimillion-dollar brine-shrimp industry. Recently, however, the lake has under-gone dramatic declines in surface area and substantial increases in salinity (Williams 2016; Wurtsbaugh et al. 2017). Between 1988 and 2016, the lake shrank by 70 percent, falling from 3,300 square miles to less than 950 square miles. As a result, large numbers of Ameri-can White Pelicans died in 2016 (Williams 2016).

Environmentalists worry that storms of alkali dust coming from the exposed lake bed will increase in frequency and will pose a health hazard to the thousands of people living nearby in the Salt Lake City area, similar to what happened to people living near Owens Lake once it dried up. Reducing dust production from des-iccated saline lake beds is costly, so the goal should be to prevent the problem from ever happening. Los Angeles will pay an esti-mated $3.6 billion to mitigate dust production from the dry bed of Owen's Lake, according to analysts (Wurtsbaugh et al. 2017). Unfor-tunately, further water developments are being planned for the Bear River, the Great Salt Lake's main tributary; if they are built, they will exacerbate the adverse effects of current water diversions and climate change.

Back in 2002, Williams asserted that governments around the world fail to value salt lakes and that this is why they allow develop-ment to damage them. He said that to stop this from happening, our first and most obvious step is to raise awareness of an array of fac-tors: (1) the lakes' many ecological and economic values; (2) threats against the lakes and probable impacts of these threats; and (3) the management required to make these natural resources ecologically viable. Fortunately, a growing number of scientists, conservation organizations, and citizens are pressuring governments to act to resolve the water crises affecting salt lakes worldwide.

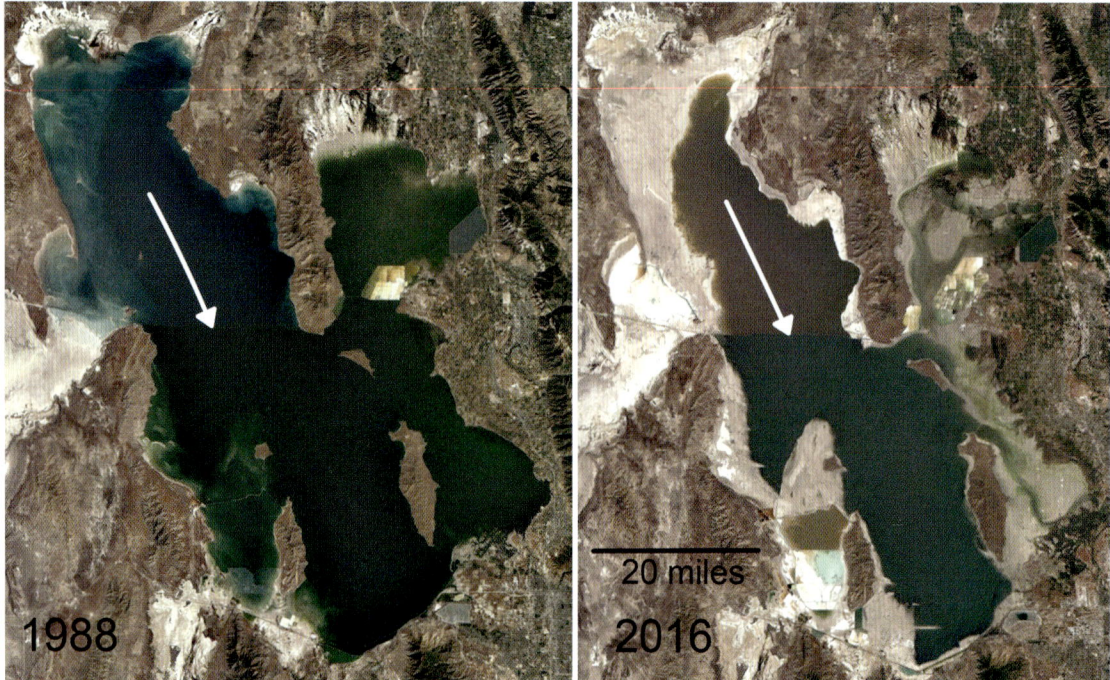

Changes in the surface area of the Great Salt Lake since 1988. The causeway running east and west across the lake (marked by the arrows) isolates Gunnison Bay in the north from Gilbert Bay in the south, which changes the water quality on each side of the causeway. Gunnison Bay has smaller inflows, so it has undergone the greatest decrease in area and the greatest increase in salinity. Google Earth images.

Future Effects of Climate Change

Based on climate-change predictions in Oregon, the drought that smacked Lake Abert in 2014 will recur often, and very likely with greater frequency. In fact, for WY 2020, the lake's elevation dropped precipitously once again because regional precipitation decreased about 50 percent from the normal average. Those conditions continued through 2021 and even into 2022.

There can be no doubt that much of Oregon and much of the West are warming. In fact, some researchers now use the term *aridification* to describe what is occurring (Overpeck and Udall 2020). The extent of this problem was brought home in June 2021, when a heat dome formed over much of the Pacific Northwest and caused many deaths, and large wildfires erupted, scorching nearly a half-million acres in southern Oregon alone.

Worries about the impact of climate change have been growing. Following the passage of a bill in 2007, the Oregon Climate Research Institute (OCRI) began conducting biennial assessments in conjunction with greenhouse-gas reduction goals. In their sixth assessment (Fleishman 2023), the report's authors stated that Oregon's average temperature had already warmed by more than 2°F in the past century, that the state would warm up by as much as 5°F by the 2050s, and that summer temperatures would be even higher. Climate change may lead to more rainfall in winter, but the benefits will be negated by a smaller snowpack and by drier summers resulting in a reduction in stream flows.

Higher temperatures will have multiple adverse effects on Lake Abert and other Great Basin terminal lakes. These effects may well include: (1) reduced stream flows, due to increased *evapotranspiration* (water loss from the soil caused by evaporation and transpiration from upland and riparian plants) and to greater evaporation from streams and wetlands; (2) reduced water flows to the lakes because upstream irrigators will demand more water to offset evapotranspiration from crops and the longer growing season; and (3) greater evaporation from lakes that will further reduce water levels, causing lake ecosystems to face extremely high salinities or even complete desiccation.

Higher water temperatures will put further stress on aquatic ecosystems by reducing dissolved oxygen concentrations (DOC) and by increasing the oxygen demands of aquatic life. Great Basin saline lake ecosystems are especially vulnerable to heat waves because: most of these lakes are at an elevation above 4,000 feet, where there is 15 percent less oxygen than there is at sea level; higher temperatures will reduce DOC—with each 10°C increase in temperature, there is a 20 percent reduction in DOC; the presence of salt lowers DOC even more substantially; and, finally, oxygen demands by aquatic organisms nearly double with each 10°C increase in temperature. Thus, the combined effects of climate change on temperature and of salinity on DOC and respiration will stress aquatic organisms until they eventually die. This is apparently what happened in Lake Abert during August 2010, when windrows of dead and dying brine shrimp turned the nearshore water bright red (see chapter 4).

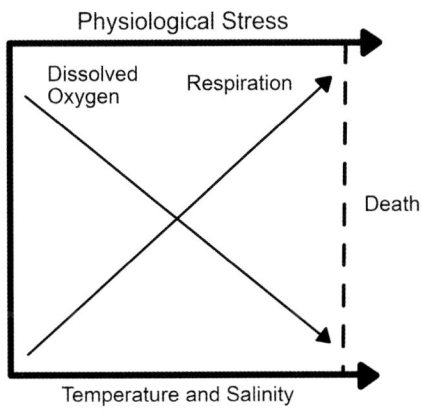

Potential effects of climate change on aquatic organisms in salt lakes. Increased temperatures and higher salinities result in reduced concentrations of dissolved oxygen and cause the organisms' respiration rates to increase. The combined effect of these factors increases physiological stress and will eventually cause death. This environmental stress likely decimated *Artemia* in Lake Abert in August 2010.

As Abert and other Great Basin lakes overheat, they could experience even longer periods of reduced water levels and higher salinities. Dangerously low DOC events could also become more frequent. Also, salinities reach toxic levels—as they did at Lake Abert in 2010—by midsummer or even sooner. Consequently, when most shorebirds typically arrive at the lake in July to feed, there could be nothing for them to eat. Under these conditions, the Lake Abert ecosystem could become increasingly unstable, and its productivity might decrease, due to greater variability in salinity, DOC, and temperature. These conditions could take the entire ecosystem past a tipping point, into longer periods of reduced productivity and low diversity for area plant and animal species and perhaps ever closer to a state of irreversible harm. Sadly, the low numbers of birds seen at Lake Abert in recent years may suggest that we are already at this tipping point.

Although ecologists cannot precisely say how climate change will affect any particular region, they agree that we must maintain an ecosystem's natural resilience in order to minimize the harm that it faces (Holling 1986; Gunderson 2000; Walker and Salt 2006; Hixon et al. 2010). What this means in practice is that we must take actions to preserve the natural variability under which an ecosystem and its species evolved before humans interfered. Figuring out nature's pattern for Lake Abert is tricky because people began altering the upstream watershed a century ago; they have substantially reduced inflows to the lake by diverting its water; and recent wildfires have further damaged the watershed. However, there is

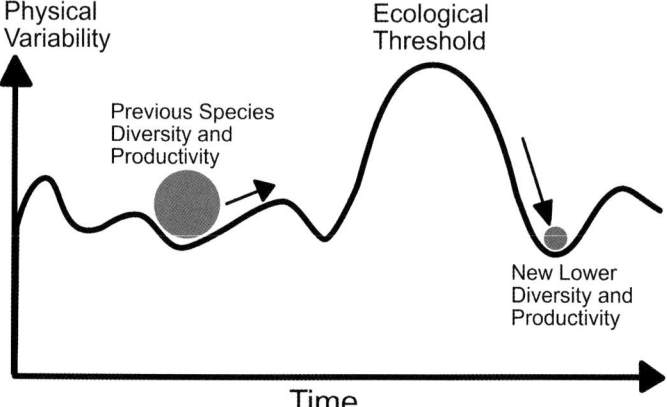

When increased physical variability pushes a habitat across an ecological threshold, species diversity and productivity decrease substantially. Species are resilient to low levels of physical variability, but as conditions worsen, they can no longer adapt, they begin to die off, and their environment becomes less productive, as is shown here by the smaller size of the ball.

little doubt that irrigation diversions have lowered the lake's level more frequently, have lowered the level more significantly, and have exacerbated high salinity events above and beyond anything that could be caused by natural climate variability (Larson et al. 2016; Moore 2016). Scientists don't know exactly when these adverse effects will take the ecosystem past its tipping point, but they do warn us that the longer salinities stay above 15% percent, the higher the risk will be of severe damage to the lake's ecosystem.

Possible Effects of Climate Change on Waterbirds

Some researchers have already gathered evidence that climate change is impacting the Great Basin's aquatic ecosystems (Donnelly et al. 2020; Haig et al. 2019; Hall et al. 2023). For example, they have documented that recent warming has reduced inflows and has also changed their timing, so the inflows come earlier. They maintain that these changes have impacted migratory waterbird populations there and are likely to result in the degradation of the interior portion of the Pacific Flyway. Scientists also emphasize just how critical these wetlands are to a variety of waterbird species: the Great Basin seasonally supports the entire North American population of Eared Grebes, up to 90 percent of all Wilson's Phalaropes, half of the global population of American White Pelicans, and over half of all American Avocets (Haig et al. 2019).

Although I am very concerned about the future of Lake Abert as an essential habitat for migratory waterbirds, I take some solace from studies in Australia, which show that some birds are amazingly resilient to short-term changes in the distribution of suitable habitats and can cross long distances to find better conditions (Halse et al. 1998; Roshier et al. 2001, 2006; Timms 2009, Pedler et al. 2014). Nevertheless, we know that long migrations cost the birds a great deal of energy and pose increased predation risks (Kersten and Piersma 1987; Lank et al. 2003). In the northern Great Basin in 2014, and again in 2021 and 2022, habitat for waterbirds was very limited, and the closest similar saline habitats to Lake Abert's were those at Mono Lake (three hundred miles away) and at the Great Salt Lake (four hundred miles distant). Crossing such distances is risky for any waterbird, but it is especially so for young birds with no experience, and many likely died trying to find a suitable environment.

Scientists don't know what the long-term effects of Lake Abert's desiccation events will be on waterbirds in the region because current waterbird monitoring in the Great Basin is inadequate. However, we can predict general outcomes based on studies done on shorebirds elsewhere. One example is a study conducted on the Red

Knot (*Calidris canutus rufa*), which is a small sandpiper known to experience higher mortality during migration when its fat reserves are too low (Niles et al. 2008). Perhaps this is an extreme example, because Red Knots fly nearly ten thousand miles, from Tierra del Fuego, near the tip of South America, to the Canadian Arctic, and then back again every year, and thus they have exceptional energy demands. To be able to fly those distances, the Red Knot must feed at multiple stopover points.

Several long-distant migrants also use Lake Abert, such as Red-necked and Wilson's Phalaropes, whose migrations include flying six thousand to ten thousand miles round trip every year, between breeding sites in North America and wintering sites in South America, and consequently their survival also depends on finding adequate food before migrating.

To make matters worse for shorebirds, many of their populations are already small and declining. In fact, one-third of 70 Arctic-nesting shorebird populations number less than 25,000 birds, and therefore, climate change is expected to further stress them (Wauchope et al. 2016). Additionally, based on another threat assessment, nearly 90 percent of the shorebirds breeding in the United States and Canada will face a greater risk of extinction in the near future from climate change. Based on this latter analysis, the Snowy Plover, Western Sandpiper, and Wilson's Phalarope are most at risk (Galbraith et al. 2014).

In 2015, the National Audubon Society also assessed climate-change impacts to birds, based on threats to their current summer and winter habitats. The conservation nonprofit's evaluation included twenty-eight species at Lake Abert threatened by habitat changes. The list comprised fifteen waterbird species that they categorized as climate endangered and three that they listed as climate threatened. Among these birds are nine species that have relatively large seasonal populations at the lake, including: American Avocets, American Wigeons, California Gulls, Eared Grebes, Mallards, Northern Shovelers, Ring-billed Gulls, and Wilson's Phalaropes. The list also included eight upland birds that they assessed as climate endangered and two that they listed as climate threatened, so their report showed that the lake's upland birds are also at risk.

Managing These Valuable Ecosystems

We can prevent catastrophic harm to migratory waterbirds if we properly manage their habitats. Fortunately, there is still some hope for the Great Basin lakes. Several recent efforts give me reason to be optimistic that these salt lakes are finally getting some much-deserved attention.

One of these efforts—the Saline Lake Ecosystems in the Great Basin States Program Act—was introduced as a bill by Senators Jeff Merkley (D-OR) and Mitt Romney (R-UT) in 2021 and was passed by Congress later that year. The act has funded and established a program to study the Great Basin salt lakes in support of coordinated management and conservation actions "to benefit those ecosystems, migratory birds, and other wildlife." Perhaps even more significant, in 2019 the Utah Legislature passed, and then Governor Gary Herbert signed, a Concurrent Resolution to Address Declining Water Levels of the Great Salt Lake. Their resolution recognized "the critical importance of ensuring adequate water flows to Great Salt Lake and its wetlands to maintain a healthy and sustainable lake system."

As a result, the 2020 Great Salt Lake Resolution Steering Group produced a report identifying sixteen strategic opportunities and making sixty recommendations for ways to protect the lake and its water. One aspect of their work may serve as a model: they approached the problem of limitations on water availability by focusing on "strategies that share finite water supplies in ways that support multiple uses." While such an approach overlooks the reality that the demand for water in the West greatly exceeds supply, any solution requires that stakeholders come together. When they are willing to do so, they may find that they have a lot in common.

California, in an effort to protect lake ecosystems, has recognized the government's role in protecting essential public resources like air and water, as well as the belief that these resources are so valuable that they belong to everyone. Historians have traced this idea—that everyone owns such critical resources—all the way back to the Roman Emperor Justinian in the sixth century CE. He asked a commission to consolidate existing laws in the empire, and in their compendium, they wrote, "By the law of nature, these things are common to mankind—the air, running water, the sea, and consequently the shores of the sea" (Ryan 2015).

What gradually became known as the Public Trust Doctrine (PTD) in English and then in American jurisprudence is now an instrumental legal concept. California courts have relied on the theory in their decisions to protect essential inflows to Mono Lake and to reflood areas of Owens Lake. Nevada's legal system has also applied the PTD to Walker Lake. The local government in Mineral County, Nevada, sought to ensure that more water reached the lake and to protect the ecosystem from rising levels of salinity. In 2020, Nevada's Supreme Court ruled that the PTD applies to all waters of the state and also to water rights, but it did not go so far as to reallocate water to the lake (Blumm and Smith 2022).

In Oregon, the courts have not yet tested the full scope of the PTD (Blumm and Doot 2012). However, what is clear is that the state holds its waters in trust for the public and that this responsibility "burdens the state with protective duties." Therefore, it is certainly logical to believe that the state has to ensure that its water-use decisions don't harm wildlife or the ecosystems on which they depend.

Another legal theory, the Rights of Nature Doctrine, asserts that a natural ecosystem like a river or lake is entitled to "exist, flourish, regenerate its vital cycles, and naturally evolve without human-caused disruption" (Challe 2021). Under this paradigm, an ecosystem has the right to be represented in court by a guardian, a higher level of protection than our existing laws provide. So far, seventeen countries have enacted rights of nature laws.

In Oregon, the Burns Paiute Tribal Council adopted a resolution in 2020 establishing the "Rights of the Malheur River Ecosystem to exist, flourish, and naturally evolve." Although it might not have the force of law, this resolution is a clear statement that emphasizes how critical the river ecosystem is to the tribe's well-being.

The failure of existing laws to protect ecosystems has shifted attention to such alternative strategies. Lake Abert is just one of many valuable habitats that remain vulnerable, due to a lack of protection from current law. I take hope from the fact that we are much better educated now than in past decades, when most environmental laws were enacted, so we are aware that threats from climate change are real. Our awareness means, perhaps, that we can make real progress toward ensuring that our ecosystems remain resilient.

There are no easy fixes for maintaining the ecological integrity of Lake Abert, or for protecting any of the other salt lakes, because the threats have been around for a century or more. Furthermore, state legislatures have been unwilling to make needed reforms in water laws because special interest groups that benefit from the overuse and misuse of our water oppose these reforms. As we work toward solutions, we certainly have to prioritize people's essential needs. Is it really in our best interest, however, to have multinational corporations like Simplot using most of the Chewaucan Basin's water?

We can't continue to ignore the region-wide needs of migratory waterbirds. As a nation, we have vowed to protect these flocks under the international Migratory Bird Treaty. Based on what is happening at Lake Abert and at other waterbird habitats in the West, can we say to our fellow signatories—our neighbors in Canada and Mexico, who have already made great strides in protecting migratory birds— that we have lived up to our sworn responsibility? And if we haven't done so, can we say that we still expect them to continue to conserve birds without our help? To survive, migratory birds must have protection throughout their range, not just in parts of it.

Current debates over how best to save the Great Salt Lake, Mono Lake, Walker Lake, and the Salton Sea may eventually offer solutions to help Lake Abert. However, the birds that depend on the Lake Abert ecosystem can't wait, so Oregon needs to step up now to develop a workable plan to protect this incredible place and to save other terminal lakes in the state. Critics will say that the lake has dried up in the past and has refilled, so there is no serious problem. We did, of course, see this cycle in 2014–2016 and again in 2021 and 2022, but the year-to-year variability in Lake Abert's salinity is likely greater now than at any time in the past. Approximately half the water is diverted upstream and never reaches the lake, and so the ecosystem struggles to adjust (Moore 2016).

Indeed, unlike most salt lakes in Australia, which lie in remote locations and are therefore largely pristine (Timms 2009), nearly all the Great Basin salt lakes are heavily impacted by water diversions, which threaten the resiliency of their ecosystems. As some scientists have taken pains to point out, "desert aquatic ecosystems and the unique species they support are fragile. While both the physical and ecosystem components and associated flora and fauna have withstood thousands of years of climatic and geologic change, their long histories have not conditioned them to the sudden man-made change [they now face]" (Sigler and Sigler 1992).

How can we protect Lake Abert and boost its resiliency? After all, its integrity and beauty are threatened not only by climate change, but also by an ongoing lack of responsible management by the agencies entrusted with its care. The Oregon Water Resources Commission (OWRC) has already tried to grapple with these complex problems. It adopted its first Integrated Water Resources Strategy in 2012, "to understand Oregon's water needs and articulate a strategy to meet those needs into the future." The far-reaching plan called for the state's Water Resources Department to create an integrated policy with four primary objectives: (1) to gather, process, and share water-resource data; (2) to improve knowledge of current and future water needs; (3) to analyze factors affecting water resources and needs; and (4) to integrate water resources management and planning among all levels of government.

Lake Abert provides a perfect opportunity for the state's water resources officials to follow through on these commitments, and they have a legal mandate to do so, but they have made little if any progress. In fact, the OWRC updated the Integrated Water Resources Strategy in 2017 (Mucken and Bateman 2017), but they did not cite any measurable successes, so it seems that the plan is largely being ignored. In a related 2018 effort, state agencies drafted Oregon's 100-year Water Vision, a plan for investing in and managing both manmade and natural water infrastructure. Although the agencies developed laudable goals, they did not provide any guidelines for how officials can achieve these objectives.

It is becoming clearer with each year of continued drought that water has become a major issue in Oregon. State-level officials have drawn up several documents that spell out what needs to be done to ensure that the state preserves enough water to meet the needs of Oregon's residents and ecosystems. What is lacking is action on the part of the governor, the state legislature, and the state agencies to begin addressing these concerns. Up to now, the state's only noticeable response to our water dilemma is the governor's annual drought declaration, and this statement might be actually harmful because it loosens restrictions on groundwater use in some years. What we really need are concrete actions that will make Oregon more resilient to drought.

How Can We Save the Lake Abert Ecosystem?

So, how can we save Lake Abert? I wish that I had a simple answer to that question, but I don't. However, I do know that a solution won't just happen: it will require a change in the way that Oregon values water and the ecosystems that water supports. As Rick Bastasch, a former OWRD employee, starkly warns: "Oregon has turned its back to its water future." Unfortunately, based on the climate of the past several years, that future is looking much drier than it did just a decade ago.

As the first step toward finding a solution to Oregon's water problems, Bastasch recommends a very simple action: we must value water as a scarce commodity and properly charge everyone who uses it, just as we do for such other resources as timber, minerals, fish, and wildlife (Bastasch 1998). Charging everyone for their water use won't create more water, but the policy could reduce waste and improve efficiency. Furthermore, the revenue could be put to a variety of good uses: installing gauges at all water-diversion points to enhance monitoring of surface water and groundwater; providing cost-share funding to improve water-use efficiency; and establishing environmental purchasing practices—those that reduce harmful impacts for products and services related to the use of Oregon's streams and lakes and which promote sustainability. All these ideas could help Lake Abert.

The free use of water—a public resource—has been a sacred cow for a century, so changing this practice will require legislative action. Moving the legislature on this issue won't be easy because, historically, agriculture, hydropower, and the timber industry have largely controlled Oregon's water. Thus, even getting a seat at the table to begin addressing the inequities in who gets this resource will be a challenge. However, as the popular adage states, "If you don't have a seat at the table, you will be on the menu."

So, how do we get to the table and become part of a solution that ensures that water is valued for all its uses, including the ecological ones? One way to do this is by supporting conservation organizations like the Oregon Natural Desert Association (ONDA), which is already working hard to ensure that we protect the state's arid landscapes and their ecosystems. Thousands of like-minded people who value these habitats already champion ONDA. The nonprofit has committed staff and financial resources to tackling such difficult issues as how to get more water into the lake.

I also recommend supporting WaterWatch of Oregon, a nonprofit that has already achieved some success in restoring and protecting the natural flows of streams and rivers throughout the state. A third worthy organization is the Oregon Lakes Association (OLA), a nonprofit that strives to "be a voice for quiet waters." OLA uses its expertise to ensure that all Oregonians have access to clean water and to healthy lake ecosystems. A fourth effective nonprofit is the National Audubon Society. Audubon's Western Water team focuses on Birds and Water in the Arid West. The group has a dedicated staff working to align the protection and restoration of bird habitats with more-reliable water supplies for all communities, while addressing historic injustices. Thankfully, all four of these nonprofits are already actively seeking ways to protect the Lake Abert ecosystem, and they are making progress. This gives me hope that before too long, the Lake Abert ecosystem will be sustainably managed.

Another contribution you can make is by joining your local Audubon chapter and helping to organize volunteer-run waterbird surveys, just as volunteers do at the East Cascades Audubon Society in Bend. You can't manage a resource if you are not sure that it exists, a problem that has plagued tracking efforts at Lake Abert. Existing surveys, such as the Christmas Bird Count and the Breeding Bird Survey, have not effectively counted many of the lake's shorebirds, so more helpers are needed, birders who are willing to use their identification skills and who want to help protect bird populations. Finally, you can write elected officials and express your concern for the future of Lake Abert and other Great Basin lakes and wetlands.

In the summer of 2021, the Chewaucan Basin was a terrible scene of drought, and the air was filled with acrid smoke from nearby wildfires, which made breathing difficult. This environmental disaster was a wakeup call, and it alerted us to the fact that change is occurring. Just a few years earlier in 2017, ample snow and rain had filled the river and canals, and grass in the pastures had grown up high enough to touch the bellies of cows. In the marshes, Sandhill Cranes had danced and bugled in the early dawn, and snipe had performed aerial displays, winnowing in the evening. And in the lake, an abundance of alkali flies and brine shrimp had attracted avocets, grebes, phalaropes, and sandpipers in vast numbers.

Conditions in 2017 showed us what is still possible, and I think we could even sustain such an environment with less water. To achieve this balancing act, however, we need people who are willing to listen to diverse viewpoints and who are willing to work together in a spirit of compromise and cooperation. It is my fervent hope that we can find solutions to our complex water issues in the Chewaucan Basin, so that family farms and ranches thrive, freshwater wetlands remain healthy, and Lake Abert can continue to offer a productive home to the avocets and other waterbirds that grace its waters and shores.

A glimmer of hope for Lake Abert came in 2022, when conservationists and Chewaucan Basin residents began meeting to discuss solutions to water problems facing the lake and people living in the area. Amazingly, that group, consisting of folks who would be expected to have wildly disparate viewpoints, found common ground, and in early 2023, they helped write a bill that, if funded, would provide facilitation toward finding an equitable solution, as well as support for more data collection and analysis. While it's too early to declare this effort a success, it makes me more optimistic that a solution will be found for the lake.

References

Abele, D. "Toxic Oxygen: The Radical Life-Giver." *Nature,* vol. 420, 2002, p. 27.

Abuduwaili, J., D. W. Liu, and G. Y. Wu. "Saline Dust Storms and Their Ecological Impacts in Arid Regions." *Journal of Arid Land,* vol. 2, no. 2, 2010, pp. 144-50.

Ackerman, J. T., C. A. Hartman, M. P. Herzog, J. Y. Takekawa, J. A. Robinson, L. W. Oring, J. P. Skorupa, and R. Boettcher. "American Avocet (*Recurvirostra americana*)," version 1.0. *Birds of the World.* Edited by A. F. Poole. Cornell Lab of Ornithology, 2020.

Adovasio, J. M., and D. Pedler. *Strangers in a New Land: What Archaeology Reveals About the First Americans.* Firefly Books, 2016.

Aikens, C. M., T. J. Connolly, and D. L. Jenkins. *Oregon Archaeology.* Oregon State UP, 2011.

Aikens, M., and M. Couture. "The Great Basin." *The First Oregonians,* pp. 274-83. Edited by L. Berg. Oregon Council for the Humanities, 2007.

Aladin, N., I. Plotnikov, P. Micklin, and T. Ballatore. "Aral Sea: Water Level, Salinity and Long-term Changes in Biological Communities of an Endangered Ecosystem—Past, Present and Future." *Saline Lakes Around the World: Unique Systems with Unique Values,* pp. 179-83. Edited by A. Oren, D. L. Naftz, P. Palacios, and W. A. Wurtsbaugh. *Natural Resources and Environmental Issues,* vol. 15, 2009.

Allison, I. S. *Fossil Lake, Oregon, Its Geology and Fossil Faunas.* Oregon State UP, 1966.

———. *Pluvial Fort Rock Lake, Lake County, Oregon, Special Paper 7.* State of Oregon, Department of Geology and Mineral Industries, 1979.

———. "Geology of Pluvial Lake Chewaucan, Lake County, Oregon." *Studies in Geology,* no. 11. Oregon State UP, 1982.

Andres, B. A., P. A. Smith, R. I. G. Morrison, C. L. Gratto-Trevor, S. C. Brown, and C. A. Friis. "Population Estimates of North American Shorebirds." *Wader Study Group Bulletin,* vol. 119, no. 3, 2012, pp. 178-94.

Asem, A., A. Eimanifar, M. Djamali, P. De Los Rios, and M. Wink. "Biodiversity of the Hypersaline Urmia Lake National Park (NW Iran)." *Diversity,* vol. 6, 2014, pp. 102-32.

Asem, A. N., Rastegar-Pouyani, and P. De Los Rios. "The Genus *Artemia* Leach, 1819 (Crustacea: Branchiopoda): True and False Taxonomical Descriptions." *Latin American Journal of Aquatic Research,* vol. 38, 2010, pp. 501-6.

Austin, M. *The Land of Little Rain.* Penguin Books, 1988 reprint (1903 original date).

Avery, S. "What Future for Lake Turkana?" *African Studies Centre.* U of Oxford, 2013. https://www.africanstudies.ox.ac.uk/sites/default/files/africanstudies/documents/media/whatfuturelaketurkana-_update_0.pdf.

Badger, T. C., and R. J. Watters. "Gigantic Seismogenic Landslides of Summer Lake Basin, South-Central Oregon." *Geological Society of America Bulletin,* 2004, pp. 687-97.

Bailey, V. US Department of Agriculture, Bureau of Biological Survey. "The Mammals and Life Zones of Oregon." *North American Fauna,* no. 55, 1936.

Bartruff, A. L. "A Characterization of Lake Abert Tufa Mounds." MS thesis. Portland State U, 2013.

Bastasch, R. *The Oregon Water Handbook: A Guide to Water and Water Management.* Oregon State UP, 2006.

Batchelor, J. L., W. J. Ripple, T. M. Wilson, and L. E. Painter. "Restoration of Riparian Areas Following the Removal of Cattle in the Northwestern Great Basin." *Environmental Management,* vol. 55, 2015, pp. 930-42.

Bauld, J. "Occurrence of Benthic Microbial Mats in Saline Lakes." *Hydrobiologia,* vol. 81, 1981, pp. 87-111.

Beauchamp, D. "Material and Remembrance: Three Fractured Petroglyph Boulders, Lake County, Oregon." 2013. https://www.academia.edu/7672593/Material_and_Remembrance_Three_Fractured_Petroglyph_Boulders_Lake_County_Oregon_A_Photo_Essay.

Benson, L. V. "The Tufas of Pyramid Lake, Nevada." *US Geological Survey, Circular 1267,* 2004.

Benson, L. V., E. M. Hattori, J. Southon, and B. Aleck. "Dating North America's Oldest Petroglyphs, Winnemucca Lake Subbasin, Nevada." *Journal of Archaeological Science,* vol. 40, no. 12, 2013, pp. 4466-76.

Bent, A. C. *Life Histories of North American Birds.* 21 vols. United States National Museum Bulletin (republished by Dover), 1919–1968.

Betancourt, J. L., T. R. Van Devender, and P. S. Martin. *Packrat Middens: The Last 40,000 Years of Biotic Change.* U of Arizona P, 1990.

Bettinger, R. L., and M. A. Baumhoff. "The Numic Spread: Great Basin Cultures in Competition." *American Antiquity,* vol. 47, no. 3, 1982, pp. 485-503.

Blank, R. R., J. A. Young, and F. L. Allen. "Aeolian Dust in a Saline Playa Environment, Nevada." *U.S.A. Journal of Arid Environments,* vol. 41, 1999, pp. 365-81.

Blumm, M. C., and E. Doot. "Oregon's Public Trust Doctrine: Public Rights in Waters, Wildlife, and Beaches." *Environmental Law,* vol. 42, 2012, pp. 375-414.

Blumm, M. C., and M. B. Smith. "Walker Lake and the Public Trust in Nevada's Waters." *Virginia Environmental Law Journal,* vol. 40, 2022, pp. 1-39.

Bodaker, I. O. Beja, I. Sharon, R. Feingersch, M. Rosenberg, A. Oren, M. Y. Hindiyeh, and H. I. Malkawi. "Archaeal Diversity in the Dead Sea: Microbial Survival Under Increasingly Harsh Conditions." *Saline Lakes Around the World: Unique Systems with Unique Values,* pp. 141-47. Edited by A. Oren, D. L. Naftz, P. Palacios, and W. A. Wurtsbaugh. *Natural Resources and Environmental Issues,* vol. 15, 2009.

Bone, M., S. Johnson, P. Kelaidis, and M. Kintgen. *Steppes: The Plants and Ecology of the World's Semi-Arid Regions.* Timber Press, 2015.

Boula, K. M. "Foraging Ecology of Migrant Water Birds, Lake Abert, Oregon." MS thesis. Oregon State U, 1986.

Boyd, W. S., and J. R. Jehl, Jr. "Estimating the Abundance of Eared Grebes (*Podiceps nigricollis*) at Mono Lake, California, by Aerial Photography." *Colonial Waterbirds,* vol. 21, 1998, pp. 236-41.

Buchanan, J. B., C. T. Schick, L. A. Brennan, and S. G. Herman. "Merlin Predation on Wintering Dunlins: Hunting Success and Dunlin Escape Tactics." *Wilson Bulletin,* vol. 100, no. 1, 1988, pp. 108-18.

Bucher, E. H., and E. D. Curto. "Managing Salt Lakes in the Neotropics: Challenges and Alternatives—the Case of Mar Chiquita, Argentina." *Saline Lakes Around the World: Unique Systems with Unique Values,* p. 135. Edited by A. Oren, D. L. Naftz, P. Palacios, and W. A. Wurtsbaugh. *Natural Resources and Environmental Issues,* vol. 15, 2009.

Buechner, H. K. "The Bighorn Sheep in the United States: Its Past, Present, and Future." *Wildlife Monographs,* no. 4, 1960.

Burrows, M. "Jumping from the Surface of Water by the Long-Legged Fly *Hydrophorus* (Diptera, Dolichopodidae)." *Journal of Experimental Biology,* vol. 216, 2013, pp. 1973–1982.

Buseck, R. S., D. A. Keinath, and M. Geraud. "Species Assessment for Great Basin Spadefoot Toad (*Spea intermontana*) in Wyoming." US Department of the Interior Bureau of Land Management, Wyoming State Office. 2005. https://www.uwyo.edu/wyndd/_files/docs/reports/speciesassessments/greatbasinspadefoottoad-jan2005.pdf.

Butler, V. L. "Tui Chub Taphonomy and the Importance of Marsh Resources in the Western Great Basin of North America." *American Antiquity,* vol. 61, 1996, pp. 699-717.

Byers, J. A. *American Pronghorn Social Adaptations and Ghosts of Predators Past.* U of Chicago P, 1997.

Camp, V. E., M. E. Ross, R. A. Duncan, N. A. Jarboe, R. S. Coe, and J. A. Johnson. "The Steens Basalt: Earliest Lavas of the Columbia River Basalt Group." *Geological Society of America, Special Paper,* vol. 497, 2013, pp. 87-116.

Cannon, W. J. "Cultural Resource Survey of the West Shore of Lake Abert." Bureau of Land Management, Lakeview, Oregon. Unpublished report, 1977.

Castro-Mejía, J., G. Castro-Mejía, R. De Lara-Andrade, M. C. Monroy-Dosta, D. I. Orozco-Rojas, J. Á. Torrez-Ramírez. "Biometry Characteristics Comparison of *Artemia franciscana* Inland Waters Strains from México with 'Originally' Species from San Francisco Bay (SFB) Population." *Revista de Biologia Tropical,* vol. 59, no. 1, 2011, pp. 199-206.

Caumette, P. "Ecology and Physiology of Phototrophic Bacteria and Sulphate-Reducing Bacteria in Marine Salterns." *Experientia,* vol. 49, 1993, pp. 473-81.

Caumette, P., and S. J. Lucas. *Microbial Mats: Structure, Development, and Environmental Significance.* Springer-Verlag, 1994.

Challe, T. "The Rights of Nature—Can an Ecosystem Bear Legal Rights?" Columbia Climate School, State of the Planet. 22 April 2021. https://news.climate.columbia.edu/2021/04/22/rights-of-nature-lawsuits/.

Christensen, L., and H. Crimmel, editors. *Teaching About Place: Learning from the Land.* U of Nevada P, 2008.

Cohen, A. S., M. R. Palacios-Fest, R. M. Negrini, P. E. Wigand, and D. B. Erbes. "A Paleoclimate Record for the Past 250,000 Years from Summer Lake, Oregon, USA: II. Sedimentology, Paleontology and Geochemistry." *Journal of Paleolimnology,* vol. 24, 2000, pp. 151-82.

Collopy, M. W., B. Woodbridge, and J. L. Brown. "Golden Eagles in a Changing World." *The Journal of Raptor Research,* vol. 51, no. 3, 2017, pp. 193-96.

Colwell, M. A., and J. R. Jehl, Jr. "Wilson's Phalarope (*Phalaropus tricolor*)," version 1.0. *The Birds of North America.* Edited by A. Poole and F. Gill. Cornell Lab of Ornithology, 2020.

Connolly, T. J., and P. Baker. "Great Basin Sandals." *The Great Basin: People and Place in Ancient Times,* pp. 68-73. Edited by C. S. Fowler and D. D. Fowler. School for Advanced Research Press, 2008.

Connolly, T. J., P. Hlavacek, and K. Moore. *10,000 Years of Shoes.* U of Oregon, 2011.

Connolly, T. J., N. P. Jew, M. E. Swisher, W. J. Cannon, K. J. Sullivan, and M. Waller. "Analyses of Household Artifacts from Rattlesnake Cave (35LK1295): A Site in the Chewaucan Basin of Southeast Oregon." *Journal of California and Great Basin Anthropology,* vol. 36, no. 2, 2016, pp. 293-310.

Conte, F. P, and P. A. Conte. "Abundance and Spatial Distribution of *Artemia salina* in Lake Abert, Oregon." *Hydrobiologia,* vol. 158, 1988, pp. 167-72.

Corl, A., A. R. Davis, S. R. Kuchta, and B. Sinervo. "Selective Loss of Polymorphic Mating Types Is Associated with Rapid Phenotypic Evolution During Morphic Speciation." *Proceedings of the National Academy of Sciences of the United States of America,* vol. 107, no. 9, 2010, pp. 4254-59.

Couture, M. D., M. F. Ricks, and L. Housley. "Foraging Behavior of a Contemporary Northern Paiute Population." *Journal of California and Great Basin Anthropology,* vol. 8, no. 2, 1986, pp. 150-160.

Coville, F. V. "Notes on the Plants Used by the Klamath Indians of Oregon." *US National Museum Contributions, US National Herbarium,* vol. 5, 1897, pp. 87-108.

Cressman, L. S. "Petroglyphs of Oregon." U of Oregon Monographs, *Studies in Anthropology, no. 2,* 1937.

———. "Archaeological Researches in the Northern Great Basin." Carnegie Institution of Washington, Publication 538, 1942.

———. "Klamath Prehistory: The Prehistory of the Culture of the Klamath Lake Area, Oregon." *Transactions of the American Philosophical Society,* vol. 46, no. 4, 1956, pp. 375-513.

———. *The Sandal and the Cave: The Indians of Oregon.* Oregon State UP, 1981 (originally published in 1962 by Beaver Books).

———. *A Golden Journey: Memoirs of an Archaeologist.* U of Utah P, 1988.

Cronquist, A., A. H. Holmgren, N. H. Holmgren, and J. L. Reveal. *Intermountain Flora, Vascular Plants of the Intermountain West, U.S.A., Volume 1.* New York Botanical Garden, 1972.

Currens, K. P., C. B. Schreck, and H. W. Li. "Evolutionary Ecology of Redband Trout." *Transactions of the American Fisheries Society,* vol. 138, 2009, pp. 797-817.

Dambacher, J. M., K. K. Jones, and H. W. Li. "The Distribution and Abundance of Great Basin Redband Trout: An Application of Variable Probability Sampling in a 1999 Status Review." Oregon Department of Fish and Wildlife, 2001.

Darby, M. "Wapato for the People: An Ecological Approach to Understanding the Native American Use of *Sagittaria latifolia* on the Lower Columbia." MA thesis. Portland State U, 1996.

DasSarma, P., G. Klebahn, and H. Klebahn. Translation of Henrich Klebahn's "Damaging Agents of the Klippfish—a Contribution to the Knowledge of the Salt-Loving Organisms." *Saline Systems,* vol. 6, 2010, p. 7.

DasSarma, P., K. Tuel, S. D. Nierenberg, T. Phillips, W. T. Pecher, and S. DasSarma. "Inquiry-Driven Teaching and Learning Using the Archaeal Microorganism *Halobacterium* NRC-1." *The American Biology Teacher,* vol. 78, no. 1, 2016, pp. 7-15.

DasSarma, S. "Extreme Microbes." *American Scientist,* vol. 95, 2007, pp. 224-31.

DasSarma, S., and P. DasSarma. *Halophiles.* Wiley, 2012.

Davis, R. "Oregon's Lake Abert Is 'in Deep Trouble.' The State Shut Down Its Effort to Figure Out Why." *The Oregonian,* 1 Jan. 2022.

D'Azevedo, W. L. *Handbook of North American Indians, Volume 11: Great Basin.* Smithsonian Institution. 1986.

DeCourten, F. L. *The Broken Land: Adventures in Great Basin Geology.* The U of Utah P, 2003.

De Deckker, P. "Australian Salt Lakes: Their History, Chemistry, and Biota—a Review." *Hydrobiologia,* vol. 105, 1983, pp. 231-44.

Deocampo, D. M., and B. F. Jones. "Geochemistry of Saline Waters." *Treatise on Geochemistry,* second ed. Edited by H. D. Holland and K. K. Turekian. Elsevier Science, 2014, pp. 437-69.

Desender, K. "A Wingless Intertidal Ground Beetle, New to Belgian Fauna, in the River Ijzer Estuary Nature Restoration Site: *Bembidion nigropiceum* Marsham, 1802." *Belgium Journal of Zoology,* vol. 135, no. 1, 2005, pp. 95-96.

De Vera, J., D. Schulz-Makuch, A. Khan, A. Lorek, A. Koncz, D. Möhlmann, and T. Spohn. "Adaptation of an Antarctic Lichen to Martian Niche Condidtions Can Occur Within 34 Days." *Planetary and Space Science,* vol. 98, 2014, pp. 182-90.

DiGiacomo, P. M., W. M. Hammer, P. P. Hammer, and R. M. A. Caldeira. "Phalaropes Feeding at a Coastal Front in Santa Monica Bay." *California Journal of Marine Systems,* vol. 37, 2002, pp. 199-212.

Donnelly, J. P., S. L. King, N. L. Silverman, D. P. Collins, E. M. Carrera-Gonzalez, A. Lafon-Terrazas, and J. N. Moore. "Climate and Human Water Use Diminish Wetland Networks Supporting Continental Waterbird Migration." *Global Climate Change Biology,* vol. 26, no. 4, 2020, pp. 2041-59.

Dugas, D. P. "Late Quaternary Variations in the Level of Paleo-Lake Malheur, Eastern Oregon." *Quaternary Research,* vol. 50, 1998, pp. 276-82.

Dupraz, C., R. P. Reid, O. Braissant, A. W. Decho, R. S. Norman, P. T. Visscher. "Processes of Carbonate Precipitation in Modern Microbial Mats." *Earth-Science Reviews,* vol. 96, 2008, pp. 141-62.

Egger, A. E., D. E. Ibarra, R. Widden, R. M. Langridge, M. Marion, and J. Hall. "Influence of Pluvial Lake Cycles on Earthquake Recurrence on the Northwest Basin and Range, USA." *Geological Society of America Special Paper,* vol. 536, 2018, pp. 1-28.

Eilers, J., and S. Walker. "Lakes Winnemucca and Pyramid: One Gone, One Saved." *Lakeline,* vol. 34, no. 3, 2014, pp. 39-42.

Einarsen, A. S. *The Pronghorn Antelope and Its Management.* The Wildlife Management Institute, 1948.

Eiselt, B. S. "Defining Ethnicity in Warner Valley: An Analysis of House and Home." U of Nevada-Reno, Department of Anthropology, Technical Reports 97-2, 1997.

El-Maarry, M. R. "Desiccation Crack Polygon." *Encyclopedia of Planetary Landforms.* Springer, 2014.

El-Maarry, M. R., W. J. Markiewicz, M. T. Mellon, W. Goerz, J. M. Dohm, and A. Pack. "Crater Floor Polygons: Desiccation Patterns of Ancient Lakes on Mars?" *Journal of Geophysical Research,* vol. 115, 2010, p. E10006.

Emerson, D. O., and W. E. Hoffman. "Mineralogy of Woodrat, *Neotoma cinerea,* Urine Deposits from Northeastern California." *Journal of Mammalogy,* vol. 59, no. 2, 1978, pp. 424-25.

Erlandson, J. M., M. H. Graham, B. J. Bourque, D. Corbett, J. A. Estes, and R. S. Steneck. "The Kelp Highway Hypothesis: Marine Ecology, the Coastal Migration Theory, and the Peopling of the Americas." *The Journal of Island and Coastal Archaeology,* vol. 2, no. 20, 2007, pp. 161-74.

Eshleman, J. A., R. S. Malhi, J. R. Johnson, F. A. Kaestle, J. Lorenz, and D. G. Smith. "Mitochondrial DNA and Prehistoric Settlements: Native Migrations on the Western Edge of North America." *Human Biology,* vol. 76, 2004, pp. 55-75.

Fansler, V. A., and J. M. Mangold. "Restoring Native Plants to Crested Wheatgrass Stands." *Restoration Ecology,* vol. 19, 2010, pp. 16-23.

Federal Register. "Nt Hydro; Notice of Application Accepted for Filing and Soliciting Comments, Protests, and Motions to Intervene." 30 Jan. 2008, pp. 6953-54.

———. "Endangered and Threatened Wildlife and Plants; Removing the Borax Lake Chub from the List of Endangered and Threatened Wildlife." 13 July 2021, pp. 35574-94.

Finley, R. B., Jr. "Formation and Occurrence of Calcium Oxalate Deposits on Rocks at Wood Rat (*Neotoma*) Dens." Annual Meeting of the American Society of Mammalogists, June 1992, pp. 1-17.

Fleishman, E., editor. "Sixth Oregon Climate Assessment." Oregon Climate Change Research Institute, Oregon State U, 2023. https://blogs.oregon state.edu/occri/oregon-climate-assessments.

Fowler, C. S. "Subsistence." *Handbook of North American Indians, Volume 11: Great Basin,* pp. 64-97. Edited by W. L. D'Azevedo. Smithsonian Institution, 1986.

———. "Willard Z. Park's Ethnographic Notes on the Northern Paiute of Western Nevada, 1933–1944." University of Utah Anthropology Papers 114, 1989.

———. "In the Shadow of Fox Peak: An Ethnography of the Cattail-Eater Northern Paiute People of Stillwater Marsh." Cultural Resource Series 5. US Fish and Wildlife Service, 1992.

Fowler, C. S., and J. E. Bath. "Pyramid Lake Northern Paiute Fishing: The Ethnographic Record." *Journal of California and Great Basin Anthropology,* vol. 3, 1981, pp. 176-86.

Fowler, C. S., and S. Liljeblad. "Northern Paiute." *Handbook of North American Indians, Volume 11: Great Basin,* pp. 435-65. Edited by W. L. D'Azevedo. Smithsonian Institution, 1986.

Fremont, J. C. "Report of the Exploring Expedition to the Rocky Mountains in the Year 1842, and to California in the Years 1843–1844." US Senate executive document, 28th Congress, 2nd session, no. 174, 1845.

Frest, T. J., and E. J. Johannes. "Freshwater Mollusks of the Upper Klamath Drainage, Oregon." Deixis Consultants, 1998, unpublished yearly report.

Friedel, D. "Pleistocene Lake Chewaucan: Two Short Pieces on Hydrological Connections and Lake-Level Oscillations." *Quaternary Studies near Summer Lake, Oregon: Friends of the Pleistocene, Ninth Annual Pacific Northwest Cell Field Trip,* pp. DF1-3. Edited by R. Negrini., S. Pezzopane, and T. Badger. 28-30 Sept. 2001. http://www.fop .cascadiageo.org/pacific_northwest_cell/2001 /2001FOPSummerLakeGB.pdf.

Fuller, R. E., and A. C. Waters. "The Nature and Origin of the Horst and Graben Structure of Southern Oregon." *Journal of Geology,* vol. 37, 1929, pp. 204-38.

Galat, D. L., M. Coleman, and R. Robinson. "Experimental Effects of Elevated Salinity on Three Benthic Invertebrates in Pyramid Lake, Nevada." *Hydrobiologia,* vol. 158, no. 1, 1988, pp. 133-44.

Galbraith, H., D. W. DesRochers, S. Brown, and

J. M. Reed. "Predicting Vulnerabilities of North American Shorebirds to Climate Change." *PLOS One,* vol. 9, no. 9, 2014, p. e108899.

Garcia, C. M., and F. X. Niell. "Burrowing Beetles of the Genus *Bledius* (Staphylinidae) as Agents of Bioturbation in the Emergent Areas and Shores of an Athalassic Inland Lake (Fuente de Piedra, Southern Spain)." *Hydrobiologia,* vol. 215, 1991, pp. 163-73.

Gatschet, A. S. "The Klamath Indians of Southwestern Oregon." *Contributions to North American Ethnography,* vol. 2, no. 2, 1890.

Gerdes, G. H. Porada, and H. Bouougri. "Biosedimentary Structures Evolving from the Interaction of Microbial Mats, Burrowing Beetles and the Physical Environment of Tunisian Coastal Sabkhas." *Senckenbergiana Maritima,* vol. 38, no. 1, 2008, pp. 5-12.

Gilbert, G. K. *Lake Bonneville.* US Geological Survey. Government Printing Office, 1890.

———. "Studies of Basin-Range Structure." US Geological Survey Professional Paper 153, 1928.

Gilbert, M. T. P., D. L. Jenkins, A. Gotherstrom, N. Naveran, J. J. Sanchez, M. Hofreiter, P. F. Thompsen, J. Binladen, T. F. G. Higham, R. M. Yohe, R. Parr, L. S. Cummings, and E. Willerslev. "DNA from Pre-Clovis Coprolites in Oregon, North America." *Science,* vol. 320, no. 5,877, 2008, pp. 786-89.

Gilchrist, B. M. "Hemoglobin in *Artemia.*" *Proceedings of the Royal Society of London,* vol. 143, 1954, pp. 136-46.

Gleick, P. H., and H. Cooley. "Freshwater Scarcity." *Annual Review of Environment and Resources,* vol. 46, 2021, pp. 319-48.

Gobalet, K. W., and R. M. Negrini. "Evidence for Endemism in Fossil Tui Chubs, *Gilia bicolor,* from Pleistocene Lake Chewaucan, Oregon." *Copeia,* vol. 1, no. 2, 1992, pp. 539-44.

Grant-Hoffman, M. N., A. Clements, A. Lincoln, and J. Dollerschell. "Crested Wheatgrass (*Agropyron cristatum*) Seedings in Western Colorado: What Can We Learn?" *Management of Biological Invasions,* vol. 3, 2012, pp. 89-96.

Grayson, D. K. *The Deserts Past: A Natural Prehistory of the Great Basin.* Smithsonian Institution P, 1993.

———. *The Great Basin: A Natural Prehistory.* U of California P, 2011.

———. *Giant Sloths and Sabertooth Cats: Extinct Mammals and the Archaeology of the Ice Age Great Basin.* The U of Utah P, 2016.

Great Salt Lake Resolution (HCR-10) Steering Group. "Recommendations to Ensure Adequate Water Flows to Great Salt Lake and Its Wetlands." 2020. https://documents.deq.utah.gov/water-quality/standards-technical-services/gsl-website-docs/other-studies/DWQ-2021-035124.pdf.

Gunderson, L. H. "Ecological Resilience: In Theory and Application." *Annual Review of Ecology and Systematics,* vol. 31, 2000, pp. 425-39.

Gutiérrez, J. S. "Living in Environments with Contrasting Salinities: A Review of Physiological and Behavioral Responses in Waterbirds." *Ardeola,* vol. 61, no. 2, 2014, pp. 233-56.

Hadley, C. J. "Mr. Spud." *Range Magazine.* Summer 1998. http://www.rangemagazine.com/archives/stories/summer98/jr_simplot.htm.

Haig, S. M., D. W. Mehlman, and L. W. Oring. "Avian Movements and Wetland Connectivity in Landscape Conservation." *Conservation Biology,* vol. 12, 1998, pp. 749-58.

Haig, S. M., S. P. Murphy, J. H. Matthews, I. Arismendi, and M. Safeeq. "Climate-Altered Wetlands Challenge Waterbird Use and Migratory Connectivity in Arid Landscapes." *Scientific Reports 9,* article 4666, 2019.

Haig, S. M., L. W. Oring, P. M. Sanzenbacher, and O. W. Taft. "Space Use, Migratory Connectivity, and Population Segregation Among Willets Breeding in the Western Great Basin." *The Condor,* vol. 104, no. 3, 2002, pp. 620-30.

Hall, D. K., J. S. Kimball, R. Larson, N. E. DiGirolamo, K. A. Casey, and G. Hulley. "Intensified Warming and Aridity Accelerate Terminal Lake Desiccation in the Great Basin of the Western United States." *Earth and Space Science,* vol. 10, no. 1, 2023, pp. 1-20.

Halse, S. A., G. B. Pearson, and W. R. Kay. "Arid Zone Networks in Time and Space: Waterbird Use of Lake Gregory in North-Western Australia." *International Journal of Ecology and Environmental Sciences,* vol. 24, 1998, pp. 207-22.

Hamilton, R. B. "Comparative Behavior of the American Avocet and Black-Necked Stilt (Recurvirostridae)." *Ornithological Monographs 17,* 1975.

Hammer, U. T. *Saline Lake Ecosystems of the World.* W. Junk Publishers, 1986.

Hargrave, J. E. "Lithostratigraphy and Fossil Avifaunas of the Pleistocene Fossil Lake Formation, Fossil Lake, Oregon, and the Oligocene Etadunna Formation, Lake Palankarinna, South Australia." PhD Dissertation. U of Oklahoma, 2009.

Harris, R. C. "Giant Desiccation Cracks in Arizona." *Arizona Geological Survey Open-File Report,* 1 April 2004.

Hartson, T. *Squirrels of the West.* Lone Pine Publishing, 1999.

Hassan, R., R. Scholes, and N. Ash, editors. *Ecosystems and Human Well-Being: Current State and Trends.* Island Press, 2005.

Heinrich, B. *Racing the Antelope.* Harper Collins, 2001.

Hemphill, B. E. "Wear and Tear: Osteoarthritis as an Indicator of Mobility Among Great Basin Hunter-Gatherers." *Prehistoric Lifeways in the Great Basin Wetlands: Bioarchaeological Reconstruction and Interpretation,* pp. 241-89. Edited by B. E. Hemphill and C. S. Larsen. U of Utah P, 1999.

Hemphill-Haley, M. A. "Quaternary Stratigraphy and Late Holocene Faulting Along the Base of the Eastern Escarpment of Steens Mountain, Southeastern Oregon." MS thesis. Humboldt State, 1987.

Herbst, D. B. "Ecological Physiology of the Larval Brine Fly *Ephydra (Hydropyrus) hians,* an Alkaline-Salt Inhabiting Ephydrid (Diptera)." MS thesis. Oregon State U, 1980.

———. "Comparative Studies of the Population Ecology and Life History Patterns of an Alkaline Salt Lake Insect *Ephydra (Hydropyrus) hians* Say (Diptera: Ephydridae)." Dissertation. Oregon State U, 1986.

———. "Comparative Population Ecology of *Ephydra hians* Say (Diptera: Ephydridae) at Mono Lake (California) and Lake Abert (Oregon)." *Hydrobiologia,* vol. 158, 1988, pp. 145-66.

———. "Aquatic Ecology of the Zone of Lake Abert: Defining Critical Lake Levels and Optimum Salinity for Biological Health." Report prepared for the Oregon Department of Fish and Wildlife and the US Bureau of Land Management, August 1994.

Herbst, D. B., and T. J. Bradley. "A Malpighian Tubule Lime Gland in an Insect Inhabiting Alkaline Salt Lakes." *Journal of Experimental Biology,* vol. 145, 1989, pp. 63-78.

———. "Salinity and Nutrient Limitations on Growth of Benthic Algae from Two Alkaline Salt Lakes of the Western Great Basin (USA)." *Journal of Phycology,* vol. 25, no. 4, 2004, pp. 673-78.

Herbst, D. B., and R. W. Castenholz. "Growth of the Filamentous Green Alga *Ctenocladus circinatus* (Chaetophorales, Chlorophyceae) in Relation to Environmental Salinity." *Journal of Phycology,* vol. 30, 1994, pp. 588-93.

Herbst, D. B., S. W. Roberts, and R. R. Medhurst. "Defining Salinity Limits on the Survival and Growth of Benthic Insects for the Conservation Management of Saline Walker Lake, Nevada, USA." *Journal of Insect Conservation,* vol. 17, 2013, pp. 877-83.

Hernández-Pérez, M., R. M. Rabanal, A. Arias, and B. Rodríguez. "Aethiopinone, an Antibacterial and Cytotoxic Agent from *Salvia aethiopis* Roots." *Pharmaceutical Biology,* vol. 37, no. 1, 2008, pp. 17-21.

Hershler, R., D. B. Madsen, and D. R. Currey. *Great Basin Aquatic Systems History: Smithsonian Contributions to the Earth Sciences 33,* 2002.

Hershler, R., and D. W. Sada. "Biogeography of Great Basin Aquatic Snails of the Genus *Pyrgulopsis.*" *Great Basin Aquatic Systems History: Smithsonian Contributions to the Earth Sciences 33,* pp. 255-76. Edited by R. Hershler, D. B. Madsen, and D. R. Currey, 2002.

Heydari, N., and H. Jabbari. "Worldwide Environmental Threats to Salt Lakes." *International Journal of Design and Nature and Ecodynamics,* vol. 7, no. 3, 2012, pp. 292-99.

Hixon, M. A., S. V. Gregory, and W. D. Robinson. "Oregon's Fish and Wildlife in a Changing Climate." *The Oregon Climate Change Assessment Report.* Edited by K. D. Dello and P. W. Mote. Oregon Climate Change Research Institute, College of Oceanic and Atmospheric Sciences, Oregon State U, 2010.

Hoehler, T. M., B. M. Bebout, and D. J. Des Maras. "The Role of Microbial Mats in the Production of Reduced Gases on the Early Earth." *Nature,* vol. 412, 2001, pp. 324-27.

Hoffman, R. *Birds of the Pacific States.* Houghton Mifflin, 1927.

Holling, C. S. "The Resilence of Terrestrial Ecosystems: Local Surprise and Global Change." *Sustainable Development of the Biosphere,* pp. 292-317. Edited by W. C. Clark and R. E. Munn. Cambridge UP, 1986.

Hudson, A. M., J. Quade, G. Ali, D. Boyle, S. Bassett, K. W. Huntington, M. G. De los Santos, A. S. Cohen, K. Lin, and X. Wang. "Stable C, O and Clumped Isotope Systematics and ^{14}C Geochronology of Carbonates from the Quaternary Chewaucan Closed-Basin Lake System, Great Basin, USA: Implications for Paleoenvironmental Reconstructions Using Carbonates." *Geochimica et Cosmochimica Acta,* vol. 212, 2017, pp. 274-302.

Hurlbert, S. H., M. Lopez, and J. O. Keith. "Wilson's Phalarope in the Central Andes and Its Interaction with the Chilean Flamingo." *Revista Chilena de Historia Natural,* vol. 57, 1984, pp. 47-57.

Ibarra, D. E., A. E. Egger, K. L. Weaver, C. R. Harris, and K. Maher. "Rise and Fall of Late Pleistocene Pluvial Lakes in Response to Reduced Evaporation and Precipitation: Evidence from Lake Surprise, California." *Geological Society of America Bulletin,* vol. 126, no. 11-12, 2014, pp. 1387-1415.

Intermountain West Joint Venture 2013 Implementation Plan. http://iwjv.org/resource/iwjv-2013-implementation-plan-entire-plan.

Isaacs, F. B. "Golden Eagles (*Aquila chrysaetos*) Nesting in Oregon, 2011–2017." Draft Annual Report, 2018. Oregon Eagle Foundation.

Jackman, E. R., and R. A. Long. *The Oregon Desert.* Caxton Printers, 1964.

Jarboe, N. A., R. S. Coe, P. R. Renne, M. G. Jonathan, and E. A. Mankinen. "Quickly Erupted Volcanic Sections of the Steens Basalt, Columbia River Basalt Group, Secular Variation, Tectonic Rotation, and Steens Mountain Reversal." *Geochemistry Geophysics Geosystems,* vol. 9, no. 11, 2008, pp. 1-24.

Jehl, J. R., Jr. "Biology of the Eared Grebe and Wilson's Phalarope in the Non-breeding Season: A Study of Adaptations of Saline Lakes." *Studies in Avian Biology,* vol. 12, 1988, 1-74.

———. "Changes in Saline and Alkaline Lake Avifaunas in Western North America in the Past 150 years." *Studies in Avian Biology,* vol. 15, 1994, pp. 258-72.

———. "Fat Loads and Flightlessness in Wilson's Phalaropes." *The Condor,* vol. 99, 1997, pp. 538-43.

———. "Population Studies of Wilson's Phalaropes at Fall Staging Areas, 1980–1997: A Challenge for Monitoring Waterbirds." *International Journal of Waterbird Biology,* vol. 22, no. 1, 1999, pp. 37-46.

———. "Why Do Eared Grebes Leave Hypersaline Lakes in Autumn?" *Waterbirds,* vol. 30, no. 1, 2007, pp. 112-15.

Jehl, J. R. Jr., H. Ellis, and A. E. Henry. "Optimizing Migration in a Reluctant and Inefficient Flier: The Eared Grebe." *Avian Migration,* pp. 199-209. Edited by P. Berthold, E. Gwinner, and E. Sonnenschein. Springer-Verlag, 2003.

Jellison, R., W. D. Williams, B. Timms, J. Alcocer, and N. Aladin. "Salt Lakes: Values, Threats, and Future." *Aquatic Ecosystems,* pp. 94-110. Edited by N. V. C. Polunin. Cambridge UP, 2008.

Jellison, R., Y. S. Zadereev, P. A. DasSarma, J. M. Melack, M. R. Rosen, A. G. Degermendzhy, S. DasSarma, and G. Zambrana. "Conservation and Management Challenges of Saline Lakes: A Review of Five Experience Briefs." Lake Basin Management Initiative Briefs, 2004. https://www.researchgate.net/publication/228885117_Conservation_and_management_challenges_of_Saline_lakes_A_review_of_five_experience_briefs.

Jenkins, D. L., T. J. Connolly, and C. M. Aikens. "Early and Middle Holocene Archaeology in the Northern Great Basin: Dynamic Natural and Cultural Ecologies." *Early and Middle Holocene Archaeology of the Northern Great Basin,* pp. 1-20. Edited by D. L. Jenkins, T. J. Connolly, and C. M. Aikens. University of Oregon Anthropological Papers 62, 2004.

Jenkins, D. L., and nineteen coauthors. "Clovis Age Western Stemmed Projectile Points and Human Coprolites at the Paisley Caves." *Science,* vol. 337, 2012, pp. 223-28.

Johnson, D. M., R. R. Petersen, D. R. Lycan, J. W. Sweet, M. E. Neuhaus, and A. L. Schaedel. *Atlas of Oregon Lakes.* Oregon State UP, 1985.

Jones, B. F., A. S. VanDenburgh, A. H. Truesdell, and S. L. Rettig. "Interstitial Brines in Playa Sediments." *Chemical Geology,* vol. 4, 1969, pp. 253-62.

Kaden, H., F. Peeters, A. Lorke, R. Kipfer, Y. Tomonaga, and M. Karabiyikoglu. "Impact of Lake Level Change on Deep-water Renewal and Oxic Conditions in Deep Saline Lake Van, Turkey." *Water Resources Research,* vol. 46, 2010, pp. 1-14.

Kallenbach, E. A. "The Chewaucan Cave Cache: A Specialized Tool Kit from Eastern Oregon." *Journal of California and Great Basin Anthropology,* vol. 33, no. 1, 2013, pp. 72-87.

Karban, R., W. C. Wetzel, K. Shiojiri, E. Pezzola, and J. D. Blande. "Geographic Dialects in Volatile Communication Between Sagebrush Individuals." *Ecology,* vol. 97, no. 11, 2016, pp. 2917-24.

Keen, F. P. "Climatic Cycles in Eastern Oregon Indicated by Tree Rings." *Monthly Weather Review,* vol. 65, no. 5, 1937, pp. 183-88.

Keil, K. "Geological History of Asteroid 4 Vesta: The 'Smallest Terrestrial Planet.'" *Asteroids III,* pp. 573-84. Edited by W. Bottke, A. Cellino, P. Paolicchi, and R. P. Binzel. U of Arizona P, 2002.

Keister, G. P., Jr. "The Ecology of Lake Abert: Analysis of Further Development." Special Report, Oregon Department of Fish and Wildlife, April 1992.

Kelly, I. "Ethnography of the Surprise Valley Paiute." *University of California Publications in American Archaeology and Ethnology,* vol. 31, no. 3, pp. 67-210. U of California P, 1932.

Kerr, A. *Oregon Desert Guide: 70 Hikes.* Mountaineers Books, 2000.

Kersten, M., and T. Piersma. "High Levels of Energy Expenditure in Shorebirds: Metabolic Adaptations to an Energetically Expensive Way of Life." *Ardea,* vol. 75, 1987, pp. 175-87.

Kessler, M. A., and B. T. Werner. "Self-Organization of Sorted Patterned Ground." *Science,* vol. 299, no. 5,605, 2003, pp. 380-83.

Kittredge, W. *Hole in the Sky: A Memoir.* Random House, 1993.

Koriche, S. A., S. D. Nandini-Weiss, M. Prange, J. S. Singarayer, K. Arpe, H. L. Cloke, M. Schulz, P. Bakker, S. A. G. Leroy, and M. Coe. "Impacts of Variations in Caspian Sea Surface Area on

Catchment-Scale and Large-Scale Climate." *Journal of Geophysical Research: Atmospheres,* vol. 126, 2021, pp. 1-17.

Kristensen, K., M. Stern, and J. Morawski. "Birds of North Lake Abert, Lake County, Oregon." *Oregon Birds,* vol. 17, 1991, pp. 67-77.

KTVB. "Monday's Wind, Rain Leaves Treasure Valley Cars in Need of a Wash." 2022. https://www.ktvb.com/article/weather/mondays-wind-rain-leaves-cars-in-need-of-a-wash/277-8dd57ca6-26e4-4ee7-b07b-103d84f39433.

Kuehn, S. C., and R. M. Negrini. "A 250 k.y. Record of Cascade Arc Pyroclastic Volcanism from Late Pleistocene Lacustrine Sediments near Summer Lake, Oregon, USA." *Geosphere,* vol. 6, no. 4, 2010, pp. 397-429.

Lank, D. B., R. W. Butler, J. Ireland, and R. C. Ydenberg. "Effects of Predation Danger on Migration Strategies of Sandpipers." *Oikos,* vol. 103, 2003, pp. 303-19.

Larson, R. "Oregon Plants, Oregon Places: Gearhart Mountain Wilderness." *Kalmiopsis,* vol. 14, 2007, pp. 17-23.

Larson, R., J. Eilers, K. Kreuz, W. T. Pecher, S. DasSarma, and S. Dougill. "Recent Desiccation-Related Ecosystem Changes at Lake Abert, Oregon." *Western North American Naturalist,* vol. 76, no. 4, 2016, pp. 389-404.

Lesterhuis, A. J., and R. P. Clay. "Conservation Plan for Wilson's Phalarope (*Phalaropus tricolor*)." Version 1.1. Manomet Center for Conservation Sciences, 2010.

Licciardi, J. M. "Chronology of Latest Pleistocene Lake-Level Fluctuations in the Pluvial Lake Chewaucan Basin, Oregon, USA." *Journal of Quaternary Science,* vol. 16, no. 6, 2001, pp. 545-53.

Lima, S. L. "Ecological and Evolutionary Perspectives on Escape from Predatory Attack: A Survey of North American Birds." *Wilson Bulletin,* vol. 105, 1993, pp. 1-47.

Littlefield, C. D. *Birds of the Malheur National Wildlife Refuge, Oregon.* Oregon State UP, 1990.

Liu, T., and W. S. Broecker. "Millennial-Scale Varnish Microlamination Dating of Late Pleistocene Geomorphic Features in the Drylands of Western USA." *Geomorphology,* vol. 187, 2013, pp. 38-60.

Lorenz, R. D., B. K. Jackson, J. W. Barnes, J. Spitale, and J. M. Keller. "Ice Rafts Not Sails: Floating the Rocks at Racetrack Playa." *American Journal of Physics,* vol. 79, no. 1, 2011, pp. 37-42.

Loring, J. M, and L. Loring. *Pictographs & Petroglyphs of the Oregon Country, Parts I & II: Monographs 21/23.* Institute of Archaeology, U of California, Los Angeles, 1996.

Lund, N. "Wilson's Phalarope: The Rebel." The Sketch. *Audubon Magazine,* September–October 2015.

Ma, L., J. Wu, and J. Abuduwalli. "Distinguishing Between Anthropogenic and Climatic Impacts on Lake Size: A Modeling Aapproach Using Data from Ebinur Lake in Arid Northwest China." *Journal of Limnology,* vol. 73, no. 2, 2014, pp. 148-55.

Madsen, D. B., R. Hershler, and D. R. Currey. "Introduction." *Great Basin Aquatic Systems History,* pp. 1-10. Edited by R. Hershler, D. B. Madsen, and D. R. Currey. Smithsonian Institution P, 2002.

Mahood, G. A., and T. R. Benson. "Using 40Ar/39Ar Ages of Intercalated Silicic Tuffs to Date Flood Basalts: Precise Ages for Steens Basalt Member of the Columbia River Basalt Group." *Earth and Planetary Science Letters,* vol. 459, 2017, pp. 340-51.

Maloney, A. B. "Fur Brigade to the Bonaventura: John Work's California Expedition of 1832–33 for the Hudson's Bay Company." *California Historical Society Quarterly,* vol. 22, no. 3, 1943, pp. 193-222.

Mansfield, D. H. *Flora of Steens Mountain.* Oregon State UP, 1999.

Markle, D. F. *A Guide to Freshwater Fishes of Oregon.* Oregon State UP, 2016.

Marshall, D. B, M. G. Hunter, and A. L. Contreras, editors. *Birds of Oregon: A General Reference.* Oregon State UP, 2003.

Maser, C. *Mammals of the Pacific Northwest: From the Coast to the High Cascades.* OSU P, 1998.

Mathewson, W. *William L. Finley: Pioneer Wildlife Photographer.* Oregon State UP, 1986.

Mathis, W. N., T. Zatwarnicki, and M. G. Krivosheina. *Studies of Gymnomyzinae (Diptera: Ephydridae), V: A Revision of the Shore-Fly Genus Mosillus Latreille.* Smithsonian Institution P, 1993.

Matthiessen, P. *The Wind Birds: Shorebirds of North America.* Chapters Publishing, 1994.

Melack, J. M., R. Jellison, and D. B. Herbst, editors. *Saline Lakes.* Developments in Hydrobiology series. Kluwer Academic Publishers, 2001.

Merola, M. "Observations on the Nesting and Breeding Behavior of the Rock Wren." *Condor,* vol. 97, 1995, pp. 585-87.

Micklin, P. P. "Desiccation of the Aral Sea: A Water Management Disaster in the Soviet Union." *Science,* vol. 241, no. 4870, 1988, pp. 1170-75.

Miller, A. W., K. D. Kohl, and M. D. Dearing. "The Gastrointestinal Tract of the White-Throated Woodrat (*Neotoma albigula*) Harbors Distinct Consortia of Oxalate-Degrading Bacteria." *Applied Environmental Microbiology,* vol. 80, no. 5, 2014, pp. 1595-1601.

Miller, M. B. *Roadside Geology of Oregon* (second edition). Mountain Press, 2014.

Miller, R. F., and P. E. Wigand. "Holocene Changes in Semiarid Pinyon-Juniper Woodlands." *Bioscience,* vol. 44, no. 7, 1994, pp. 465-74.

Millsap, B. A., G. S. Zimmerman, J. R. Sauer, R. M. Nielson, M. Otto, E. Bjerre, and R. Murphy. "Golden Eagle Population Trends in the Western United States: 1968–2010." *Journal of Wildlife Management,* vol. 77, 2013, pp. 1436-48.

Minckley, T. A., C. Whitlock, and P. J. Bartlein. "Vegetation, Fire, and Climate History of the Northwestern Great Basin During the Last 14,000 Years." *Quaternary Science Reviews,* vol. 26, 2007, pp. 2167-84.

Mladen, Z., R. M. Negrini, and P. E. Wigand. "Evidence of Synchronous Climate Change Across the Northern Hemisphere Between the North Atlantic and the Northwestern Great Basin, United States." *Geology,* vol. 30, no. 7, 2002, pp. 635-38.

Moore, J. "Recent Desiccation of Western Great Basin Saline Lakes: Lessons from Lake Abert, Oregon, U.S.A." *Science of the Total Environment,* vols. 554-55, 2016, pp. 142-54.

Morrison, R. B. "Quaternary Stratigraphic, Hydrologic, and Climatic History of the Great Basin, with Emphasis on Lakes Lahontan, Bonneville, and Tecopa." *The Geology of North America: Quaternary Nonglacial Geology: Conterminous U.S.,* vol. K-2, pp. 283-320. Edited by R. B. Morrison. Geological Society of America, 1991.

Mozingo, H. N. *Shrubs of the Great Basin: A Natural History.* U of Nevada P, 1987.

Mucken, A., and B. Bateman, editors. "Oregon's 2017 Integrated Water Resources Strategy." Oregon Water Resources Department, 2017.

Munoz, J. F. Amat, A. J. Green, J. Figuerola, and A. Gomez. "Bird Migratory Flyways Influence the Phylogeography of the Invasive Brine Shrimp *Artemia franciscana* in Its Native American Range." *PeerJ,* vol. 1, 2013, p. e200.

NASA. "Phoenix Mars Lander: Exploring the Arctic Plain of Mars." Missions, 13 May 2008. https://www.nasa.gov/mission_pages/phoenix/multimedia/5302-20080513.html.

———. "Saltiest Pond on Earth." 2014. https://earthobservatory.nasa.gov/images/84955/saltiest-pond-on-earth.

———. "Red Lake Urmia." 2016. http//:earthobservatory.nasa.gov.

———. "Shrinking Aral Sea." 2016. http://earthobservatory.nasa.gov.

———. "A Salt Bath in Bolivia." 2022. https://earthobservatory.nasa.gov/images/149502/a-salt-bath-in-bolivia.

National Audubon Society. G. Langham, J. Schuetz, C. Soykan, C. Wilsey, T. Auer, G. Le Baron, C. Sanchez, and T. Distler, contributors. "Audubon's Birds and Climate Change Report." 2014. http://climate.audubon.org/sites/default/files/Audubon-Birds-Climate-Report-v1.2.pdf.

NatureServe Explorer. https://explorer.natureserve.org/Taxon/ELEMENT_GLOBAL.2.797081/Pyrgulopsis_robusta.

Nebert, D. D. "Development and Application of a Water Budget Model for Lake Fluctuation, Goose Lake Basin, Oregon-California." MS thesis. Portland State U, 1985.

Neel, L. A., and W. G. Henry. "Shorebirds of the Lahontan Valley, Nevada, USA: A Case History of Western Great Basin Shorebirds." *International Wader Studies,* vol. 9, 1996, pp. 15-19.

Negrini, R. M. "Pluvial Lake Sizes in the Northwestern Great Basin Throughout the Quaternary Period." *Great Basin Aquatic Systems History,* pp. 11-53. Edited by R. Hershler, D. B. Madsen, and D. Currey. Smithsonian Institution P, 2002.

Negrini, R., D. B. Erbes, K. Faber, A. M. Herrera, A. P. Roberts, A. S. Cohen, P. E. Wigand, and F. F. Foit. "A Paleoclimate Record for the Past 250,000 Years from Summer Lake, Oregon, USA: I. Chronology and Magnetic Proxies for Lake Level." *Journal of Paleolimnology,* vol. 24, 2000, pp. 125-49.

Negrini, R., S. Pezzopane, and T. Badger, editors. "Quaternary Studies near Summer Lake, Oregon." Friends of the Pleistocene, Ninth Annual Pacific Northwest Cell Field Trip, 28-30 Sept. 2001. http://www.fop.cascadiageo.org/pacific_northwest_cell/2001/2001FOPSummerLakeGB.pdf.

Nehls, H. "Oregon Shorebirds: Their Status and Movements." Oregon Department of Fish and Wildlife, Wildlife Diversity Program Report, 2 Jan. 1994.

Newbold, T. A., and J. A. MacMahon. "Consequences of Cattle Introduction in a Shrub-Steppe Ecosystem: Indirect Effects on Desert Horned Lizards (*Phrynosoma platyrhinos*)." *Western North American Naturalist,* vol. 68, no. 3, 2008, pp. 291-302.

Niles, L. J., and twenty-one coauthors. "Status of the Red Knot, *Calidris canutus rufa,* in the Western Hemisphere." *Studies in Avian Biology,* vol. 36, 2008, pp. 1-185.

Norris, R. D., J. M. Norris, R. D. Lorenz, J. Ray, and B. Jackson. "Sliding Rocks on Racetrack Playa, Death Valley National Park: First Observation of Rocks in Motion." *PLOS One,* vol. 9, no. 8, 2014, p. e105948.

Northup, D. E., J. R. Snider, M. N. Spilde, M. L. Porter, J. L. van de Kamp, P. J. Boston, A. M. Nyberg, and J. R. Bargar. "Diversity of Rock Varnish Bacterial Communities from Black

Canyon, New Mexico." *Journal of Geophysical Research,* vol. 115 (GO2007), 2010, pp. 1-19.

O'Brien, M., R. Crossley, and K. Karlson. *The Shorebird Guide.* Houghton Mifflin, 2006.

Obst, B. S., W. M. Hamner, P. P. Hamner, E. Wolanski, M. Rubega, B. Littlehales. "Kinematics of Phalarope Spinning." *Nature,* vol. 384, 1996, p. 121.

Oetting, A. C. "Villages and Wetland Adaptations in the Northern Great Basin: Chronology and Land Use in the Lake Abert-Chewaucan Marsh Basin, Lake County, Oregon." University of Oregon *Anthropological Papers,* vol. 41, 1989, pp. 1-351.

———. "Aboriginal Settlement in the Lake Abert-Chewaucan Marsh Basin, Lake County, Oregon." *Wetlands Adaptations in the Great Basin: Museum of Peoples and Cultures Occasional Papers No. 1,* pp. 183-206. Edited by J. Janetski and D. B. Madsen. Brigham Young U, 1990.

———. "Chronology and Time Markers in the Northwestern Great Basin: The Chewaucan Basin Cultural Chronology." *Archaeological Researches in the Northern Great Basin: Fort Rock Archaeology Since Cressman,* pp. 41-62. Edited by C. M. Aikens and D. L. Jenkins. University of Oregon Anthropological Papers 50, 1994.

———. "Early Holocene Rabbit Drives and Prehistoric Land Use Patterns on Buffalo Flats, Christmas Lake Valley, Oregon." *Archaeological Researches in the Northern Great Basin: Fort Rock Archaeology Since Cressman,* pp. 155-70. Edited by C. M. Aikens and D. L. Jenkins. University of Oregon Anthropological Papers 50, 1994.

Oregon Department of Water Resources. "Goose and Summer Lakes Basin Report," 1989. https://www.oregon.gov/owrd/wrdreports/Goose_and _Summer_Lakes_Basin_1989.pdf.

Oren, A., J. Gavrieli, M. Kohen, J. Lati, M. Aharoni, and I. Gavrieli. "Long-term Mesocosm Simulation of Algal and Archaeal Blooms in the Dead Sea Following Dilution with Red Sea Water." *Saline Lakes Around the World: Unique Systems with Unique Values,* pp. 149-55. Edited by A. Oren, D. L. Naftz, P. Placios, and W. A. Wurtsbaugh, *Natural Resources and Environmental Issues,* vol. 15, 2009.

Oring, L. W., L. Neel, and K. E. Oring. "U.S. Shorebird Conservation Plan: Intermountain West Regional Shorebird Plan," 2009. https://www .shorebirdplan.org/wp-content/uploads/2013/01 /IMWEST4.pdf.

Oring, L. W., and J. M. Reed. "Shorebirds of the Western Great Basin of North America: Overview and Importance to Continental Populations." *International Wader Studies,* vol. 9, 1997, pp. 6-12.

Orme, A. R. "Pleistocene Pluvial Lakes of the American West: A Short History of Research." *Geological Society, London, Special Publications,* vol. 301, 2008, pp. 51-78.

Orosei, R., and twenty-one coauthors. "Radar Evidence of Subglacial Liquid Water on Mars." *Science,* vol. 361, no 6401, 2018, pp. 490-93.

Overpeck, J. T., and B. Udall. "Climate Change and the Aridification of North America." *PNAS,* vol. 117, no. 22, 2020, pp. 11856-58. https://doi.org /10.1073/pnas.2006323117.

Oyewusi, H. A., R. A. Wahab, M. F. Edbeib, M. A. N. Mohamad, A. A. A. Hamid, Y. Kaya, and F. Huyop. "Functional Profiling of Bacterial Communities in Lake Tuz Using 16S rRNA Gene Sequences." *Biotechnology & Biotechnological Equipment,* vol. 35, no. 1, 2021, pp. 1-10.

Paige, C., and S. A. Ritter. "Birds in a Sagebrush Sea: Managing Sagebrush Habitats for Bird Communities." Partners in Western Flight working group, 1999. https://partnersinflight.org/resources/birds -in-a-sagebrush-sea/.

Parsons, A. J., and A. D. Abrahams. *Geomorphology of Desert Environments.* Springer, 2009.

Patton, M. A., G. McCaskie, and P. Unitt. *Birds of the Salton Sea.* U of California P, 2003.

Paul, D. S., and A. E. Manning. "Great Salt Lake Waterbird Survey, Five-Year Report (1997–2001)." Utah Division of Wildlife Resources publication number 08-38, 2002.

Paulson, D. *Shorebirds of the Pacific Northwest.* U of Washington P, 1993.

Pay, D., R. Weldon, S. McClure, and K. Schumacher. "Three Occurrences of Oregon Sunstones." *Gems and Gemology,* vol. 49, no. 3, 2013.

Pedler, R. D., R. F. H. Ribot, and A.T. D. Bennett. "Extreme Nomadism in Desert Waterbirds: Flights of the Banded Stilt." *Biology Letters,* vol. 10, issue 10, 2014, p. 20140547.

Pellant, M. "Cheatgrass: The Invader That Won the West." Interior Columbia Basin Ecosystem Management Project, BLM Idaho State Office, 1996. http://www.icbemp.gov/science/pellant.pdf.

Perreault, T. "Climate Change and Climate Politics: Parsing the Causes and Effects of the Drying of Lake Poopó, Bolivia." *Journal of Latin American Geography,* vol. 19, no. 3, 2019, pp. 26-46.

Perry, R. S., and M. A. Sephton. "Solving the Mystery of Desert Varnish with Microscopy." *Infocus,* vol. 11, 2008, pp. 62-76.

Peters, T. J., M. F. Diggles, and D. S. Kostick. "Brine Mineral Occurrence in the Diablo Mountain Wilderness Study Area, Oregon, and Its Possible Significance to Pacific Rim Trade." *Oregon Geology,* vol. 56, no. 51, 1994, pp. 51-61.

Pettigrew, R. M. "The Ancient Chewaucanians: More on the Prehistoric Lake Dwellers of Lake Abert, Southwestern Oregon." *Association of*

Oregon Archaeologists Occasional Papers, vol. 1, 1980, pp. 49-67.

———. *Archaeological Investigations on the East Shore of Lake Abert, Lake County, Oregon.* University of Oregon Anthropological Papers 32, 1985.

Pezzopane, S. "Pluvial Lake Chewaucan Shoreline Elevations." Quaternary Studies near Summer Lake, Oregon: Friends of the Pleistocene, Ninth Annual Pacific Northwest Cell Field Trip, pp. SP1-4. Edited by R. Negrini, S. Pezzopane, and T. Badger. 28-30 Sept. 2001. http://www.fop.cascadiageo.org/pacific_northwest_cell/2001/2001FOPSummerLakeGB.pdf.

Pezzopane, S. K., and R. J. Weldon. "Tectonic Role of Active Faulting in Central Oregon." *Tectonics,* vol. 12, 1993, pp. 1140-69.

Phillips, K. N., and A. S. Van Denburgh. "Hydrology and Geochemistry of Abert, Summer, and Goose Lakes, and Other Closed-Basin Lakes in South-Central Oregon." Closed-Basin Investigations: *US Geological Survey Professional Paper* 502-B, 1971.

Pianka, E. R., and W. S. Parker. "Ecology of Horned Lizards: A Review with Special Reference to *Phrynosoma platyrhinos.*" *Copeia,* 1975, pp. 141-62.

Pillod, D. S., J. L. Welty, and R. S. Arkle. "Refining the Cheatgrass-Fire Cycle in the Great Basin: Precipitation Timing and Fine Fuel Composition Predict Wildfire Trends." *Ecology and Evolution,* vol. 7, 2017, pp. 8126-51.

Pintarich, R. M. *The Swamp Land Act in Oregon, 1870–1895.* Portland State U, 1980.

Plissner, J. H., S. M. Haig, and L. W. Oring. "Within- and Between-Year Dispersal of American Avocets Among Multiple Western Great Basin Wetlands." *Wilson Bulletin,* vol. 111, no. 3, 1999, pp. 314-20.

———. "Post-breeding Movements of American Avocets and Implications for Wetland Connectivity in the Western Great Basin." *The Auk,* vol. 117, no. 2, 2000, pp. 29098.

Pohl, K. A., K. S. Hadley, and K. B. Arabas. "A 545-Year Drought Reconstruction for Central Oregon." *Physical Geography,* vol. 23, 2002, pp. 302-20.

Prakash, M., D. Quéré, and J. W. M. Bush. "Surface Tension Transport of Prey by Feeding Shorebirds: The Capillary Ratchet." *Science,* vol. 320, 2008, pp. 931-34.

Pratt, J. "Truckee-Carson River Basin Study: Final Report." Report to the Western Water Policy Review Advisory Commission. Clearwater Consulting Corporation, 1997.

Purdue, J. R. "Adaptations of the Snowy Plover on the Great Salt Plains, Oklahoma." *Southwestern Naturalist,* vol. 21, 1976, pp. 347-57.

Ramsey, J. *Coyote Was Going There: Indian Literature of the Oregon Country.* U of Washington P, 1977.

Raymond, A. W., and E. Sobel. "The Use of Tui Chub as Food by Indians of the Western Great Basin." *Journal of California and Great Basin Anthropology,* vol. 12, no. 1, 1990, pp. 2-18.

Reheis, M. C., "Extent of Pleistocene Lakes in the Western Great Basin." *USGS Miscellaneous Field Studies Map Mf-2323,* 1999.

Reheis, M. C., K. D. Adams, C. G. Oviatt, and S. N. Bacon. "Pluvial Lakes in the Great Basin of the Western United States—a View from the Outcrop." *Quaternary Science Reviews,* vol. 97, 2014, pp. 33-57.

Reheis, M. C., J. R. Budahn, and P. Lamothe. "Geochemical Evidence for Diversity of Dust Sources in the Southwestern United States." *Geochimica et Cosmochimica Acta,* vol. 66, 2022, pp. 1569-87.

Reichert, K. L., J. M. Licciardi, and D. S. Kaufman. "Amino Acid Racemization in Lacustrine Ostracods, Part II: Paleothermometry in Pleistocene Sediments at Summer Lake, Oregon." *Quaternary Geochronology,* vol. 6, 2011, pp. 174-85.

Reidel, S. P., V. E. Camp, M. E. Ross, J. A. Wolff, and B. S. Martin. "The Columbia River Flood Basalt Province: Stratigraphy, Areal Extent, Volume, and Physical Volcanology." *The Columbia River Flood Basalt Province,* pp. 1-43. Edited by S. P. Reidel, V. E. Camp, M. E. Ross, J. A. Wolff, B. S. Martin, T. L. Tolan, and R. E. Wells. Geological Society of America, vol. 497, 2013.

Reisner, M. D., J. B. Grace, D. A. Pyke, and P. S. Doescher. "Conditions Favoring *Bromus tectorum* Dominance of Endangered Sagebrush Steppe Ecosystems." *Journal of Applied Ecology,* vol. 50, 2013, pp. 1039-49.

Remple, S. "Taxonomy and Systematic Relationships of Tui Chubs (*Siphateles*: Cyprinidae) from Oregon's Great Basin." MS thesis. Oregon State U, 2013.

Risatti, J. B., W. C. Capman, and D. A. Stahl. "Community Structure of a Microbial Mat: The Phylogenetic Dimension." *Proceedings of the National Academy of Sciences,* vol. 91, no. 21, 1994, pp. 10173-77.

Roberts, A. J., M. R. Conover, J. Luft, and J. Neil. "Population Fluctuations and Distribution of Staging Eared Grebes (*Podiceps nigricollis*) in North America." *Canadian Journal of Zoology,* vol. 92, 2013, pp. 906-13.

Robinson, A. *The Spoils of Dust: Reinventing the Lake That Made Los Angeles.* Applied Research and Design Publishing, 2018.

Robinson, J. A., and L. W. Oring. "Long-Distance Movements by American Avocets and

Black-Necked Stilts." *Journal of Field Ornithology,* vol. 67, no. 2, 1996, pp. 307-20.

Rogler, G. A., and R. J. Lorenz. "Crested Wheatgrass—Early History in the United States." *Journal of Range Management,* vol. 36, 1983, pp. 91-93.

Rom, W. N., W. Greaves, K. M. Bang, M. Holthouser, D. Campbell, and R. Bernstein. "An Epidemiologic Study of the Respiratory Effects of Trona Dust." *Archive of Environmental Health,* vol. 38, no. 2, 1983, pp. 86-92.

Rosenbaum, J. G., and R. L. Reynolds. "Record of Late Pleistocene Glaciation and Deglaciation in the Southern Cascade Range" and "Flux of Glacial Flour in a Sediment Core from Upper Klamath Lake, Oregon." *Journal of Paleolimnology,* vol. 31, 2004, pp. 235-52.

Rosentreter, R. "Sagebrush Identification, Ecology, and Palatability to Sage-Grouse." USDA Forest Service Proceedings RMRS-P-38, 2005, pp. 3-15.

Roshier, D. A., N. I. Klomp, and M. Asmus. "Movements of a Nomadic Waterfowl, Grey Teal *Anas gracilis,* Across Inland Australia—Results from Satellite Telemetry Spanning Fifteen Months."*Ardea,* vol. 94, no. 3, 2006, pp. 461-75.

Roshier, D. A., A. I. Robertson, R. T. Kingsford, and D. G. Green. "Continental-Scale Interactions with Temporary Resources May Explain the Paradox of Large Populations of Desert Waterbirds in Australia." *Landscape Ecology,* vol. 16, 2001, pp. 547-56.

Rubega, M. A., and C. Inouye. "Prey Switching in Red-Necked Phalaropes *Phalaropus lobatus*: Feeding Limitations, the Functional Response and Water Management at Mono Lake, California, U.S.A." *Biological Conservation,* vol. 70, 1994, pp. 205-10.

Rubega, M. A., and J. A. Robinson. "Water Salinization and Shorebirds: Emerging Issues." *International Wader Studies,* vol. 9, 1996, pp. 45-54.

Russell, I. C. "A Geological Reconnaissance in Southern Oregon." *US Geological Survey Fourth Annual Report,* 1884, pp. 435-62.

———. "Geological History of Lake Lahontan, a Quaternary Lake in Northwest Nevada." *US Geological Monographs,* no. 11, 1885.

———. "Present and Extinct Lakes of Nevada." *National Geographic Monographs,* vol. 1, no. 4, 1895, pp. 101-36.

Ryan, E. "The Public Trust Doctrine, Private Water Allocation, and Mono Lake: The Historic Saga of the National Audubon Society v. Superior Court." *Environmental Law Review,* vol. 45, no. 2, 2015, pp. 561-640.

Ryser, F. A., Jr. *Birds of the Great Basin.* U of Nevada P, 1985.

Saban, C. "Palynological Perspectives on Younger Dryas to Early Holocene Human Ecology at Paisley Caves, Oregon." MA thesis. U of Oregon, 2015.

Sada, D. W., and G. L. Vinyard. "Anthropogenic Changes in Biogeography of Great Basin Aquatic Biota." *Smithsonian Contributions to the Earth Sciences,* vol. 33, 2002, pp. 277-93.

Santi, L., A. Arnold, D. E. Ibarra, C. Whicker, J. Mering, C. G. Oviatt, and A. Tripati. "Lake Level Fluctuations in the Northern Great Basin for the Last 25,000 Years." *2019 Desert Symposium,* pp. 176-86. https://eartharxiv.org/repository/view /870/.

Savvaitova, K., and T. Petr. "Lake Issyk-Kul, Kirgizia." *International Journal of Salt Lake Research,* vol. 1, no. 2, 1992, pp. 21-46.

Scarberry, K. C. "Extension and Volcanism: Tectonic Development of the Northwestern Margin of the Basin and Range Province in Southern Oregon." Dissertation. Oregon State U, 2007.

Scarberry, K. C., A. J. Meigs, and A. L. Grunder. "Inception and Style of Faulting at the Edge of a Propagating Continental Rift System: Insight from the Development of the Abert Rim Fault, Southern Oregon." *Tectonophysics,* vol. 488, 2010, pp. 71-86.

Schultz, B. W., R. J. Tausch, P. T. Tueller. "Size, Age, and Density Relationships in Currlleaf Mountain Mahogany (*Cercocarpus ledifolius*) Populations in Western and Central Nevada: Competitive Implications." *Great Basin Naturalist,* vol. 51, 1990, pp. 183-91.

Schultz, L. M. "The Genus *Artemisia* (Asteraceae: Anthemidae)." *Flora of North America North of Mexico, Volume 19: Asterales,* pp. 503-34. Oxford UP, 2006.

Senner, N. R., J. N. Moore, S. T. Seager, S. Dougill, and S. E. Senner. "A Salt Lake Under Stress: Relationships Among Birds, Water Levels, and Invertebrates at a Great Basin Saline Lake." *Biological Conservation,* vol. 220, 2018, pp. 320-29.

Shaver, S. A, A. P. Rose, R. F. Steele, and A. E. Adams. *An Illustrated History of Central Oregon, Embracing Wasco, Sherman, Gilliam, Wheeler, Crook, Lake and Klamath Counties, State of Oregon.* Western Historical Publishing, 1905.

Shiojiri, K., S. Ishizaki, R. Ozawa, and R. Karban. "Airborne Signals of Communication in Sagebrush: A Pharmacological Approach." *Plant Signal and Behavior,* vol. 10, 2015, p. e1095416.

Shuford, W. D., N. Warnock, K. C. Molinda, and K. K. Strum. "The Salton Sea as Critical Habitat to Migratory and Resident Waterbirds." *Hydrobiologia,* vol. 473, 2002, pp. 255-74.

Sigler, J. W., and W. F. Sigler. "Aquatic Resources of the Arid West: Perspectives on Fishes and Wilderness Management." *Wilderness Issues in the Arid Lands of the Western United States,* pp. 69-83. Edited by S. I. Zeveloff and C. M. McKell. U of New Mexico P, 1992.

Sigler, W. F., and J. W. Sigler. *Fishes of the Great Basin.* U of Nevada P, 2014.

Smith, G. R., T. E. Dowling, K. W. Goblet, T. Lugaski, D. K. Shiozawa, and R. P. Evans. "Biogeography and Timing of Evolutionary Events Among Great Basin Fishes." *Great Basin Aquatic Systems History,* pp. 175-233. Edited by R. Hershler, D. B. Madsen, and D. R. Currey. Smithsonian Institution P, 2002.

Snyder, J. O. "Relationships of the Fish Fauna of the Lakes of Southeastern Oregon." *Bureau of Fisheries Document,* vol. 636, 1908, pp. 71-102.

Soucie, M. T. "Burns Paiute Tribe." *The First Oregonians,* pp. 44-59. Edited by L. Berg. Oregon Council for the Humanities, 2007.

Spier, L. "Klamath Ethnography." *University of California Publications in American Archaeology and Ethnography,* vol. 30, 1930, pp. 1-338.

Spigel, R. H., and G. W. Coulter. "Comparison of Hydrology and Physical Limnology of the East African Great Lakes: Tanganyika, Malawi, Victoria, Kivu and Turkana (with References to Some North American Great Lakes)." *The Limnology, Climatology, and Paleoclimatology of the East African Lakes,* pp. 103-35. Edited by T. C. Johnson and E. O. Odada. Gordon and Breach Publishers, 1996.

The State of North America's Birds 2016. http://www.stateofthebirds.org/2016/resources/species-assessments/.

Stein, N., and nine coauthors. "Desiccation Cracks Provide Evidence of Lake Drying on Mars, Sutton Island Member, Murray Formation, Gale Crater." *Geology,* vol. 46, no. 6, 2018, pp. 515-18.

Stern, M. A., J. F. Morawski, and V. Marr. "Distribution, Abundance and Movements of Snowy Plovers in Southeast Oregon, 1990." Final Report submitted to Oregon Department of Fish & Wildlife Nongame Program and Lakeview District Bureau of Land Management, 1991.

Stevenson, A. E., and V. I. Butler. "The Holocene History of Fish and Fisheries of the Upper Klamath Basin, Oregon." *Journal of California and Great Basin Anthropology,* vol. 35, no. 2, 2016, pp. 169-88.

Stewart, O. C. "The Northern Paiute Bands." *Anthropological Records,* vol. 2, no. 3, 1939, pp. 127-49.

Stone, R. "Saving Iran's Great Salt Lake." *Science,* vol. 349, no. 6252, 2015, pp. 1044-47.

Strauss, E. A., C. A. Ribic, and W. D. Shuford. "Abundance and Distribution of Migratory Shorebirds at Mono Lake, California." *Western Birds,* vol. 33, no. 4, 2002, pp. 222-40.

Sullivan, P. T. "Field Notes: Eastern Oregon, Fall 1995." *Oregon Birds,* vol. 22, no. 2, 1996, pp. 59-63.

Swarth, C. W. "Foraging Ecology of Snowy Plovers and the Distribution of Their Prey at Mono Lake, California." MS thesis. California State U, 1983.

Sylvestre, P. D. "The Art and Science of Natural Discovery: Israel Cook Russell and the Emergence of Modern Environmental Exploration." MA thesis. Colorado State U, 2008.

Temeles, E. J., J. S. Miller, and J. L. Rifkin. "Evolution of Sexual Dimorphism in Bill Size and Shape of Hermit Hummingbirds (Phaethornithinae): A Role for Ecological Causation." *Philosophical Transactions of the Royal Society of London, Biological Sciences,* vol. 12, no. 1543, 2010, pp. 1053-63.

Thomas, S. M., J. E. Lyons, B. A. Andres, E. E. Smith, E. Palacios, J. F. Cavitt, J. A. Royle, S. D. Fellows, K. Maty, W. H. Howe, E. Mellink, S. Melvin, and T. Zimmerman. "Population Size of Snowy Plovers Breeding in North America." *Waterbirds,* vol. 35, no. 1, 2012, pp. 1-14.

Timms, B. V. "Waterbirds of the Saline Lakes of the Paroo, Arid-Zone Australia: A Review with Special Reference to Diversity and Conservation." *Saline Lakes Around the World: Unique Systems with Unique Values,* pp. 226-33. Edited by A. Oren, D. Naftz, P. Palacios, and W. A. Wurtsbaugh. *Natural Resources and Environmental Issues,* vol. 15, 2009.

Tinniswood, W. R. "Adfluvial Life History of Redband Trout in the Chewaucan and Goose Lake Basins." *Redband Trout: Resilience and Challenge in a Changing Landscape,* pp. 99-112. Edited by R. Schroeder and James D. Hall. Oregon Chapter, American Fisheries Society, 2007.

Tourtlotte, G. I. "Studies on the Biology and Ecology of the Northern Scorpion." *The Great Basin Naturalist,* vol. 34, no. 3, 1974, pp. 167-79.

Tremble, S. *The Sagebrush Ocean: A Natural History of the Great Basin.* U of Nevada P, 1999.

Umek, J., S. Chandra, and J. Brownstein. "Limnology and Food Web Structure of a Large Terminal Ecosystem, Walker Lake (NV, USA)." *Saline Lakes Around the World: Unique Systems with Unique Values,* pp. 94-98. Edited by A. Oren, D. Naftz, P. Palacios, and W. A. Wurtsbaugh. *Natural Resources and Environmental Issues,* vol. 15, 2009.

US Bureau of Land Management. "High Desert Management Framework Draft Plan Amendment and Environmental Impact Statement for the Proposed Lake Abert Area of Critical Environmental Concern (ACEC) in Lake County, Oregon." 1995.

US Department of Agriculture. "Rabbit Brush: A New High Value Rubber Crop for Nevada." 2006. https://reeis.usda.gov/web/crisprojectpages/0204073-rabbit-brush-a-new-high-value-rubber-crop-for-nevada.html.

US Fish and Wildlife Service. *Endangered and Threatened Wildlife and Plants:12-Month Finding for a Petition to List the Great Basin Redband Trout as Threatened or Endangered. Federal Register* 54, 2000, pp. 14932-36. https://www.govinfo.gov/content/pkg/FR-2000-03-20/pdf/00-6864.pdf.

———. *Walker Lake Ecosystem: Research and Monitoring Summary Report, 2006–2013.* Lahontan National Fish Hatchery Complex, 2013. https://static1.squarespace.com/static/550a1fc8e4b0e1de27f15703/t/59533409b11be18d82b0b4b5/1498625059456/FWS-Walker-Lake-Ecosystem-Report-2013.pdf.

———. *Bald and Golden Eagles: Population Demographics and Estimation of Sustainable Take in the United States.* Division of Migratory Bird Management, 2016 update. https://www.fws.gov/media/population-demographics-and-estimation-sustainable-take-united-states-2016-update.

Van Denburgh, A. S. "Solute Balance at Abert and Summer Lakes, South-Central Oregon: Closed-Basin Investigations." US Geological Survey Professional Paper 502-C, 1975.

Van den Hout, P. J., K. J. Mathot, L. R. M. Maas, and T. Piersma. "Predator Escape Tactics in Birds: Linking Ecology and Aerodynamics." *Behavioral Ecology,* vol. 21, 2010, pp. 16-25.

Van Dyke, W. A., A. Sands, J. Yoakum, A. Polenz, and J. Blaisdell. *Wildlife Habitats in Managed Rangelands—The Great Basin of Southeastern Oregon.* USDA General Technical Report PNW-159, 1983.

Van Winkle, W. *The Quality of Surface Waters of Oregon.* US Geological Survey Water-Supply Paper 363, 1914.

Verts, B. J., and L. N. Carraway. *Land Mammals of Oregon.* U of California P, 1998.

Wagner, T. W. *Lake Abert Half Graben: Estimate of Basin Fill from Gravity Profiling and Geologic Mapping.* Abstract, Paper No. 4-5, Geological Society of America, 103 Annual Meeting, 2007.

Walker, B., and D. Salt. *Resilience Thinking: Sustaining Ecosystems and People in a Changing World.* Island Press, 2006.

Waring, G. A. *Geology and Water Resources of a Portion of South-Central Oregon.* US Geological Survey Water-Supply Paper 220, 1908.

Warnock, N., S. M. Haig, and L. W. Oring.
"Monitoring Species Richness and Abundance of Shorebirds in the Western Great Basin." *The Condor,* vol. 100, no. 4, 1998, pp. 589-600.

Washington State University 2015. "Rare 'Milky Rain' Most Likely Traveled from Ancient Lake." *WSU Insider.* https://news.wsu.edu/press-release/2015/02/10/rare-milky-rain-most-likely-traveled-from-ancient-lake/.

Wauchope, H. S., J. D. Shaw, O. Varpe, E. G. Lappo, D. Boertmann, R. B. Lanctot, and R. A. Fuller. "Rapid Climate-Driven Loss of Breeding Habitat for Arctic Migratory Birds." *Global Change Biology,* vol. 23, no. 3, 2016, pp. 1085-94.

Weide, M. L. "Cultural Ecology of Lakeside Adaptation in the Western Great Basin." Dissertation. U of California, Los Angeles, 1968.

Welch, B. "The ZX Ranch: The Story Behind the Iconic Brand." *American Cowboy,* 2017. https://americancowboy.com/lifestyle/zx-ranch-oregon/.

Weppner, K. N., J. L. Pierce, and J. L. Betancourt. "Holocene Fire Occurrence and Alluvial Responses at the Leading Edge of Pinyon–Juniper Migration in the Northern Great Basin, USA." *Quaternary Research,* vol. 80, no. 2, 2013, pp. 143-57.

Werschkul, D. F. "Species-Habitat Relationships in an Oregon Cold-Desert Lizard Community." *Great Basin Naturalist,* vol. 42, no. 3, 1982, pp. 380-84.

Wheat, M. M. *Survival Arts of the Primitive Paiutes.* U of Nevada P, 1967.

Whitaker, J. O., and C. Maser. "Food Habits of Seven Species of Lizards from Malheur County, Southeastern Oregon." *Northwest Science,* vol. 55, no. 3, 1981, pp. 202-8.

White, D. E. "Memorial to Gerald Ashley Waring, 1883-971." *GSA Memorials—Geological Society of America,* vol. 3, 1971, pp. 205-10.

Whitley, D. S. "Rock Art Dating and the Peopling of the Americas." *Journal of Archaeology 2013,* article ID713159.

Williams, T. T. *Refuge: An Unnatural History of Family and Place.* Vintage Books, 1991.

———. "The Lost Daughter of the Oceans." *High Country News,* vol. 48, no. 19, 2016, pp. 12-15.

Williams, W. D. "Lake Corangamite, Australia, a Permanent Saline Lake: Conservation and Management Issues." *Lakes and Reservoirs: Research and Management,* vol. 1, no. 1, 1995, pp. 55-64.

———. "Environmental Threats to Salt Lakes and the Likely Status of Inland Saline Ecosystems in 2025." *Environmental Conservation,* vol. 29, no. 2, 2002, pp. 154-67.

Wirston, T., and G. M. Smith. "Late Pleistocene to Holocene History of Lake Warner and Its Prehistoric Occupations, Warner Valley, Oregon (USA)."

Quaternary Research, vol. 88, no. 3, 2017, pp. 1-23 and 491-513.

Wurtsbaugh, W. A., C. Miller, S. E. Null, R. J. DeRose, P. Wilcock, M. Hahnenberger, F. Howe, and J. Moore. "Decline of the World's Saline Lakes." *Nature Geoscience,* vol. 10, 2017, pp. 816-23.

Young, J. A., and C. D. Clements. *Cheatgrass, Fire and Forage on the Range.* U of Nevada P, 2009.

Zeveloff, S. I., and F. R. Collett. *Mammals of the Intermountain West.* U of Utah P, 1988.

Zhang, G., H. Xie, S. Duan, M. Tian, and D. Yi. "Water Level Variation of Lake Qinghai from Satellite and In Situ Measurements Under Climate Change." *Journal of Applied Remote Sensing,* vol. 5, 2011, pp. 3532-47.

Zolá, R. P., and L. Bengtsson. "Long-Term and Extreme Water Level Variations of the Shallow Lake Poopó, Bolivia." *Hydrological Sciences Journal,* vol. 51, 2006, pp. 98-114.

Index

Page numbers in **bold** indicate illustrations.

effects on biota, **95**, 208. *See also* hypersaline, Lake Abert, salt lakes

hypersaline, 47, 51, 207. *See also* hypersaline lake, Lake Abert, salt lakes

Ibis. *See* White-faced Ibis

ice rafting, 38–39, **39**. *See also* drop stone, Lake Abert, Summer Lake

International Phalarope Working Group, 256. *See also* phalaropes, Ryan Carle

J. R. Simplot Inc., 82, 298. *See also* ZX Ranch

jackrabbit. *See* mammals

Juniper Creek, 10, 185

Killdeer (*Charadrius vociferus*), 190; natural history 217–18, **217**; nesting in Great Basin, 214. *See also* shorebirds

Klamath Basin Refuges, 210, 263. *See also* William Finley

Klamath Basin: John Frémont exploration, 12; lakes, 256; Native American presence, 281

Klamath Marsh, 206, 285

Klamath Reservation, 277, 279. *See also* Klamath Tribe

Klamath Tribe, 4, 19. *See also* Klamath Reservation, Native American, Yahooskin

Kreuz, Keith and Lynn, 106–7, **107**

Lahontan cutthroat trout. *See* fish

Lahontan Valley wetlands, 185, 227, 245

Lake Abert, **5, 9, 43, 56**; algae, 98–100; alkali dust, 62; Area of Critical Environmental Concern, 20; birding, 261–262; Chewaucan River inflows, 80–83; climate change threats, 292–95; conservation, 299–302; desiccation polygons, 54; fauna, 92–118; geochemistry, 55–60; geology, 27–28, **28**, 31; historical background, 11–21; hydrology, 78–87, **84–87**; ice cover, 55–56, **56**; Indian

villages near, 264–85, **279**; limnology, 41–60; maps, **6, 8, 10, 16, 17**; microbiota, 92–98; playa, 38; red water, **94, 108**; salinity, 46–47; shore plants, 125–27; storm surges, 55; tufa deposits, 64; water diversion threats, 87, 89, 294–295; waterbirds, 187–88, 203–61

landslides and rockfalls, 34–35. *See also* Abert Rim, Summer Lake

lava and lava flow, 23, 28–31, 33–34. *See also* basalt, flood basalt

Least Sandpiper (*Calidris minutilla*), **213, 224**, 253. *See also* shorebirds, waterbirds

lichens, 128–31, **130**; cobblestone lichen (*Pleopsidium flavum*), 129, 131; sunburst lichen (*Xanthoria elegans*), 131; woodrat signpost association, 171–73, **172, 173**. *See also* bushy-tailed woodrat

lizards, 161–64; desert horned lizard (*Phrynosoma platyrhinos*), 162–63, **162**; leopard lizard (*Gambelia wislizenii*), 163, **164**; pygmy short-horned lizard (*Phrynosoma douglasii*), **162**; sagebrush lizard (*Sceloporus graciosus*), **161**; side-blotched lizard (*Uta stansburiana*), 163–64; western fence lizard (*Sceloporus occidentalis*), **161**

Loggerhead Shrike (*Lanius ludovicianus*), 185, 195–96, **196**, 262. *See also* upland birds

Long-billed Curlew (*Numenius americanus*), 5, **9, 213**. *See also* shorebirds

long-legged fly (*Hydrophorus plumbeus*), 115–16, **116**

Malheur National Wildlife Refuge, 210, 231, 263

mammals, 166–82; bighorn sheep (*Ovis canadensis*), 166–67, 178–81, **179**; black-tailed jackrabbit (*Lepus californicus*), 166–68, **167**; bushy-tailed woodrat (*Neotoma cinerea*), 130–131, 166, 170–74; coyote

(*Canis latrans*), 166, 181–82, **181**; mountain cottontail (*Sylvilagus nuttallii*), 166–68, **167**; pronghorn (*Antilocapra americana*), 174–78, **175, 176**, 182–83; Sierra Nevada bighorn sheep (*Ovis canadensis sierrae*), 179; white-tailed antelope squirrel (*Ammospermophilus leucurus*), 166, 168–69, **169**; yellow-bellied marmot (*Marmota flaviventris*), 166, 169–170, **169**

maps: Abert Rim Wilderness Study Area, **153**; American Avocet, Western Sandpiper and Wilson's Phalarope breeding and winter range, **205**; Basin and Range, **23**; Columbia River Flood Basalt, **31**; Great Basin, **23**; Great Basin paleolakes, **67**; High Lakes Plateau, **69**; important waterbird habitats near Lake Abert, **206**; Lake Abert, **6, 16, 17**; Lake Abert playa, **10**

Mar Chiquita, Argentina, 49, 245, 287–88

Mediterranean sage (*Salvia aethiopis*), 155–56, **156**

Mile Post 74 Springs, **125**; waterbird use, 204, **208**, 250, 261

mollusks, 95, 118–20; Great Basin ramshorn snail (*Helisoma newberryi*), 119–20, **119, 120**; robust (Jackson Lake) springsnail (*Pyrgulopsis robusta*), 44, 118–19, **119**

Mono Lake: importance for Eared Grebe and Wilson's Phalarope, 205, 211, 245; Public Trust Doctrine protections for, 297; waterbird abundance, 258–259

Mosillus bidentatus fly, 115, **116**

mosses, 131–32

Mozingo, Hugh, 142–43

National Audubon Society: American Avocet impacted by climate change, 228; California Gull and Eared Grebe as climate endangered, 211, 250; conservation efforts for waterbirds, 301; Lake Abert as

Important Bird Area, 185; Lake Abert waterbirds as climate threatened, 296; Willet and Wilson's Phalarope as climate endangered, 235, 247

Native Americans: Chewaucan Cave use, 282; disappearance from Chewaucan Basin, 285; drought impacts, 282; graves uncovered, 18; marsh resource use, 281; native plant use, 4, 135, 143, **276**, 277; petroglyphs, **271**, **284**; presence in Chewaucan Basin, 264–85; rabbit drives, 168; Rattlesnake Cave use, 65; rock rings, 272–273, **273**; sagebrush ceremonial use, 138; stone kitchen tools, **272**, **273**. *See also* Chewaucan culture, Klamath Indians, Northern Paiute, Yahooskin

Northern Paiute, 19; alkali fly pupae harvesting, 112; forced onto reservations, 287; marmots as food source, 170; preadapted to dry climate, 283; presence in South Central Oregon, 285; tribal bands named after foods, 277. *See also* Native Americans, Yahooskin

Northern Pintail (*Anas acuta*), **252**, 253. *See also* waterfowl

Northern Shoveler (*Anas clypeata*): abundance at Lake Abert, 251, 253; climate change threats, 296; presence at Lake Abert, 212–13, **212**; salt removal from food, 207; vortex feeding, 236. *See also* waterfowl

Oregon Lakes Association, 301

Oregon Natural Desert Association, 301

ostracod (*Candona* sp.), 70–71, **71**, 73; value for dating lake sediments, 70, 73

Owens Lake: dust hazards, 52, 61–62, 291; Public Trust Doctrine protections, 297; water-diversion impacts, 52

paleolakes and Pleistocene lakes, 22, 51, 66–69, **67**; fossil fishes and snails, 118–20; Lake

Alvord, 60; Lake Bonneville, 22, 26, 66–67, **67**; Lake Chewaucan, 66–68, **67**, 71–77, **72**; Lake Lahontan, 65–67, **67**, 69, 122; maps, **67**, **69**; Winter Lake, **72**, **75**; ZX Lake, **72**, **75**. *See also* Ira Allison

Peregrine Falcon (*Falco peregrinus*), 194–95, **216**. *See also* upland birds

petroglyphs: dated at Winnemucca Lake, 66; desert varnish association, 36–37, **37**; Lake Abert area locations, **37**, 148, 152, **271**, **284**; Luther Cressman surveys, 270–271. *See also* Richard Pettigrew

Pettigrew, Richard: Lake Abert archaeological surveys, 264, 270–75, 277. *See also* Chewaucan culture

phalaropes: freshwater spring use, 207; natural history, 235–38. *See also* Red-necked Phalarope, Red Phalarope, shorebirds, waterbirds, Wilson's Phalarope

pit-houses, 266, 272, 279. *See also* Chewaucan culture, Native Americans

playa lakes, 45; American Avocet presence, 223; branchiopod presence, 109; drought effects, 221–22; on Abert Rim, 154; on High Lakes Plateau, **69**. *See also* playa

playa: alkali salt crust, 59; clay content, 62; desiccation patterns, 62–64; drop stones and ice rafted boulders, 38–40, **39**; dust impacts, 61–62; evolution, 51; prevalence in Great Basin, 51–54, **53**, 68; sediments, 57; storm surge effects, **55**. *See also* Lake Abert, playa lake, Summer Lake

Pleistocene epoch: defined, 22, 51; lake levels, 74–76; mammalian predators, 177; paleolake presence, 65–69, 72; pronghorn relatives, 178. *See also* Pleistocene lakes

Poison Creek, 10, 34–35, 143, 147–148

Prairie Falcons (*Falco*

mexicanus), 185, 194–95, **195**. *See also* upland birds

Prior Appropriation Doctrine, 81. *See also* water rights

pronghorn, and pronghorn antelope. *See* mammals

Public Trust Doctrine, 297

pygmy short-horned lizard. *See* lizards

rabbitbrush (Ericameria nauseosa and Chrysothamnus viscidiflora), 126, 138–40, **139**

rabbits and hares. *See* mammals

radiocarbon dating: Lake Abert archaeological sites, 273–74; Lake Chewaucan elevations, 73; Paisley Caves coprolites, 267; tui chub bones, 121; Winnemucca Lake petroglyphs, 66; woodrat middens, 174

Red Phalarope (*Phalaropus fulicarius*), 235–36, 251

Red-necked Phalarope (*Phalaropus lobatus*), 203, **204**, **224**, **239**; abundance at Lake Abert, 241, 251, 253; climate change threats, 240; natural history at Lake Abert, 239–41; population at Lake Abert, 260. *See also* phalaropes, shorebirds, waterbirds

Rights of Nature Doctrine, 298

Ring-billed Gull (*Larus delawarensis*), bathing at springs, **208**; climate change threats, 250; natural history at Lake Abert, 247–50, **248**, **249**; nesting at Summer Lake, 205. *See also* gulls

River's End Reservoir, 4, 7, 18–19, **19**, 90

Rock Wren (*Salpinctes obsoletus*), 185, 198–99, **199**, 202. *See also* upland birds

rocks and minerals: basalt, 30–34, **34**; Oregon sunstone, **33**; Steens basalt, 30–34, **31**; tuff, 31–32, **32**. *See also* salt, salt crystals and minerals, trona, tufa

Russell, Israel Cook, 24–28, **24**, 42, 67

Ryser, Fred A., Jr., 184, 190

About the Author

Ron Larson earned a PhD in marine sciences in Canada and completed his post-doctoral work in Florida while studying deep-sea animals. He was employed by the U.S. Fish and Wildlife Services, where some of his work focused on water development and endangered species. His research interests include the ecology of waterbirds and shorebirds. Larson currently works with a group of scientists who are studying the effects of climate change on the Great Basin. He is the author of *Swamp Song*.